T0073772

The Curious Human Knee

THE
CURIOUS
HUMAN KNEE

Han Yu

Columbia University Press
New York

Columbia University Press
Publishers Since 1893
New York   Chichester, West Sussex
cup.columbia.edu

Library of Congress Cataloging-in-Publication Data
Names: Yu, Han, 1980– author.
Title: The curious human knee / Han Yu.
Description: New York : Columbia University Press, [2023] |
Includes bibliographical references and index. |
Identifiers: LCCN 2022044171 | ISBN 9780231207027 (hardback) |
ISBN 9780231556774 (ebook)
Subjects: LCSH: Knee—Anatomy. | Knee—Aging.
Classification: LCC QM549 .Y8 2023 | DDC 612.9/8—dc23/eng/20220920
LC record available at https://lccn.loc.gov/2022044171

Columbia University Press books are printed on permanent and
durable acid-free paper.
Printed in the United States of America

Cover design and illustration: Philip Pascuzzo

Contents

The Curious Human Knee

1

Knees Before the Brain

Man became distinct from the apes when he became a biped.
—SHERWOOD WASHBURN (1911–2000),
LEGENDARY AMERICAN ANTHROPOLOGIST

Biped /ˈbīped/
noun: an animal that uses two legs for walking

Related words:
bipedal (*adjective*); bipedally (*adverb*); bipedalism (*noun*)

Walking upright on two hind legs, all the time and with grace, is
a unique feature of modern humans. Yes, other things make us
humans and separate us from apes, such as a big brain, dexterous hands, language, and material culture.[1] But bipedalism is a turning point, some say *the* turning point, in evolution. On the dusty hills of Hadar, Ethiopia, if you were to kick loose a rock and reveal an animal fossil, and if that fossil showed anatomical features of bipedalism, like a knee with a certain look, then congratulations, because you had chanced upon not any animal but a *hominin*.

"Hominin" is not the same as "human." This taxonomy group includes modern humans (i.e., us), our ancient ancestors after they split from chimpanzees, and our extinct relatives (like the Neanderthals). Really, it is a pretty diverse group. It's just that we are the only ones who, as of now, still exist. So, rather than saying "*man* became distinct from the apes when *he* became a biped," it's more accurate to say "hominins became distinct from the apes when they became bipeds." After all, extinct hominins walked the Earth bipedally long before we did. And, after all, a female human can get around bipedally just as well as a man can.

Why is bipedalism so special? Well, for one thing, no other mammals do it. Birds can move about on two feet, but no other mammals habitually stand, walk, or run on two legs.[2] Cats, dogs, and horses are all quadrupedal—that is, they move about on four feet. Our close relatives the chimpanzees and gorillas are also quadrupeds. Their form of quadrupedal walking is called knuckle walking, which they accomplish by curling the fingers of their two hands and then walking on the knuckles as well as the flat of their two feet. You and I can accomplish this form of movement too, but we won't be going very fast. Our arms are too short to comfortably reach our knuckles to the ground, and our knees would be constantly in the way. Chimpanzees and gorillas don't have this problem because they have longer arms and shorter legs.

At times, chimpanzees and gorillas do stand up and walk bipedally. They waddle and look a tad clumsy, taking a sort of "bent-hip, bent-knee" posture. They feel this awkwardness too, which is why they are not fans of bipedalism and perform it only occasionally. If an exceptional ape comes along that seems to embrace bipedalism, it becomes an instant celebrity, like Louis, a 6-foot, 450-pound gorilla at the Philadelphia Zoo. In 2018, Louis was caught on film walking swiftly across his yard, snacks in hands strutting his stuff.[3] The video went viral on social media, getting countless likes, laughs, and comments about how human-like Louis looks. According to the zoo, one reason Louis walks upright is to avoid getting his hands and snacks dirty when the ground is muddy.[4] Now that's practically human.

Off social media and in real life, bipedalism is more often a source of pain than amusement. By standing and walking upright, we put a tremendous amount of weight—that of our head, upper torso, and upper limbs, as well as whatever we happen to be carrying—through our spine onto our knees.[5] When we power walk or run, the force on our lower limbs increases further to several times our body weight.[6] If we then proceed to do something fancy like rapidly changing directions or going up or down hills, the lower limbs lose their ability to absorb shock, further endangering the knees.[7] No wonder so many people have torn menisci, torn anterior cruciate ligaments, and inflamed or arthritic knees.

Annoying and painful as these conditions are to us, they could be deadly to our ancient ancestors living in a primitive world. Indeed, even when their knees *were* healthy, early hominins would have traded some dire consequences for becoming bipeds.

For an animal to run, it not only needs to thrust its body forward but also to hold that body upright.[8] By standing up on two legs, our ancestors (as do we) consumed a significant amount of energy. Think about it: as soon as we use one leg to generate energy and push our bodies forward, the other leg must be planted in front to stop that forward motion and prop our bodies up. We can try to lean forward to increase forward motion, but there is a limit to how far we can lean before we fall on our face.

This is why a quadruped—a lion, a dog, and even a lowly rabbit—can surpass us in running, always.[9] With four feet on the ground, they lean so far forward that they are already parallel to the ground. Most of the energy generated by their hind legs is thus put into the forward direction. The front legs, rather than being used to stop forward motion and hold the body up, function as a pivot point for the body to rest momentarily before the hind legs push forward again. This is why we can never catch Fido when he's running away with a stolen sock (or worse) in his mouth. More soberly, this is why our ancestors had no chance of outrunning hungry saber-toothed cats on the African savannah.

Of course, our ancestors, before they split from apes, were not ground dwellers. They were arboreal, or tree living. A glimpse of their abilities can be seen in the mad skills of modern apes. Gibbons, for example, get around trees by swinging from branch to branch like a pendulum, negotiating gaps up to fifty feet with a single swinging leap and reaching speed up to thirty-five miles per hour.[10] Orangutans may be slower, but they can flex and twist those arms and legs like acrobats in hairy, orange suits, grabbing branches every which way using any foot or hand. (Incidentally, the name "orangutan" has nothing to do with the orange color. In the Malay language, it means "person of the forest." Quite poetic, isn't it?)

But forests aren't forever. Starting about seven to five million years ago, during the late Miocene epoch and the subsequent Pliocene epoch, the Earth underwent a period of cooling and reduced rainfall, which turned widespread lush forests to a mixture of woodlands and grasslands.[11] As forests shrank, fruits, leaves, and other vegetarian food diminished. As a result, our ancestors had to come down from the trees and expand into new territories to eke out a living.

And once they did come down (if only occasionally at first), the wheels of bipedalism started turning. The familiar story goes like this: A bipedal posture freed our ancestors' hands, allowing them to make and use tools. This, in turn, spurred brain and hand development. Endowed with bigger brains and more dexterous hands, our ancestors made more complex tools, carried them, and used them, which further required that they stand up. Suppose they needed to jab an animal or hurl stones. They couldn't do that very well by squatting on the ground, unless their victims were very small.[12]

An upright posture, the brain, hands, and tools, each piece reinforced the other, creating a positive loop that helped our ancestors to survive.

As bipedalism became the new survival strategy, anatomical features that facilitated this type of movement were selected. These included a curved spine that keeps the head and torso in a vertical line and absorbs stress; a short, wide pelvis that brings hip muscles to the side of the body for balance; and arched feet with the big toes adducted in line with the other toes (instead of sticking out to the side) for ground push off and shock absorption.

The knee, in particular, saw important changes. In apes, the thigh bone and shinbone meet more or less straight on to form the knee. In humans, the knee grew angulated (figure 1.1A). If you draw imaginary lines to extend the shafts of your two thigh bones, the lines will meet somewhere below your feet, forming a very long V shape. Each half of that V constitutes an angle, known as the bicondylar angle, which measures about eight to eleven degrees in adult humans.[13] This angle allows us to place each foot directly below our body so we can momentarily balance on one foot between every two steps.[14]

This angled knee arose as a direct response to human bipedalism: it is missing not only in apes but also in human newborns. It starts to form only when human babies learn to walk, becoming noticeable in two-year-olds, and reaching lower adult values in four-year-olds.[15] However, this trait cannot be cultivated in other animals. Monkeys who are trained to walk upright from a young age never develop an angled knee, nor do birds such as ostriches or penguins.[16]

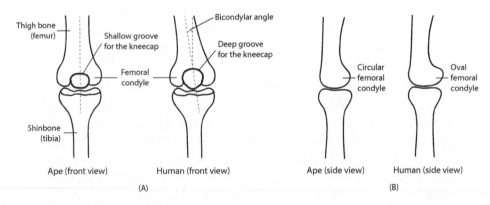

FIGURE 1.1 Changes to the human knee

Another distinct feature of our knees is that our femoral condyles—the round knobs at the lower end of the thigh bone (or femur)—grew elongated and oval in shape rather than in circular fashion as in apes (figure 1.1B). With this change, when the thigh bone and shinbone meet at the knee, there is an increased area of contact between the bones to reduce bipedal stress.[17] A deep groove also formed between the two condyles, which is largely flat in apes (figure 1.1A).[18] In this groove, our kneecap can slide up and down as we repeatedly bend and straighten knees during bipedal movement.

A knee alone can set us apart from apes. A knee alone can help us identify hominin ancestors.

In the fall of 1973, with $43,000 in his pocket, Donald Johanson headed for Hadar, Ethiopia.[19] Barely thirty years old, Don had just finished his doctorate at the University of Chicago, where he majored in paleoanthropology—that is, the study of ancient (*paleo*) humans (*anthropo*).

In the caravan with him, fighting sheep, goats, and donkeys out of the capital city Addis Ababa, were ten other scientists. Each brought to the trip a different focus in studying ancient lives: animal fossils, plants, and rocks. As the caravan began its journey toward Hadar, some two hundred miles northeast of Addis Ababa in a region of the country called the Afar, all on board dreamed the same dream: their destination held the secret to the origin of Earth and its lives.

Don dreamed of something even bigger. He was not just hoping to find ancient lives; he was looking for ancient hominins. This promise landed him the $43,000 grant money from the National Science Foundation. The money was to last him two years, but it didn't look like it was going to. He had already spent $10,000 to outfit a Land Rover that could move over the rough terrain. Another alarming amount went to buying tents, mosquito nets, and other equipment—for Don and for others on the team who didn't come with funds.

With its heavy load of ambition and money spent, the caravan drove past extinct volcanos, dark lava fields, and millions of years of sand and rocks into Hadar. Upon arrival, the team swiftly set up camp on a low bluff by the Awash River that runs through the region. Under the hundred-degree sunshine, it was going to be hot no matter what, but the bluff at least allowed some wind, and the river offered precious, if not always clean, water. Camp was minimal, consisting of a large tent for dining and mosquito nets for sleeping. Long tables were set up for cleaning and sorting all of the fossils and artifacts the teams were sure they would find.

After settling in, everyone eagerly dove into the field to do whatever they came for: dissecting geological strata, looking for fossil pollens, and, in Don's case, finding those hominins. Don had reasons to be optimistic. Last year, before committing to this expedition, he had made a quick survey trip to the area and, during three short days, had already found some well-preserved mammal skulls and pig teeth. Judging by the fossils' anatomical features, they dated to three million years ago. It was only a matter of time, and patience, Don thought, before the Hadar hominins would reveal their presence.

Hands behind his back, and the scorching sun behind that, Don did some dogged surveying, gluing his eyes on every nook in the sand, every rock, and every stone. Hours grew into days, then weeks. Time slipped away like the sand that surrounded him, without a hominin rising to the surface. One always sweats in Hadar, but Don was sweating from more than the heat. His grant was running out, and he had nothing to show for it. He wondered whether he had acted rashly by organizing such a large trip. Would he get a reputation for irresponsibility? Would his career crash before it even launched?

These thoughts preoccupied Don when he was out surveying one afternoon on October 30. Dejected, he kicked what looked like a hippo rib sticking out the sand. It came loose. Not a rib. It was the proximal tibia—or the upper end of a shinbone, *proximal* meaning close to the center of the body. The bone looked like it belonged to a primate—perhaps a monkey.

Don bent down to collect it and proceeded to record the find and its location in his notebook. While writing, he noticed two more pieces of bones lying a few yards away. One was a distal femur—or the lower end of a thigh bone, *distal* meaning being away from the center of the body. The bone was broken. Only one of its condyles—again, the round knobs at the end—remained attached. The other condyle was lying in the sand.

Don picked up these two and put them together with the shin bone to see if they would form a single knee joint. Almost unbelievably, they did. The pieces were of the same size and color and fit together perfectly. More remarkably, as Don now realized, the thigh bone and shinbone joined at an angle. Don tried to bring them in line, but they wouldn't. They *naturally* formed an angle. This was no monkey. It was a hominin!

Suddenly, the burden that had laid heavily on Don lifted. "I felt light-headed," Don recalled, "floating with relief—and at the same time incredulous at my luck. . . . I wondered if I weren't dreaming. No one in the history of anthropology had ever seen the knee joint of a three-million-year-old [hominin] before. If this was indeed one, it was unique in the world."[20] If this was indeed one, it would be solid evidence that something—some*one*—walked the Earth bipedally before us. If this was indeed one, Don would need to hold a press conference so he could borrow the fossils and take them back to the United States for further study.

The thought of a press conference stopped Don in his tracks. Suppose he was wrong? Suppose he made a mistake about the knee? With no anatomist on the staff to consult, Don needed a human knee to compare the fossils. But where to find a human knee in the middle of nowhere? Well, not far from camp, Don thought, was a local burial mound where the Afar people laid their dead to rest.

Ignoring the wrath and guns of the Afar people—and I should add the ethical protocols of scientific research—Don snuck into the burial mound: a loosely built, partially collapsed dome of boulders. As Don looked in, bones were piled high and, *supposedly*, a thigh bone was just "lying on the top, almost asking to be taken."[21]

Back at camp, under the feeble butane lamp, Don compared his ancient find with his illicit one: other than the fossils being smaller, they were identical. Whoever owned that ancient knee walked on two hind legs, several million years ago.

If conventional wisdom about the rise of bipedalism is correct, then the owner of Don's knee should have had a decent-sized brain and a fair ability to make tools. Remember? Bipedal movement, brain development, and tool making should form a positive feedback loop in which each triggers and reinforces the other.

Well, evidently, such a loop does not exist.

After finding the knee joint in 1973, Don Johanson went back to Hadar the next year hoping to find its owner.[22] Not *the* owner of the knee, but someone just like the owner, in a whole fossil skeleton.

In life, some of us are luckier than others. Or, maybe we are each lucky in different facets of life. As far as paleoanthropology goes, Don was *really* lucky. On November 24, 1974, thirteen months after he had stumbled on the knee joint, he found Lucy. That day, Don was supposed to stay in camp to do paperwork, but, feeling lucky, he ventured out. "I'm superstitious," he recalled. "Many of us are, because the work we do depends a great deal on luck. The fossils we study are extremely rare. . . . When I got up that morning I felt it was one of those days when you should press your luck. One of those days when something terrific might happen."[23] Two hours after feeling lucky, he found Lucy.

Lucy was scattered on the slope of a little gully: a bit of an arm here, a skull piece there, then part of a thigh bone, a couple of vertebrae, part of a pelvis, altogether several hundred pieces, many of which fragments. Collecting Lucy took three weeks. When the last piece was picked up and assembled, 40 percent of a skeleton emerged. It belonged to one individual: everything fit and looked the same with no extras, like two left legs.

The night Lucy was found, a tape recorder in the camp was blasting the Beatles song "Lucy in the Sky with Diamonds." That's how Lucy got her name. We know she is a "she" because her pelvis has the telltale signs and anatomical features that allow females to give birth. As an Ethiopian, Lucy also has an Ethiopian name: Dinkinesh, which means "you are marvelous."[24] Scientifically, Lucy falls under the genus *Australopithecus*, which means "southern (Latin, *australo-*) ape (Greek, *-pitheco*)." This genus was named in the first quarter of the twentieth century based on fossil discovery. It designates hominins that are too primitive to belong in the same genus as us humans, which is *Homo*, meaning "man" (naturally, we are told the word is scientifically neutral and includes modern women).

By the time Don found Lucy, other species had been unearthed and classified under the *Australopithecus* genus, each with different anatomical features. Lucy doesn't look quite the same as any of them, so she was given a new species name: *afarensis*, meaning "from Afar," the larger region surrounding Hadar where Lucy was found. Lucy's full scientific name is thus *Australopithecus afarensis*: "southern ape from Afar." By contrast, you and I belong to the species *sapiens*, meaning "wise," so our full name is *Homo sapiens*: "wise man." The name was coined in 1758 by Swedish scientist Carl Linnaeus—a man, known as the father of modern taxonomy.

Judging by her wisdom teeth, Lucy had been full grown. Standing at 3.5 feet, she was quite short, most definitely shorter than the actual owner of the knee found the year before. But, anatomically, her knee was identical with the earlier find: it was angled, had a groove for the kneecap, and featured an elongated condyle.[25] Other parts of her body whispered bipedalism, too:[26] her spine was curved to support a habitually upright stance; her pelvis was wide to stabilize the body during upright walking; and the connection between her thigh bone and pelvis was elongated, which brought hip muscles to the side of her body to facilitate bipedal movement.

Lucy may be the most famous *Australopithecus afarensis*, but she's not the only one. More of her people (I know, *Australopithecus afarensis* are not exactly *people*, a word usually reserved to describe *Homo sapiens*, but you know what I mean. It just feels unneighborly to say "Lucy and her kind.") Anyway, more of her people turned up in Hadar and elsewhere in the Afar region and Ethiopia in following years. Notably, bones that belong to at least thirteen *Australopithecus afarensis* men, women, and children—nicknamed the First Family—were discovered at a single location in Hadar. These assorted bones likewise showed unmistakable signs of bipedalism from hip to toe.[27]

At the same time, however, Lucy and her people *also* retained primitive, ape-like features.[28] Notably, they had longer arms and shorter legs, and their fingers and toes were curved. Of course, these features could simply be evolutionary baggage that was still in the process of being selected against. But, they could also signal that Lucy and her people spent time in trees climbing and grasping branches. Perhaps, when they did walk on the ground, they walked less like us and more like apes with a bent hip and bent knees?[29]

Luckily, Lucy and her people left footprints.

Some 1,050 miles southwest of Hadar, within the borders of Tanzania, lies another ancient site: Laetoli. Despite the distance, hominin fossils unearthed at Laetoli look just like Lucy and belong to the same species of *Australopithecus afarensis*. Laetoli isn't exactly famous for these fossils, though. It was put on the map for something more elusive: footprints.

Standing 9,400 feet tall in Laetoli is Sadiman, a volcano active some four million years ago.[30] Back then, it periodically blew massive amounts of volcanic ash that covered the area in half-inch layers. If rain immediately followed, ash would dampen, like newly laid cement. If *Australopithecus afarensis* happened to venture out and walk upon this wet ash, they would leave behind footprints. At this juncture, if the sun came out, the ash would harden and lock the prints in place. Before the delicate prints could be damaged, Sadiman must spew again, fresh ash covering up and preserving the prints. Stars must align for us to see trails of hominins from millions of years ago.[31] At Laetoli, stars did.

In 1978, an eighty-eight-foot-long trail containing about seventy *Australopithecus afarensis* footprints was discovered in Laetoli.[32] The trail includes two separate tracks. The first was made by one individual with clear prints. The second was made by two—possibly even three—individuals walking in single file with overlapping prints.[33] Based on pressure patterns, the print makers were walking at a medium, comfortable speed.[34] Were they having a morning stroll after a refreshing rain, like we do on a beach vacation? Because if you look at these prints, you probably won't be able to tell them apart from your own on the beach. They have a well-shaped heel, a good ball of the foot, and big toes in line with the other four toes.[35] Statistical and comparative analyses further confirmed that the prints were made by creatures who walked essentially like we do with extended knees, and not in an ape-like, bent-knee posture.[36] You see, walking with bent knees puts more pressure on the front of the feet, which would have created deeper toe impressions and shallower heel impressions. This was not the case in the Laetoli prints.

So, everywhere we look, Lucy and her people had human-like knees and other lower limb functions that enabled bipedalism. What about their brain development and tool-making skills?

Their brains, judging by their skulls, are *tiny*, comparable in size to that of a chimpanzee.[37] Presumably, a brain this size would not generate enough creative juice for complex tool making. They might be able to, like chimpanzees, use twigs to fish termites out of nests, but can they fracture stones into meat cutters or scrapers? Fossil dating says no.

By measuring the decay of naturally occurring, radioactive elements in the sediments where the various fossils were discovered, we can determine that Lucy was 3.18 million years old, the First Family was slightly older at 3.2 million years of age, and the Laetoli footprint makers were older yet at 3.66 million years of age.[38] By contrast, the oldest stone tools found in the Afar region were made 2.5–2.6 million years ago,[39] some 580,000 years *after* Lucy had died.[40]

If Lucy and her people didn't and couldn't make tools, then why did they stand up? If conventional wisdom about the rise of bipedalism vis-à-vis a big brain is wrong, what pressured our ancestors to evolve those bipedal knees?

Before we turn eagerly to these questions, I must make two quick clarifications.

Clarification 1: Lucy and her people are *not* the first bipeds. Older hominins have walked the earth on two feet, though probably less effectively. As we push further back into history, evidence also becomes less conclusive.

Notably, in 2009, after fifteen years of digging, sifting, and reconstructing, Ardi was presented to the world. A partial adult female skeleton, Ardi was found in Aramis, Ethiopia, and belongs to the species *Ardipithecus ramidus*, which means "ground-dwelling ape at the root of human ancestry." Dated 4.4 million years old, Ardi was a good 1.2 million years older than Lucy. Her foot featured a big toe stuck to the side, suggesting branch grasping, but her other toes formed a rigid forefoot that could be pushed off as she walked.[41] Ardi's pelvis also had mixed features that suggest both bipedal walking and tree climbing.[42] The general view is that Ardi walked upright on the ground but was not as adapted as Lucy was to long, strenuous treks. When it comes to climbing and clambering in the trees, Ardi was more effective than Lucy but less so than modern-day chimpanzees.[43]

Tantalizing evidence of still older bipedal hominins also exists. In 2002, some 5.7-million-year-old footprints were discovered in Trachilos, Greece,[44] preceding Lucy and her people's Laetoli prints by more than two million years. The Trachilos footprints have a big toe in line with the other four toes, but they also feature an ape-like sole, so we can't be sure if the print makers habitually walked upright.[45]

Still older at six million years of age is an upper thigh bone discovered in the Tugen Hills of central Kenya and attributed to the species *Orrorin tugenensis* (meaning "original man in the Tugen region"). The shape and structure of the bone resembles that of Lucy, suggesting bipedalism.[46] But a single thigh bone can tell us only so much. Without a telltale knee, the verdict isn't conclusive.[47]

Clarification 2: The rise of bipedalism, like human evolution in general, was not a straight and narrow one-way street. Multiple forms of moving about on two feet, before and after Lucy, have appeared (and disappeared) along the way.

A younger cousin of Lucy, for example, is *Australopithecus sediba* ("sediba" means "fountain" or "wellspring"), who lived about 1.9 million years ago, more than a million years after Lucy. Fossils of this species were discovered in Malapa, South Africa, and among them was a more-or-less complete knee joint. The knee shows a bicondylar angle that, together with hip, foot, and ankle features, suggests bipedalism.[48] But the fossils also show something peculiar: these hominins were hyperpronators—that is, when they walked, they put body weight on the inside of the foot with the ankle collapsed inward. In modern humans, hyperpronation is a pathological condition that can cause knee problems, among other ailments.[49] But, in *Australopithecus sediba*, it seems that hyperpronation was the "norm," a way to accommodate both bipedal walking and some kind of tree-living lifestyle.[50]

Even our closest relatives, the extinct *Homo* species, did not necessarily move about as we do.[51] *Homo luzonensis*, which were discovered in Luzon, the Philippines, despite being very, very young (dated only 50,000–67,000 years ago), had hand and foot features that suggest an adaptation to both bipedal walking and tree living.[52] For example, their feet have more mobility in the middle, which would have made *Homo luzonensis* better climbers but compromised their ability to push off the ground when walking, causing them to walk as if wearing floppy slippers.[53]

So, long story short, we don't know which precise species led to *us* in which precise lineage. Similarly, we don't know whose precise bipedalism gave rise to our way of walking and running. What we do know is that, for some reason or reasons, millions of years ago, our ancestors developed bipedal knees *before* they developed big brains. Why that should happen is a question of much intrigue and heated debate.

Quadrupeds may be agile and fast, but they are stumped when it comes to carrying. All four of their limbs touch the ground and have evolved to carry their bodies, not other stuff. When the situation calls for it, their best bet is to put stuff in their mouths and either hold it or drag it on the ground. Either way, they can't carry much, or go very far. This is a blessing for scavengers in the wild: when lions can't carry their kill back to the den, hyenas get to sneak in and feast on it. It is also a blessing in modern domestic life: Fido may get away with one stolen sock in his mouth, but he sure can't loot a whole armful at once.

Knuckle-walking gorillas and chimpanzees fare the carrying challenge a little better. By curling their hands inward and walking on the knuckles, they can carry small objects in those hands—an apple, for example, or a banana. But on a good day when they come across several apples or bananas, knuckle walking becomes cumbersome. This is when gorillas and chimpanzees may stand up and walk bipedally, to free their hands and carry more food. Indeed, this is one theory explaining why our ancestors stood up and became habitual bipeds millions of years ago.

American anthropologist Gordon Hewes proposed this theory in the early 1960s. According to Hewes, our ancestors scavenged before they developed the weapons and skills to hunt big game.[54] Imagine that Lucy and her people had just found a lion's kill site. Without powerful teeth, they could not devour the meat then and there and would need considerable time to gnaw and chew. So, they had to act fast and move the meat to the safety of their home before the lion came back or other predators came along. In a less dramatic scenario, Lucy and her people had chanced upon a bunch of wild fruit, couldn't finish at once, and wanted to carry the bounty home for a rainy day. In either case, standing up, taking an armful, and holding the food against their chest would be the way to go.

Such behaviors can be readily induced in other primates.[55] When monkeys are presented with large and not immediately consumable food, such as a large grapefruit or hard "chow" biscuits, they will rise to the bipedal position and run off with the food to consume in private. If pursued, the animals will continue to hold onto their food and run on two feet, as long as they are not so scared by the pursuer that they abandon the food and resort to a faster, four-legged retreat. If the pursuer stops, the monkeys will remain in their upright position, ready to take off at a moment's notice while trying to stuff food into their mouths.

If our ancestors constantly partook in such bipedal food maneuvers, then bones, joints, and muscles adapted to bipedalism would be given selective priority. Over time, habitual bipedalism would arise.

The theory makes sense, but Owen Lovejoy, a renowned paleoanthropologist at Kent State University, thinks the real story is far juicier. Lovejoy doesn't deny the importance of food carrying but suggests that, ultimately, what drove our ancestors to carry them food and to stand up was sex.

Biologically speaking, sex is made to produce offspring, to propagate one's genes, and to ensure the survival of one's species. In the natural world, there are two opposite strategies in sexual reproduction: (1) to produce a lot of offspring without much, if any, parental care, and pray that at least some of them make it; or (2) to produce very few offspring and care for them intensively so that most, if not all, of them make it.[56]

An extreme example of the first strategy is the oyster, which can produce millions of eggs a year. The oyster chooses this strategy because it can't take care of its offspring in any shape or form: the oyster parent can't think, move, or carry. All it can do is scale up production, way up, and hope for the best.

Clearly, a mother who is more intelligent and more capable at childcare has a leading edge in this game. To be more intelligent and capable, a mother would need a relatively large brain. A large-brained mother begets large-brained children—some of whom would one day also become intelligent mothers. Bringing large-brained babies to term, however, requires a lot of energy, which means a mother can't afford frequent pregnancies. Once born, an infant also needs considerable time to further develop its large brain before it can face the world on its own. This means the mother would be tied up in prolonged and intensive childcare and could not afford another pregnancy right away. This, then, is the second sexual reproduction strategy: fewer kids, better care.

When this strategy is pushed to the extreme, we have the great apes—gorillas, chimpanzees, and orangutans—who produce an infant every four to eight years. As Jane Goodall, the famous primate expert, wrote of the chimpanzees,

> the infant does not start to walk until he is six months old, and he seldom ventures more than a few yards from his mother until he is over nine months old. He may ingest a few scraps of solid food when he is six months, but solids do not become a significant part of his diet until he is about two years of age and he continues to nurse until he is between four-and-a-half and six years old. Moreover, while he may travel short distances . . . when he is about four years old, he continues to make long journeys riding on his mother's back until he is five or six.[57]

As one can imagine, such a reproduction system can be precarious. Not every baby makes it, no matter how hard a mother tries. According to Goodall, a major cause of death among infant chimpanzees is "injuries caused by falling from the mother."[58] It can't be helped. A mother has to be out and about foraging for food, she has to dodge predators, and she has to make sudden movements. Accidents are bound to happen—and the risk would have been even higher for our ancestors, whose babies were losing the ability to grasp with their feet. Add to that any food shortage or sickness, and a species can easily be endangered.

The way out of this reproductive dead end, Own Lovejoy believes, is enlisting the help of fathers so mothers can stay at home to take care of the young while fathers go out to forage for food for the whole family—the so-called male provisioning hypothesis. With this arrangement, mothers and children could conserve energy and stay out of harm's way, while fathers could safely pass their genes to one or more children.

How does this division of labor promote bipedalism? Well, at the home base, mothers had to *carry* their dependent children, and out there, fathers had to find food and *carry* it home. Either way, our ancestors engaged in extensive bipedal walking, which put selective pressure on bones, joints, and muscles to adapt.

For Lovejoy's theory to work, there is one prerequisite: our ancestral male had to be pretty sure that the children he was helping to raise were his own and carried his genes. Otherwise, he had no reason to help. And for him to be pretty sure, our ancestors had to practice what (many) modern humans do: monogamy. Sexual reproduction would happen between established pairs, not opportunistically among multiple partners (as with chimpanzees) or polygamously between one or more males and multiple females (as with baboons). With monogamy, a father would be reasonably sure of a child's paternity and would not worry about wanton males usurping his reproductive rights when he was out looking for food.

But how did monogamy happen in the first place? There were multiple factors, according to Lovejoy: the shrinking of forest, the disappearance of food source, the increased time and labor spent to search for food, the increased exposure to predators, and a reduced number of fertile females.[59] Under these circumstances, males had to safeguard their reproductive investment, rather than go and reproduce with multiple partners, if they wanted their children to survive.

Similarly, females would have wanted supportive, cooperative males to help bring up their young. The aggressive alpha males may fit the "hot stuff" stereotype, but our ancestors lived in a hostile environment teeming with far more aggressive predators: saber-toothed cats, giant hyenas, and large raptors, to name but a few.[60] Maintaining a close social group in which everyone looked out for each other was a key survival strategy. In such a group, friendly and amicable males were more attractive.[61]

And then, there was love—that is, the development of sexually stimulating signs that were specific to individuals and aroused brain signals of affection.[62] Rather than obvious generic signs, such as scent and swelling that announce their estrus cycles to everyone, females (and males too) started to develop distinct features—their hair, their body shapes, and their faces—that attracted specific individuals of the opposite sex. Over time, attraction gave rise to pair bonding, monogamy, nuclear families, food/infant carrying, and bipedalism.

In Lovejoy's words, "I didn't get my start by being a tool-using ape. . . . I was a socially and sexually innovative ape who became a biped and, as a result, managed to propagate my kind better than other apes. It was sheer luck that my ability to stand up and use my hands led to a later development of tools and a culture, and a still larger brain."[63]

In other words, the knees came before the tools and the brain, and sex came before it all.

Lovejoy's theory is not without criticism. Some critics question the feasibility of carrying low-calorie food—fruits, nuts, and roots—to feed a whole family.[64] Because a large amount of such food was needed to sustain a family, to carry it, our ancestral father would have needed some kind of container, a tool, which he supposedly did not yet know how to make. Then again, if our ancestral father could scavenge meat, he would be able to provide more sustenance.

Other critics question the whole idea of monogamy, for which we have no material evidence, like million-year-old marriage certificates. There *is* one clue to marriage systems, though, and that is sexual dimorphism: the male and female body differences other than sex organs. These differences evolved as a result of mating competition: males that were bigger, stronger, and better weaponized (e.g., endowed with larger canine teeth) could prevail over other males and monopolize female mates. So, generally speaking, the more intense the mating competition, the more pronounced the sexual dimorphism. Among gorillas, males fight to become the alpha and gain access to multiple females. Consequently, they show high levels of sexual dimorphism with males weighing twice as much as females.[65] By contrast, modern humans commonly practice monogamy with established pairs. Accordingly, we exhibit low sexual dimorphism with males weighing only 15 percent more than females.[66]

Lucy's people have little difference in canine teeth between the males and females, which supports a theory of monogamy.[67] However, whether or not the males have significantly larger bodies has been debated, depending on which fossil samples we compare and how. Different studies yielded sharply different verdicts: Lucy and her people either had high skeletal dimorphism, like gorillas, or low skeletal dimorphism, like us.[68] As for 4.4-million-year-old Ardi and her people, evidence is more conclusive that both canine and body size sexual dimorphism is small, hinting at a long line of monogamy in hominins.[69]

Then, there are people who dismiss Lovejoy's theory on the ground that it is, well, sexist. These critics say that Lovejoy, by portraying males foraging for food and females exchanging sex for said food, projected "contemporary gender roles, behaviour, expectations and anxieties" into our evolutionary past.[70]

As much as I share the belief that we need to watch out for sexism at every turn, I can't help feeling that in this case, the *charge of sexism* projects gender anxieties. Sexism is a modern construct, and using it to categorize ancient hominin behaviors is anachronistic. Hominin female behaviors several million years ago cannot be simplified as either "lying around in the caves" waiting for food or "[shuffling] out to get dinner."[71] And either way, it has *nothing* to do with the status of women today. Plus, I rather think that guarding fragile babies at home against dangers we can't begin to imagine today is quite an accomplishment for our ancestral mother.

All this is not to say that Lovejoy's definitely got it right. During the last century, some thirty theories have been proposed to explain the rise of bipedalism, each formulated by people not convinced by existing theories, with each new theory likewise doubted by others.[72] No one can definitely "prove" their story, as that requires anatomical, behavioral, and ecological evidence that we currently don't and may never have.[73]

One alternative theory, for example, is dubbed the watching-out hypothesis. It suggests that our ancestors became bipeds because they needed to stand up to survey their surroundings for predators or prey, which was especially useful when they came down from the trees and ventured into the open savannah with tall grass.[74] Watching-out behaviors have been reported in apes that live in captivity. Ambam, a Western lowland gorilla at the Port Lympne Wild Animal Park in England, supposedly walks upright "to get a height advantage to look over the wall when keepers come to feed him."[75]

The watching-out hypothesis reflects our old belief that bipedalism happened on the open savannah, but recent geological and fossil findings suggest that that's not true. Instead, bipedalism happened in a mixed environment: woodland extended along river courses, which were flanked by grassy plains.[76] Granted, even so, our ancestors could still keep watch, but the hypothesis faces one principal weakness: Ambam the gorilla needs to stand up only for a few seconds to spot an incoming keeper bearing food; similarly, our ancestors needed no more than a few seconds to survey their land. This occasional act is not likely to exert enough evolutionary pressure to develop the bones, joints, and muscles needed for habitual bipedalism.[77] Indeed, *after* spotting predators or prey, wouldn't a fast *four*-legged flight or pursuit be more important and evolutionary advantageous?

Another interesting hypothesis is the postural feeding hypothesis. This theory asserts that bipedal walking came from bipedal eating.[78] When trying to reach fruit on an overhead branch, chimpanzees would stand up, grab the branch for support, and then consume the food. This happens when the animals are on the ground looking for low-hanging fruit or in the trees trying to reach fruit on branches that are too small to climb. Some think that a similar feeding posture in our ancestors led to upright movement, because it would have been inefficient to change gears between two and four legs when they moved from devouring one fruit patch to another.

However, bipedal movement between food patches is rare for modern chimpanzees: during seven hundred hours of observing twenty-six chimpanzees, only four instances were recorded, and none of them occurred on the ground.[79] The fact that one has to support one's body by hanging onto branches also decreases the evolutionary pressure for true bipedalism.

Still more intriguing is the display hypothesis, which suggests that bipedalism started as a "threat display" our ancestors used to warn off others in their groups.[80] It is, in other words, a scare tactic, a bluff. By standing up and making themselves look taller and more imposing, our ancestors avoided actually fighting for food and mating partners, which reduced bodily injuries and gave them an evolutionary advantage. Similar displays can be seen in chimpanzees and gorillas when they draw themselves up, hunch their shoulders, and make mock charges.

This theory, however, is somewhat contradictory. According to its proponents, the threat display works by their "sheer novelty" and "surprising effect."[81] In other words, they are performed occasionally and for brief periods of time, which, once again, is not likely to induce habitual bipedalism. A more "disturbing fact," as one critic put it, is that we don't know how often a bluff may backfire and lead to actual fights.[82]

In addition to these theories, we also have the water-wading hypothesis, thermoregulation hypothesis, energy-efficiency hypothesis, sexual display hypothesis, and so on.[83] I could go on, but you get the idea: everyone's got a favorite theory, and every theory has its problems and unknowns. If we can agree on anything, it is that the rise of bipedalism, the advent of the bipedal knees, was a complex event and multiple factors likely acted in synergy to make it happen.

If this is so, then maybe Lovejoy has a point by trying to account for a wide range of factors in his theory—from environment to food, from birth rate to baby rearing, and from anatomy to sexual attraction. Maybe our ancestors really did, to put it bluntly, stand up so they could have more sex and raise more children.[84] If so, then the whole thing is pretty ironic for *Homo sapiens* because, in becoming bipeds to spread our species far and wide, we also doomed ourselves, and our offspring, to a pair of knees that is one of the weakest links in our body. Doomed, especially, as we will learn, are the female *Homo sapiens*.

2

Confused Anatomy

The mammalian knee is an alarmingly complex joint . . . [and]
suffers from a rather precarious design.
—OWEN LOVEJOY, AMERICAN PALEOANTHROPOLOGIST

You broke something alright, the ACL, I think.
—DR. BRADLEY, PRIMARY CARE PHYSICIAN

Our body consists of complex parts and elements. Many people will undoubtedly appreciate that our brain is vastly intricate. Modern science has a long way to go to grasp its workings and effectively treat neurological diseases. The human genome, as another commonly cited example, is not only complex but also enormous. We are just now scratching the surface of its mystery.

The knee, in contrast, appears much less intriguing. In fact, it seems downright boring compared with the enigmatic brain and genome. I mean, it is a joint. It may be the largest joint in the human body, but it is still just a joint, consisting of a few bony parts, which move, in a mechanical sort of way. If the human brain and genome are like the sexy twenty-first-century information technology and artificial intelligence, the knee is hopelessly stuck in an eighteenth-century industrial-era cotton mill.

Even the definition of the knee sounds industrial: a hinge joint, so called because it works, well, like a hinge, the kind installed in most doors. Door hinges connect a door to the doorframe, with the hinges secured in place by screws. Through the use of these hinges, a door can swing forward or backward to open or close.

In a similar fashion, the knee connects the thigh bone (or femur) to the shinbone (or tibia). The end of the thigh bone has two round knobs (the femoral condyles), while the top of the shinbone is relatively flat (called

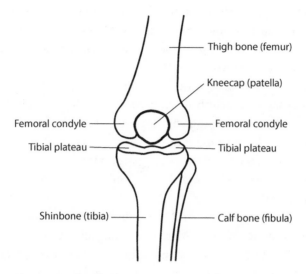

Thigh bone (femur)

Kneecap (patella)

Femoral condyle

Femoral condyle

Tibial plateau

Tibial plateau

Shinbone (tibia)

Calf bone (fibula)

FIGURE 2.1 Major bone structures in the knee

the tibial plateau) with two sections. Each of the condyles fits into each of the sections, creating the knee hinge (figure 2.1). Rather than screws, various ligaments, tendons, and connective tissues secure the hinge in place so we can swing our lower legs forward or backward to straighten or bend the knee. And this repeated movement is what allows us to walk and run.

So, there you have it: the knee. What more can one say about it?

Well, in fact, quite a bit. If the knee were this simple, it wouldn't be such a giant pain in the you-know-what.

Despite being defined a hinge joint, the human knee is so *not* a hinge. Most obviously, aside from the thigh bone and shinbone, it has a kneecap (i.e., the patella), the counterpart of which doesn't readily exist in a hinge. A flat, semi-triangular bone, the kneecap sits on top of the knee joint— like a cap. Or, more accurately, it sits atop the end of the thigh bone, in a groove (figure 2.1). Like a shield, the kneecap provides some protection to the knee, but more important, it acts like a fulcrum for the quad muscles to effectively pull on the knee and straighten it.[1]

Aside from this added component, our knees move in ways that a hinge wouldn't dream of. In addition to swinging forward and backward, the knee can rotate. Not a lot, but it does. Stand up straight, if you will, with your legs and feet facing forward. Now, turn your toes outward but continue to point your thighs forward. You will feel some tension at the knee, but you can do it—because your shinbone can rotate outward relative to your thigh bone. Similarly, turn your toes inward with legs straight, and your shinbone will rotate inward.

Your knee can also rock a little from side to side. Stand up straight again with your feet shoulder-width apart. Now, try to push your two knees apart from each other while keeping your legs straight and your feet stationary. You will feel a slight movement in the knee, as the gap between the thigh bone and shinbone opens up ever so slightly on the outer side of each knee. Similarly, push your knees toward each other, and the opening occurs on the inner side of the knee.

In fact, the human knee is said to exhibit six kinds of movements during dynamic activities. This multiangle mobility allows us to move not only in a straight line but also sideways; it allows us to not only walk but also turn, pivot, cut, to make all the dynamic movements that are needed in just about any sports—or, indeed, in everyday life.

If this description is making you feeling pretty pleased with your knees, don't. Dynamic freedom comes at a terrible cost. The knee is flexible because it is *fundamentally unstable*. It is, essentially, a few pieces of rigid, ill-fitting bones bound up by rope-like soft tissues. Nothing is fused in place, so the knee can move in just about any direction. For the same reason, it can also twist, misalign, overstretch, or simply fall apart.

The bigger ropes, or the major ligaments, are easily broken because they are the main stabilizers and take the brunt of stress. In technical terms, ligaments are stretchy, fibrous bands of tissues that connect bone to bone. The knee has four major ligaments: the anterior cruciate ligament (ACL), the posterior cruciate ligament (PCL), the medial collateral ligament (MCL), and the lateral collateral ligament (LCL) (figure 2.2). These names may seem to run together, but they make perfect sense once you know what the names mean.

The C (*cruciate*) in the ACL and PCL means cross-shaped, as in *crucifix*. These two ligaments are so named because they cross each other and form an X inside the knee as they connect the thigh bone to the shinbone. To imagine this X in three dimensions, do a "fingers-crossed" hand gesture. But rather than pointing your fingers toward the sky to appeal to a higher power, point them toward the ground. Now you have the X. Your middle finger would be the ACL, because *anterior* (A) means in the front; your index finger would be the PCL, because *posterior* (P) means in the back.

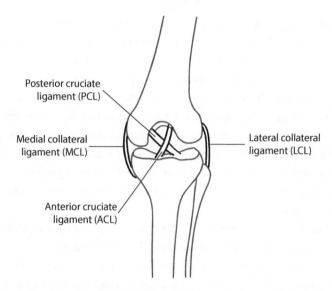

FIGURE 2.2 The four major knee ligaments: anterior and posterior cruciate ligaments and medial and lateral collateral ligaments

The ACL and PCL prevent the shinbone from swinging excessively forward or backward so that your lower leg doesn't suddenly slide out in front of or behind your thigh.[2] They also restrict the other two knee movements: rotations and side-to-side rocking.

Of the two, the ACL is more easily injured. In fact, it is the most commonly injured ligament in the knee, being sprained or torn about one to two hundred thousand times every year in the United States.[3] Some of these injuries occur as a result of external blows to the knee (such as during the game of football), but most (about 70 percent) are caused by an unfortunate movement, such as a sudden change in direction, rapid deceleration, jumping, or pivoting—movements that are common in sports such as skiing, soccer, and basketball.[4]

Given how prevalent ACL injuries are, doctors can have knee-jerk reactions to diagnosing them. When I took a tumble skiing down a Colorado mountain in the winter of 2014 and limped to my primary physician, a Dr. Bradley, he took a look at the wreckage, asked what happened, and suggested that I had probably broken my ACL. I suppose it was a good guess, but fortunately for me, he was wrong.

Compared with the ACL, the PCL is less frequently injured because it is almost twice as thick and about twice as strong.[5] The most common causes of PCL injuries are car accidents (45 percent), especially motorcycle accidents at 28 percent, and sports (40 percent), especially soccer at 25 percent.[6]

What about the other two major ligaments, the MCL and LCL? The C here, to keep us guessing, stands *not* for *cruciate* but for *collateral*, meaning parallel. These two ligaments are so named because they run parallel to the leg and to each other on each side of the knee, connecting the thigh bone to the lower leg.[7] The MCL sits on the inner side of your knee (*medial* means nearing the middle) and the LCL sits on the outer side (*lateral* means on the side).

Given their side placements, the MCL and LCL prevent excessive rotations and side-to-side rocking of the knee. For the same reason, they can be injured by excessive stress placed on the sides of the knee. A blow to the outside of the knee pushes the knee inward, thus stressing and risking the MCL. Similarly, blows delivered from inside the knee will endanger the LCL.

Because we don't usually go about daily activities or sports with legs wide open to allow inside blows, isolated LCL injuries are rare.[8] Not so for MCL injuries: in fact, 60 percent of skiing-related knee injuries involve the MCL.[9] When braking and stopping, skiers bend their knees, push their heels out, and turn the tips of the skis in, creating stress on the MCL. The same thing happens if they fall and the inside of the ski gets caught in the snow. My doctor probably didn't know this fun fact about skiing. At any event, he didn't propose that I broke my MCL and, thankfully once again, my MCL was also intact.

What, then, *did* I break? Because something definitely broke given the amount of swelling and pain. Magnetic resonance imaging (MRI) gave the verdict: the meniscus.

The word *meniscus* originally means a crescent moon, and this is where the knee meniscus got its name, for it is shaped like a crescent moon. This moon sits between the thigh bone and shinbone to absorb shock, reduce stress, and cushion otherwise rigid bones.[10] In fact, there are two moons, or two menisci, in each knee. One sits on the inner side of the knee and is thus called the medial meniscus; the other sits on the outer side and is called the lateral meniscus (figure 2.3).

Unlike the four ligaments mentioned earlier, the meniscus is not a ligament. It is cartilage, which is a flexible, rubber-like connective tissue that exists throughout our bodies. The outer ear, the part that is outside the

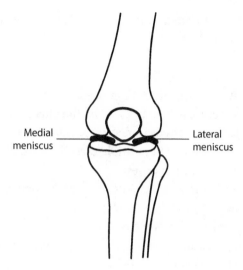

FIGURE 2.3 Medial and lateral menisci

head, for example, is made of a particularly flexible type of cartilage called elastic cartilage.

Then there is articular cartilage, which lines the ends of bones where they meet to form joints (*articular* means relating to joints). Inside the knee, the ends of the thigh bone and shinbone and the underside of the kneecap are lined with this type of cartilage. Articular cartilage is tough and smooth to help the bones glide over each other.

The kind of cartilage in the meniscus, known as fibrocartilage, is even tougher, making it the first line of defense against stress. Or, frequently enough, it fails to meet that stress: among physically active people, the rate of meniscus injury can be more than eight in one thousand; as a whole, the United States sees about one million meniscal surgeries every year.[11] Among older adults, injury is often a result of wear and tear. In younger populations, it is a result of putting a lot of downward force on the knee while rotating it—like what happens when we turn, cut, or pivot at speed.[12] I suppose that's exactly what I did: falling rapidly down the slope, while the ski, still attached to my foot, twisted—twisting my knee along with it and grinding up the moon inside.

Now, if you *have* to injure a meniscus, you would want it to be the medial one, because the lateral meniscus carries more load and provides more cushioning.[13] Injury on the lateral side is therefore more devastating, causing instability and deterioration.[14] Guess which meniscus I injured? The lateral one, naturally.

So, *now* we have it? The knee is not a hinge and not from a cotton mill. It is topped with a kneecap, is held in place by four major ligaments that are susceptible to injury, and is cushioned by menisci that are likewise easy to injure. The knee can move in multiple directions, so that we can jump, dive, and lunge to cause said injuries.

Well, I suppose this is what we know relatively well about the knee—hence, it is what we tend to talk and hear about. The problem is that there is quite a bit that we *don't* know well—things that doctors *don't* talk about and that patients *don't* ask. Frequently, it is these things that make the knee a continuous, frustrating source of pain.

Sadly, my torn meniscus is a case in point.

Menisci were once considered vestigial.[15] Like wisdom teeth, they were supposedly a useless remnant from days past, a structure that had lost its function through evolution. Surgical removal (i.e., meniscectomy) was considered to be a benign procedure. Indeed, surgeons were urged to remove injured menisci completely and not leave anything behind to cause complications.[16]

Then, in the late 1940s, studies of postsurgical patients led some to whisper that meniscectomy "is not wholly innocuous" and may cause long-term damage.[17] Still, as late as 1975, surgeons were removing menisci with gusto, regarding it as a simple procedure with fast results.[18] Today, we know without doubt that meniscectomy can do a number on our knees: it reduces the knee's ability to absorb shock by 20 percent and increases joint stress by up to 300 percent.[19] Without menisci, articular cartilage within the knee easily degenerates, causing pain, swelling, inflammation, and osteoarthritis.[20] Despite this understanding, however, the attitude that meniscectomy is a quick and simple fix hasn't entirely changed.[21]

When I tore my meniscus, I was told I needed a partial meniscectomy: removing parts of the meniscus that were torn. The meniscus, I was told, receives *no* blood supply, so it *doesn't* heal. The torn piece, being torn, provides no cushioning anyways. It just gets in the way of the movement of the joint. I was assured that meniscectomy is a common and minimally invasive procedure. Everything was to be done arthroscopically: two tiny incisions would be made in the knee, one for inserting a video camera and the other for inserting surgical instruments. The knee would be good as new afterward, and I could play whatever sports I wanted.

So, in I went for the surgery, and by all accounts, everything went fine. I was out for barely a couple of hours and was sent home the same day. The knee was heavily bandaged. I had some swelling but surprisingly little pain. After a few days, I was able to bear weight. After a week of physical therapy, I was back to walking. After two to three weeks, I had gained my full range of motion and was able to run and jump. Both the surgeon and physical therapist deemed me, for all intents and purposes, cured.

I too felt cured, compared with the condition I was in after falling off the face of the mountain. I proceeded to run, ski, do whatever my heart desired, just as the surgeon had predicted, forgetting that I even had any knee surgery.

Then, from about the fourth year on, on random days and for seemingly no apparent reasons, the knee would start feeling tight and stiff. Stairs became tricky to navigate, kneeling was awkward, and I could feel a dull ache if I sat for too long with my legs crossed.

After putting up with the discomfort for months, I finally broke down and went to the doctor, who, fearing additional meniscus tears, ordered another MRI. The MRI showed high-grade articulate cartilage loss in the lateral knee around where the meniscectomy was performed as well as a lot of inflammation. But the meniscus, or what remained of it, was fine, and nothing else was broken. The knee was, to the surgeon's credit, admirably fixed—except, as it dawned on me, it was not "good as new," and it never would be.

The doctor prescribed an anti-inflammatory drug, recommended leg-strengthening exercises to take stress off the knee, and sent me home. Resigned to the fact that I'm damaged goods, I gave up skiing and learned to favor my weak knee. I told myself that it was what it was—until, one day, when a friend happened to mention that a torn meniscus could be repaired. She had it done herself years ago. What the heck?!

Apparently, about 10 to 30 percent of the meniscus actually *have* decent blood supplies.[22] These areas around the rims of the meniscus are known as the red zone. If a tear happens in the red zone, it may be reparable. The torn piece is secured in place with sutures and other devices and is left to heal.

The inner portions of the meniscus, by contrast, lack blood supply and are known as the white zone. If a tear happens in the white zone, it is automatically removed, because, supposedly, it is unhealable and good as dead.

Presumably, my tear got into the white zone, although I was never told this one way or the other—and I didn't know to ask. I mean, who would?

And that's not all. Recent studies have shown that tears extending into the white zone may *also* be repaired and do heal. In one study,[23] twenty-nine cases of red-white zone repairs were performed and followed up some sixteen years later. Among them, 62 percent had normal or nearly normal functions and characteristics; an additional 17 percent had normal functions though exhibited some abnormalities on the X-ray or MRI.

Indeed, even injuries dead in the white zone aren't necessarily dead. At the Robert Jones and Agnes Hunt Orthopaedic Hospital in Shropshire, United Kingdom, 87 patients with white-zone tears received repair surgeries.[24] In the four-year follow up, the success rate was 68 percent—76 percent if counting patients who had additional repair surgeries.

Given these outcomes, it is speculated that aside from blood supplies at the edges of the meniscus, a cellular response happens in the white zone to break down glucose and generate energy for healing. Alternatively, the injury/surgery may cause a healthy inflammation response that spurs cell division and healing.[25] We can also resort to external measures, such as placing blood clots at the site of injury to promote healing, which had been attempted more than fifteen years before my meniscectomy.[26]

No doubt, all tears are not created equal, and depending on tear patterns, some are less reparable than others.[27] Also, repairs do fail, so one may have to endure repairs and eventually still get a meniscectomy. Plus, the rehabilitation of repair surgeries takes four to six months, which are considerably longer than that of a meniscectomy.

Still, if such repairs were even a possibility, I would have taken that chance to be "whole" again, to have better long-term outcomes than short-term relief. At the very least, I would have wanted my surgeon to give me the information before putting me under. No, I may not travel to the Robert Jones and Agnes Hunt Orthopaedic Hospital in Shropshire, but I would seek second opinions, because, clearly, when it comes to the knee, things are *not* clear-cut and experts *don't* agree.

Although the knee is more accessible than the brain or genome and can be taken completely apart for scrutiny, different dissection techniques can cause different interpretations of anatomy, which can in turn change our understandings of biomechanical functions.[28] Plus, we can't dissect live human knees just to satisfy our curiosities. We have to utilize cadavers, which are not the same thing. Embalmed knees aren't suitable for detailed, layer-by-layer dissection; even fresh frozen knees can have changed biomechanical properties.[29] Certainly, we can get a "live" view of the knee through imaging technologies, such as the X-ray or MRI, but the resolution of these images pales in comparison to the intricate—and often individual-specific—structures of the human knee.

Take, for example, the ACL, easily the most extensively studied structure in the human knee, if not the entire human body. Yet, we still puzzle over its finer anatomies and functions.

The ACL is, as mentioned earlier, a cruciate ligament, a fibrous band connecting the thigh bone and the shinbone inside the knee. The exact nature of this "band," however, is unclear. Some assert that the ACL is just one single bundle of fiber; others believe it consists of two separate bundles that work independently during knee movement.[30] Still others believe that there are three or, indeed, anywhere between six and ten bundles.[31] This debate is not only theoretical but also has important clinical implications. If we don't know the ligament's exact fiber networks and their functions, we can't hope to restore an injured ACL to its full natural function.

In fact, it is well known that even with an ACL reconstruction, up to 30 percent of patients continue to have a slipping knee, or the so-called pivot shift instability[32]—that is, when you bend and rotate your knee to pivot your body, your lower leg will shift out of the knee joint. This phenomenon was first reported in 1914, variously described as a sudden displacement of the knee or the knee giving way or crumpling up.[33] In clinical examination, a pivot shift test is one of the most accurate and specific ways to detect ACL injuries.[34]

But, curiously enough, in some knees, injured ACLs cause no slipping, not even when the entire ACL has been cut off.[35] Still more curiously, in other knees, even when the ACL is intact, the knee nonetheless slips.[36] Apparently, although the ACL provides a major restraining force to pivot shift, other structures in the knee also pull their weight.[37] The specifics of their contributions, however, are far from being clear or agreed upon.

One such structure that has drawn considerable attention—and controversy—in recent years is the anterolateral ligament (ALL). The ALL is so named because it sits on the front (*anterior*), outer (*lateral*) side of the knee, connecting the thigh bone, the lateral meniscus, and the shinbone.

The ALL was discovered a century and half ago by French surgeon Paul Segond. In 1879, Segond reported finding "a pearly, resistant, fibrous band" at the front, outer side of the knee.[38] The band "showed extreme amounts of tension" when the knee is rotated.[39] Oddly, however, in the many years since Segond's report, medical literature hardly mentioned such a band. It's like it never existed.

It was only in the past decade or so that the structure was rediscovered—well, sort of. Many agree that there is *something* in that vicinity of the knee, but what it *is* is disputed. Some think it is the mere thickening of local tissues.[40] Others are sure it is a bona fide ligament and have even measured it: thirty-four to forty-two millimeters long, five to six millimeters wide, and one to three millimeters thick.[41] And both sides have visual evidence: some dissection photos reveal an unmistakable, pearly band, while other, equally vivid photographs show nothing but flat tissues.[42]

True, no two people are identical, and we may all be differently endowed when it comes to the ALL. But, if that is all there was to the debate, then you would expect the existence of the ALL or the lack thereof to be somewhat consistent across studies. But that's not so. In one study of twenty fresh-frozen cadaver knees, not a single ALL (0 percent) was found;[43] in another study of ten fresh cadaver knees, every single one (100 percent) had an ALL.[44] In one study of forty-four embalmed cadaver knees, twenty specimens (46 percent) had an ALL;[45] in another study of forty-one embalmed knees, all but one (97 percent) had an ALL.[46] This kind of variation doesn't seem like natural differences among individuals; it seems like the result of different dissections and interpretations.

To make matters worse, even people who agree on the existence of an ALL do not agree on its function. Some researchers, through cadaver studies, concluded that the ALL had a minor stabilizing role, too minor to be worth our trouble reconstructing it.[47] Others, also studying cadavers, concluded that the ALL is an indispensable assistant to the ACL. When the ALL was cut, the knee exhibited more pivot shift, above and beyond what was produced by a severed ACL.[48] Accordingly, reconstructing the ACL alone failed to restore knee stability in cadavers if the ALLs were injured; it was only by reconstructing both ligaments that normal knee function could be restored.[49]

In a rare, "live" human study,[50] ninety-two patients were given precisely that—a combined ALL and ACL reconstruction—and were followed for two years. Before the surgery, forty-one of the patients had grade 1 (slight) pivot shift, twenty-three had grade 2 (moderate) pivot shift, and nineteen had grade 3 (severe) pivot shift. Postsurgery, only seven patients showed slight pivot shift, and the rest all tested normal. Despite these seemingly fantastic results, ALL reconstruction is far from having reached a consensus. In fact, even if we all agree to reconstruct, there is no consensus on *how*, because exactly where the ALL originates and how it is positioned on the outer side of the knee are still debated.[51]

Think that's pretty wild? Well, that's not the worst of it.

If there is one region of the knee experts are willing to admit that they don't know enough about, it is the back, outer side of the knee—that is, the posterolateral knee. This side of the knee is anatomically complex and tricky to study, earning itself the nickname "the dark side of the knee."

At least three tendons and nine ligaments crowd the dark side of the knee.[52] These structures lie below, above, in front of, behind, or across from each other in close proximity. Identification of their anatomic details is difficult, and unrecognized injuries are a common cause of failed ACL and PCL reconstruction and chronic knee instability.[53]

Part of the confusion is caused by variations in the shapes or positions of these tendons and ligaments.[54] Take, for example, the *popliteus tendon*. Unlike ligaments, which connect bone to bone, tendons connect muscles to bone. The popliteus tendon connects the popliteus muscle (which arises from the shinbone) to the thigh bone, running diagonally across the back of the knee. In some lucky knees (about 18 percent), the popliteus tendon has a strong attachment to, and thus seems to protect, the lateral meniscus; in other knees, there is little to no attachment.[55] With a shattered lateral meniscus, I doubt I was one of the lucky ones.

More confusing than these variations is the complete absence of some structures. Take, for example, the *popliteofibular ligament*, which, as its name suggests, connects the aforementioned popliteus tendon to the calf bone (the fibular). Despite being a stout ligament and a major stabilizer in the dark side of the knee, it is missing in 2 to 6 percent of the population.[56] Even flakier is the *arcuate ligament*, a Y-shaped ligament connecting the calf bone to the knee. Depending on which study you look at, it is missing in 20 to 76 percent of the population.[57]

Then, there is the *iliotibial band*, which always exists and not infrequently as a source of nagging pain for many—myself included. Pardon me as I take a deep dive into this bane of my existence.

The iliotibial (IT) band is a thick band of connective tissue (known as *fascia*) that runs down the outer side of the thigh, from just above the hip to just below the knee. It is the top layer in the dark side of the knee, lying above the assorted tendons and ligaments. The iliotibial band is so named because it connects the *ilium* (which is the broad, flaring bone that forms the upper part of your hip) to the *tibia* (shinbone). Often, it is simply called the IT band, an unfortunate acronym in this day and age of information technology.

The upper end of the IT band is connected to two muscles. One is the *gluteus maximus*, which, as the name suggests, is the largest (*maximus*) buttock (*gluteus*) muscle. If you put your hand on your tuchus, that's mostly what you are feeling. The other muscle is on the up, outer side of your thigh. It has quite a grandiose-sounding name (try saying it out loud): *tensor fascia lata*. The name literally says what the muscle does: to stretch or tense (*tensor*) the broad (*lata*) fascia, this fascia being the IT band.

The lower end of the IT band inserts into the knee area. Multiple insertion points have been reported, including the kneecap, the top of the shinbone, and several places on the thigh bone.[58]

Humans are the only animals who have an IT band. Some believe that the band, by stabilizing the hip, is part of the anatomic evolution that allows us to temporarily balance on one foot and thus move bipedally (see chapter 1).[59] Alternatively or additionally, the IT band has evolved to make us more efficient bipedal walkers and runners.[60] When you swing a leg backward, the IT band is stretched. Like a stretched rubber band, it stores energy. When you then swing the leg forward, the recoil of the band releases that energy to make for a more efficient swing. This energy-saving mechanism could have assisted our ancestors to outpace predators and preys millions of years ago.

Then, millions of years later, you and I are paying the evolutionary price, through this excruciating pain called the IT band friction syndrome, a name coined in 1975 by Lieutenant Commander James Renne.[61] At the time, Dr. Renne was serving on the medical corps of the U.S. Naval Reserve and observed a "painful, disabling condition in the region lateral to the knee" among a group of Marine Corps second lieutenants who were attending vigorous physical training.[62] Almost all of Dr. Renne's patients started to experience symptoms after running more than two miles or hiking more than ten miles.

Not finding descriptions of such a condition in the medical literature, Dr. Renne decided to call it the IT band friction syndrome, because he believed the condition is caused by painful friction between the IT band

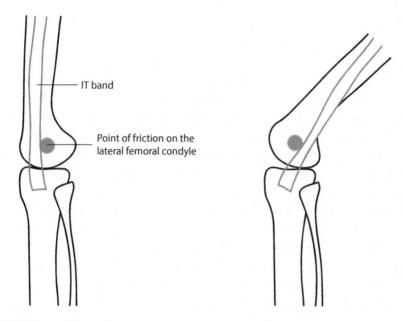

FIGURE 2.4 IT band friction

and the knee. More specifically, the friction is between the IT band and the lateral femoral condyle, or the condyle on the outer side of the thigh bone.

The process goes like this (depicted in figure 2.4): when your knee is fully straightened, the IT band sits in front of the lateral femoral condyle; when you start to bend the knee, the IT band starts to move backward, meet the condyle, and pass it as the knee continues to bend. When the knee is bent at about thirty degrees—or right when your foot makes contact with the ground during a walk or run—that's the moment when the two structures meet.

Normally, the meeting is uneventful: a smooth hello, glide, and goodbye. It's something we don't know has happened at every step we take. But if we overdid it—ran one too many laps or hiked one too many miles—the IT band can get irritated and start to rub against the condyle. Over time, this will cause inflammation and pain.[63] Walking becomes difficult. Running is even worse. When we run, the muscles at the top of the IT band will tug at the band to decelerate each foot as it hits the ground, which exerts greater tension and causes more friction.[64]

As you can tell from this description, IT band syndrome is not a traumatic injury caused by a sudden fall or blow. Rather, it is a so-called overuse injury, caused by repetitive straightening and bending of the knee. This lack of trauma makes the pain especially lame, as one doesn't get to blame a particular incident. The overuse label, in my opinion, is also annoying. It suggests, not very subtly, that one is perhaps a tad obsessive.

Upon discovering the syndrome, Dr. Renne predicted in 1975 that IT band syndrome was not unique to military personnel and "may be recognized in the future in conjunction with other kinds of athletic pursuits."[65] Sure enough, since the 1980s and with the rising popularity of running and other endurance sports, the syndrome became commonly diagnosed.[66] In the United States, an estimated thirty to thirty-four million people participate in some form of running, and among them, as many as 70 percent will sustain some injury.[67] The IT band friction syndrome is the second most common injury among these running injuries.[68] Besides runners, this syndrome also plagues cyclists, football players, downhill skiers, and weightlifters, anyone who repeatedly bends and straightens their knees.[69]

In my case, the syndrome got me during agility, a dog sport in which a human handler directs a canine partner through an obstacle course in a race for time and accuracy. The Westminster Dog Show, anyone? Even if you've never seen it, think about how fast quadrupedal dogs run and imagine running with them while attempting turns, cuts, and twists to avoid off-course obstacles—not a small physical feat. Yes, I did stop skiing because of the damaged meniscus, but some other things just can't be given up.

And so it is that IT band pain has been with me for more than two years now. It is fine on some days, bad on others, and downright terrible when I happen to twist my leg *just* so while bending it around thirty degrees. The pain is a shooting, disabling kind, the kind that makes limping down a snow-capped mountain with a broken meniscus pale in comparison.

If it seems crazy that I would sit on something this bad for this long, I agree. Well-intentioned friends have asked, "Are you *still* in pain? Did you have that looked at?" Did I ever! I have seen two sports physicians, two physical therapists, and one chiropractor and have been prescribed rest, pain medication, anti-inflammatory medication, chiropractic adjustments, and strengthening exercises.

The adjustments and exercises didn't much help—beyond making me feel better that I was doing *something*. Resting does help, but the pain would stubbornly come back, and, let's face it, one can't avoid bending those knees unless one stays in bed twenty-four/seven. Pain killers and anti-inflammatory drugs certainly help reduce pain and, as studies show, allow patients to run farther and longer.[70] But one can't keep taking those meds, which have a decent amount of side effects such as stomach bleeding.

Frustrated, I hit the medical literature to see what other remedies I could dig up. I didn't find any panacea, but at least I gained a better perspective of why I am still in pain. You would think that for a common running injury, one that has been around for forty-five years, the medical community would have a solid grasp of it. Not so. In fact, forty-five years later, we are still unsure of the exact cause of IT band pain.

As mentioned earlier, conventional wisdom blames the pain on the rubbing or friction between the IT band and the lateral femoral condyle. But lately, a new theory has arisen, suggesting that the problem is not friction; that nothing is rubbing anything; that, in fact, the IT band does not even move around the lateral femoral condyle.[71] Rather than dangling around the knee, the band encloses the entire thigh as a stocking. This stocking will press against the lateral femoral condyle when the knee is bent at about thirty degrees. It moves away when the knee is straightened. Sandwiched between the IT band stocking and the condyle is a layer of fat tissue rich in blood vessels and nerve fibers. Repeated compression causes this layer to inflame, and that's where pain comes from. In short, there is no "friction syndrome"; it's more like "compression syndrome."

As a newer discovery, if nothing else, the compression theory would seem more believable. But, I don't know. My pain doesn't *feel* like compression, like something being pushed against my knee. It is either a grinding ache at every step or stabbing pain at a wrong turn. In fact, it feels precisely like what the old theory proposes: rubbing, snapping, or scraping.

Of course, my personal feeling has no bearing on medical ruling. I am glad to hear, then, that imaging studies support how I feel. Just like we can use ultrasound to examine fetuses, we can use the same technology to track real-time knee movement. In ultrasound videos, the IT band does appear to move back and forth over the lateral femoral condyle.[72] When the knee is straightened, the rear edge of the IT band is located in front of the condyle. When the knee is bent, this edge will meet and pass over the condyle. This can explain why in most IT band patients, pain resides along the rear edge of the band.[73] At the very least, the authors of the ultrasound study conclude, the IT band may both rub *and* compress against the knee.

But—and this is a big but—either way, we don't really know how to fix the resulting pain. And it is not for lack of trying. In addition to what my doctors prescribed me, the medical literature offers a range of other possibilities.

Some suggest shoe inserts and better training regimens because, supposedly, anatomic features we were born with—anything from an imperfect foot to legs of different lengths—can predispose us to IT band syndrome. Poor training habits, such as inappropriate footwear or running on sloped roads, can also do us in. However, evidence for any such beliefs remains limited and inconsistent.[74] For example, in one study of 126 runners, both those with healthy IT bands and those with IT band syndrome had uneven legs.[75] I, per my chiropractor, also have uneven legs. One leg, apparently, is a whopping one inch shorter than the other. Curiously, it is my longer leg, not the shorter one predicted by medical literature,[76] that has IT band pain.

Another commonly mentioned remedy is to strengthen hip muscles, more precisely, hip abductor muscles. Hip abduction is the side movement of legs away from the midline of the body, like what we do when we step out of a car, or into a car. While several muscles help to abduct the leg, a particularly important one is *gluteus medius* (the middle buttock muscle). This is a fan-shaped muscle that sits right around where the back pockets of your jeans would be, originating from the top of the hip and extending to the top of the thigh bone.

You may think that side movement has nothing to do with running—unless you run sideways. But actually, running forward also engages abductor muscles. These muscles keep your body balanced so you don't fall to the side when balancing on a single leg in between steps. Indeed, it is estimated that 19 percent of total running energy is devoted to maintaining side balance.[77] If your hip abductor is weak, your thigh will move toward the midline of your body, collapse in, so to speak, and stress your knee. This is thought to put tension on the IT band and cause friction (or compression), inflammation, and pain.[78]

Looking back, my doctors obviously believed in this theory, because I was prescribed various exercises to strengthen my gluteus medius. A typical one is the side-lying leg raise. You know, you lie on your side, keep the top leg straight, slowly raise it up and lower it down, and repeat. Another exercise is the clamshell: you lie on your side with knees bent and legs stacked. Keeping your two feet together, you repeatedly lift and drop the top leg, as if a clam opening and closing its shell.

These exercises are undoubtedly effective at strengthening hip abductors, but does that help with the IT band pain? In a six-week rehabilitation program that focused on hip abductor exercises, runners with IT band syndrome increased hip abductor strength by 35 to 51 percent and, impressively, twenty-two of the twenty-four runners were able to go back to running pain free.[79] Unfortunately, in a similar six-week program, only four of sixteen patients came out pain free, despite having strengthened their hip abductors.[80]

Likewise inconsistent are studies that try to link hip abductor muscle strength to IT band pain. In one study of fifty-four distant runners, those with IT band syndrome had weak hip abductors on their injured side— about 20 percent weaker than their good side and 23 to 29 percent weaker than uninjured runners.[81] But, in other studies, no difference in hip abductor strength was found between healthy runners and injured runners, nor were there differences between the good and bad sides of injured runners.[82]

I have to admit, reading these conflicting findings make me feel more justified in neglecting my clamshell exercises.

Right up there with hip abductor strengthening is the advice to stretch the IT band. People who believe in the friction theory enthusiastically advocate stretching, which is thought to lengthen the IT band, and once lengthened, the band will no longer rub against the knee. But people who believe in the compression theory pooh-pooh the idea. The IT band, they say, is dense and tough and does not easily elongate. To achieve a permanent, significant elongation, one must apply extreme force, which is unlikely to be obtained through stretching. Sure enough, in a study of professional rugby players, stretching extended their IT bands by a measly 0.2 percent, according to ultrasound measurement.[83]

That said, in two other ultrasound studies with a total of eighty healthy young adults, stretching significantly reduced the *width* of their IT bands, which presumably meant that the bands were elongated.[84] To prove this, researchers photographed several distance runners doing IT band stretches. Reflective markers were fixed at multiple points along their IT bands. Based on the changes in these markers, stretching lengthened their IT bands by 10 to 11 percent.[85]

From 0.2 percent to 11 percent, what gives? Muscles, possibly.

Recall that the top of the IT band joins the tensor fascia lata thigh muscle. When the whole tensor fascia lata and IT band structure was peeled off cadavers and mechanically stretched, significant lengthening (4.5 percent) happened in the upper portion of the structure where the muscle lies.[86] Much less lengthening happened in the middle (1.4 percent) and lower (1.7 percent) portions, where it is just the band. This finding would suggest that when we stretch, what we are stretching and lengthening is not the IT band but rather the muscles.

This idea fueled another popular IT band treatment method: foam rolling. Through YouTube videos and exercise blogs, we are advised to roll our buttock and thigh muscles on cylindrical, high-density foams to help loosen up the IT band. If you've ever heard of, and wondered about, the very lofty sounding "self-myofascial release" treatment, that's what it is.

Myofascial refers to muscles (*myo*) as well as fascia that, like the IT band, surrounds muscles. Supposedly, knotty adhesions and tender spots known as trigger points can form in these tissues, creating pain and dysfunction. Through physical pressure, we hope to soften the tissues and release the knots. *Self-release* refers to the fact that we self-apply pressure by rolling on a foam roller or tennis ball rather than enlisting the help of a massage therapist.

Given the popularity of the foam rolling remedy on the internet, I expected to find lots of relevant medical literature on the topic. Not really. Plenty of evidence suggests that rolling leg muscles increases knee range of motion,[87] meaning you would now be able to bend your knee to a more impressive degree, but that has little to do with relieving IT band pain. One study that specifically tried foam rolling to treat IT band syndrome found the method ineffective at softening the IT band.[88] We are also unsure if foam rolling can create enough physical force to reshape or release connective tissues.[89]

Plus, wouldn't rolling, which presses the IT band into the muscles, actually create more adhesions? I'm glad I'm not the only one who wondered about this.

In 2015, two New York–based inventors filed with the U.S. patent office for a medical device specifically designed to treat IT band syndrome by breaking up adhesions and trigger points. As the inventors explained, current treatments "do not lift and separate the iliotibial band from its surrounding myofascial structures necessary for treating ITBS [IT band syndrome]."[90] The only way to lift and separate the band is by "utilizing negative pressure and suction."[91]

To do so, the inventors proposed a device pictured in figure 2.5, which consists of a round, bowl-shaped suction mechanism, with a handle attached on top. A patient is to move the device along the IT band and repeatedly press and pull the handle to create suction and thus lift the IT band.

I know I shouldn't laugh at creativity, but I couldn't help it. The proposed invention looks exactly like a toilet plunger—with a shorter handle,

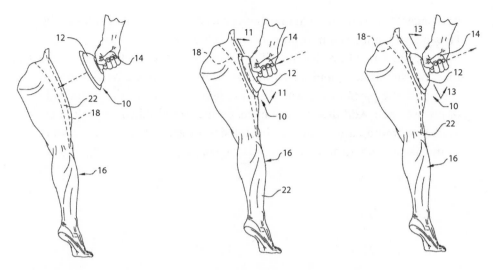

FIGURE 2.5 A proposed suction device for treating IT band syndrome
Source: Melissa Nicol Conte and Peter Kessler, "Method and device for therapeutic treatment of iliotibial band syndrome, myofascial and musculoskeletal dysfunctions," US Patent 20,150,313,788, filed November 5, 2015, http://appft.uspto.gov/netacgi /nph-Parser?Sect1=PTO1&Sect2=HITOFF&p=1&u=/netahtml/PTO/srchnum.html&r =1&f=G&l=50&d=PG01&s1=20150313788

admittedly. Even as I laughed, a small, very small, part of me wanted to go into the bathroom and try the plunger on myself.

I'm glad I didn't, because the status of the patent application, as I next saw, is "abandoned." I laughed some more at this (which again I shouldn't) and wondered what went wrong. The inability to create a sufficient seal between the device and the leg? The inability to generate enough suction to lift the tough IT band?

At this point, some of you may be wondering: "What about surgeries?" If I'm zealous enough to try toilet plungers, why not talk to a surgeon? Well, there is zeal, and then there is desperation. Current IT band surgery options are *pretty* desperate.

Surgeries can be performed to "release" the IT band, which essentially means cutting a hole in the IT band where it crosses the lateral femoral condyle so the two won't rub (or compress).[92] Surgeries can also "lengthen" the IT band, which is achieved by punching multiple small holes in the band so it, like mesh, stretches out.[93] Or, surgeons can cut the band diagonally to make flaps and then restitch the flaps so the band becomes thinner but longer.[94] As you can imagine, these procedures necessarily weaken the IT band, which may compromise knee stability.[95]

Also, did I mention that my IT band syndrome is on the same knee that had the broken meniscus, and that went on to have a partial meniscectomy? The partial meniscus that was removed was on the lateral side of my left knee. And the IT band hurts on the lateral side of my left knee. Somehow, more surgery on the lateral side of my left knee just doesn't sound like a good idea, thank you very much.

3

Bare Knees, Dicey Power

I can show my shoulders,
I can show my knees;
I'm a free-born American
And can show what I please.

—UNNAMED FLAPPERS

On August 23, 1923, in the little town of Somerset, Pennsylvania, the parent–teachers association gathered for a solemn task: to petition the school board to adopt old-fashioned school uniforms and bar "short skirts, bobbed hair, and low-necked, sleeveless dresses."[1] Catching wind of the meeting, the town's flappers came uninvited, delivered the snappy rhyme, and stormed out leaving the parents and teachers seething with indignance.

The Associated Press picked up the story, spreading it to newspapers across the country. In Wisconsin Rapids, Wisconsin, the *Daily Tribune* concluded that flappers–Puritans had declared war.[2] In Wilmington, Delaware, the *Morning News* announced that flappers had risen to defend their "near nudity."[3]

In 1920s America, it didn't take much for women to be accused of practicing nudity.

On some level, I guess, the accusation makes sense. Only thirty years earlier, skirts had been floor-length, barely showcasing women's shoes.[4] Indeed, from ancient Greece and Rome onward, the concept of "short skirts" was largely unknown to Western civilization.[5] Skirts, dresses, and gowns were meant to cover up the legs, to pretend that women didn't have ankles, much less knees. Even the *shapes* of the legs were too much to show, so skirts were made full, round, and "distant from the natural shape of the body."[6]

Interesting enough, historically, bosoms were less of a taboo. Evening gowns, including some from the prudish Victorian period, can be quite generous in lowering the neckline. Supposedly, this is because breasts are associated with breast *feeding*, a legitimate, laudable maternal activity (especially back in the day when there were no great alternatives). So, breasts can be worn as a badge of fertility. Legs, on the other hand, imply that which is between the legs—that which women, heaven forbid, may *enjoy* doing apart from their duty to reproduce.

Male sexuality, in contrast, was heartily celebrated, always. Skin-tight hose, breeches, and trousers accentuated the male lower body from the Middle Ages until the early nineteenth century.[7] When the wearer's doublet or other upper garment wasn't long enough to cover the crotch, nothing was left to the imagination.

Given this fashion history, the flapper's attire was nothing short of scandalous to her contemporaries. Head to toe, thirteen qualifications make the flapper (figure 3.1), three of which specifically concern her knee: a knee-length fringed skirt, exposed bare knees, and rolled hose to accentuate said bare knees. After centuries in hiding, the female knee had debuted—and it didn't fail to shock the onlookers.

Joining the parents and teachers in denouncing the flappers were preachers, elected officials, employers, "older fogies"[8] in general. In Newark, New Jersey, patrons complained of a particular female employee at the Fidelity Trust Company bank: "Everything about the young woman had a modern and—if the truth must be known—a provocative slant. Her hair was bobbed, her hidden ears were hung with jade earrings, her low-cut waist allowed certain exciting revelations, and suggested even more. And as she walked toward the back of her cage a pair of low-cut, flat-heeled sport shoes with champagne-colored legs springing out of them, came into view."[9]

Bank officials at Fidelity Trust agreed that short skirts and, consequently, exposed "dimpled knees" reduced business efficiency. They promptly issued a dress code, forbidding, among other items, dresses higher than twelve inches above the ground. Similar measures were made by businesses nationwide, from telephone companies in Detroit to department stores in New York.

Back in Newark, the Board of Education got involved. When board member Joseph Hauber visited a local high school, he witnessed some bare knees.[10] Believing that the "immodest dressing" was corrupting the girls,

TYPICAL FLAPPERS. You've often heard 'em called that, but did you ever really understand what it meant? This will straighten you out. It's a picture of a flapper, 100 per cent, from head to foot. Thirteen qualifications. Count 'em: No. 1, hat of soft silk or felt; No. 2, bobbed hair; No. 3, flapper curl on forehead; No. 4, flapper collar; No. 5, flapper earrings; No. 6, slip-over sweater; No. 7, flapper beads; No. 8, metallic belt; No. 9, bracelet of strung jet; No. 10, knee-length fringed skirt; No. 11, exposed, bare knees; No. 12, rolled hose with fancy garter; No. 13, flat-heeled, little-girl sandals.

FIGURE 3.1 Typical flappers: 1, hat of soft silk or felt; 2, bobbed hair; 3, Flapper curl on forehead; 4, flapper collar; 5, flapper earrings; 6, slip-over sweater; 7, flapper beads; 8, metallic belt; 9, bracelet of strung jet; 10, knee-length fringed skirt; 11, exposed bare knees; 12, rolled hose with fancy garter; 13, flat-heeled, little-girl sandals.
Source: "Typical Flappers," *Weekly Journal-Miner* (Prescott, Arizona), August 2, 1922, 1.

Hauber lodged a complaint. Alarmingly, when he came back a few days later, he was greeted by *more* bare knees. Hauber chastised the dean of the school and promptly launched a knee-cover-up crusade in high school.

Religious leaders went one step further, calling the knee-exposing flappers a menace, a symptom of anarchy. "A young woman has a perfect right to dress attractively," acknowledged Rev. Charles S. Stevens from Chicago, "but she has no right to destroy the sense of modesty of her own heart or in the minds of others. The present vogue in dress of the flapper is an appeal to the sex instinct. The flappers are simply ranging through the jungle after the beast there is in men."[11]

There you have it: the 1920s version of the contemporary myth that women "ask for" sexual advances when they dress a certain way. Seventy years later, a male juror of a rape case would put the morale in more lucid language: "She asked for it. The way she was dressed with that skirt you could see everything she had. She was advertising for sex."[12] The sexual urges in men are God-given and undeniable. Women, therefore, must take care to cover themselves up, or risk becoming the objects of male fantasies, even victims of sexual violence.[13]

No crime data, at least none that I could find, solidify this claim. In fact, evidence exists to the contrary. The What Were You Wearing exhibit shows that women were wearing anything from jeans to T-shirts to pajamas when they were assaulted. Certainly, *even* if women choose to wear something revealing, that does not equal sexual consent.[14] Unlike language, clothes have a diffuse meaning: a woman cannot control *who* will receive the message from her clothes. The meaning of clothes is also imprecise: what a woman intends to project and what an observer receives can differ by a long shot, as can the messages received by different observers.

In the case of the flappers, not every observer was bothered by the knees. "Look at the way women dress as seen on Boston's streets," said Dr. Charles Eliot, president-emeritus of Harvard University. "These young women have no desire to be indecent. They wear what they do because it is fashionable."[15] Rev. John F. Von Herrlich of Holy Rood Church concurred: "I think that the subject of dress should be left to the judgment of the fair sex themselves. I have never been disturbed by their attire and see nothing that alarms me in the matter under discussion. . . . I think that the old proverb is appropriate: 'Evil to him who evil thinks.'"[16] Or, to put it into more succinct words: "if one didn't look for bare knees one wouldn't see them."[17]

The president of the national California women's club, a Mrs. Vivian, had the sagest remark: "When men chose to wear their trouser legs prodigiously tight . . . I cannot recollect that women made them a subject for public ridicule or condemnation. . . . [W]omen undoubtedly felt that men wanted a change and were working out their own salvation in their own mysterious way. This is the case with women now—they want a change."[18]

Despite these supporting voices, in several states, bills were introduced to regulate women's skirt hemline and to hide their knees, and more.[19] In Utah, a statute was put before the legislature to ban skirts higher than three inches above the ankle. Applauding the sentiment but deeming the length immodest still, Virginia sought to ban skirts more than four inches above the *ground*. Even this could not satisfy Ohio, who proposed that women over fourteen years of age must wear skirts that completely cover their ankles.

These bills never did pass, perhaps because of their general ridiculous-ness, or because of the utter confusion they would cause when women attempted to travel across state lines. After all, a modestly dressed Utah woman could be charged of practicing near nudity in Ohio.

What *did* pass were business dress codes. No matter that the dress code at Fidelity Trust caused much indignation from female employees—who complained that it wasn't fair for them to have to buy new office clothes, and it wasn't convenient to run home and change out of the dowdy thing if they wanted to go out after work—when the day came, all women reported to work in the demure attire.[20]

It's understandable. Although flappers may have defied parents and teach-ers and turned a deaf ear to preachers, employers held their purse strings—and flapperdom wasn't cheap. The skirts, the stockings, and the bling all cost money. According to the—granted, very liberal—estimate from the New York League of Girls' Clubs, to properly clothe a flapper cost $503,[21] which is $8,000 in today's money. Even with economizing and home sewing, women had to spend a third to a half of their income on clothing if they wanted to look fashionable.[22]

With their fashion outrage, flappers were often held up as extraordinary rebels. Only thirty years before their appearance, America's archetype femininity was a girl with long hair, a thirty-six-inch bust, a precise waist, broad hips, and well-concealed legs—the Gibson girl.[23] Then 1920s roared and in came the brave, liberated flappers with bobbed hair, short skirts, and a boyish spark, flaunting their knees, legs, and freedom.

This picture is, alas, more Hollywood (and Fitzgerald) than history. In reality, seventy years of fashion rebels had preceded—and one might say, enabled—the flappers and their exposed knees.

Although skirts have always been devised to hide the female lower body, the Industrial Revolution and its attendant technological innova-tions pushed things to the extreme. Power-driven looms churned out great quantities of fabric, which allowed skirts to swell with layers of petticoats and ruffles.[24] The sewing machine then added complicated stitching and intricate trimming, tons of it.[25] By the end of the Industrial Revolution, women were more completely covered up than ever before.[26]

If lavish petticoats and decorations made for an attractive dress, they also made for a heavy one. Weighing in the neighborhood of forty pounds,[27] they effectively forced women into a sedentary lifestyle, moving as little as possible, avoiding stairs at all cost, and carrying nothing of substance. In hot summer days, layers of fabric became heat traps, creating additional discomfort to deter movement.[28] These long skirts also helped women to pick up and carry home a healthy dose of slime and germ from the floor.

Topping things off was the corset, worn to achieve a nonexistent feminine waist. A tight-laced corset can reduce a woman's waist by two to six inches—and apply twenty to eighty pounds of crushing weight on her body in the process.[29] The result is a shortage of blood supply, displaced internal organs, and ailments ranging from indigestion to miscarriage.[30]

To be fair, innovation *was* sought after to make women more comfortable in their sartorial shackles. The cage crinoline, introduced in the mid-1850s, "used concentric hoops of lightweight spring steel connected with fabric tapes . . . onto which a single petticoat and a skirt would rest."[31] This device eliminated heavy layers of petticoats and reduced the amount of fabric sweeping the ground; the steel was flexible and could be bent slightly to facilitate movement.[32]

But the cage crinoline introduced new problems.[33] Now that size was no longer restricted by the amount of fabric that could be physically hung on a hip, skirts reached a prodigious volume. Maneuvering such an enormous thing was challenging both for the wearers and those around them. The threats of entanglement and tripping were constant. And it was certain death for the wearer if the large, airy skirt swept into open flames or got caught in machinery—a real danger in the crowded, newly industrialized nineteenth-century America.

At this juncture, some women finally said enough is enough. From two rather disparate groups, the issue of dress reform was raised: water curists and women's rights activists.

Water cure is a form of alternative medicine that became massively popular in the United States around the mid-nineteenth century. Water-cure facilities, which are akin to spa villages and resorts, popped up in the hundreds across the nation. These facilities championed water therapies, such as baths, foot washing, and compresses, together with diet and exercise, to promote healing and health.[34] An inherent part of the regimen featured healthy, rational clothes for women.

In 1851, Theodosia Gilbert, a cofounder of the Glen Haven Water Cure in New York, wrote about a suit she made expressly for walking: "The suit consisted simply of a pair of cassimere pantaloons, a frock of woolen material, loose, plain waist, and sleeves, with a skirt reaching to the knees, of decent dimensions. . . . In this rig, I could just about double the distance, in the same length of time . . . and what is more, with half the fatigue."[35]

Although people were initially shocked by the rig, it was quickly adopted by women at the Glen Haven resort once they recognized the "ease, strength, agility, and freedom" it provided.[36]

The rig Gilbert described is essentially the bloomer dress named after the prominent women's rights activist Amelia Bloomer, who was one of the most earnest advocates for the dress. The signature feature of the bloomer dress was a knee-length skirt, accompanied by some sort of pants. Trying to appeal to national pride, water curists called the dress the "American costume" and contrasted it with the fancy, unhealthy French costume (figure 3.2).

FIGURE 3.2 American and French costumes contrasted. "The American and French Fashions Contrasted," *Water-Cure Journal* 12, no. 4 (1851): 96.

Furthering this nationalist rhetoric, water curists argued that American women were the cornerstone of the country's strength. Mary S. Gove Nichols, a cofounder of the American Hydropathic Institute in New York City, put it this way: "We must have free, noble, healthy mothers, before we can have men. The cramped waist, the crushed vitals, the loaded spine, the trailing skirts, the fettered limbs, the feeble, fearful being . . . can train us no Washingtons, Franklins, or Jeffersons, no wise or great men."[37]

The irony of the quote, I trust (I hope), won't be lost on today's readers. The water curists, passionate though they were about women's health, had no grand vision for their role outside the home and beyond child-rearing.

It is through women's rights activists that the bloomer dress took on a more decidedly progressive tone. These activists, the so-called bloomer-ites, donned the short dress and stood on the podium to demand women's civil and political rights: the ability to vote, to own properties, to receive education, and to work.

Sadly, on that podium the dress didn't last long. The general public, unlike fellow invalids at secluded water-cure resorts, had no sympathy for bloomerites. Men glared at them, women ostracized them, friends were ashamed of them, and strangers followed them on the street, pointing, laughing, and sneering.[38] Rather than heeding the activists' words, the public was engrossed with their dress and, hating *that*, found their words even less tolerable.

Fearing that their costume was hurting rather than helping their politi-cal cause—and tired of the constant harassment—one by one, the activists lengthened their skirts. They conceded that women's clothes were a result, not a cause, of political disempowerment. Until women had the right to vote, to be an equal of men in society and government, she stood little chance of being able to reform her clothes—or to conceive the desire to do so. And even if she succeeded in keeping her skirt short, that wouldn't do her much good. As Elizabeth Cady Stanton, a leader of the women's rights movement, wrote in 1855, "The negro slave enjoys the most unlimited freedom in his attire . . . yet in spite of his dress, and his manhood, too, he is a slave still."[39]

By 1870, the water cure fad had also passed, loosing appeal to the scien-tific and clinical approach to medicine.[40] Most water-cure facilities were closed and gone with them another voice for dress reform.

The bloomer dress, however, didn't fade out of history. It found an ally in an unexpected place: bicycles. Bicycle-riding rose to popularity in America in the 1890s, thanks to modifications made to the vehicle that improved its safety and utility. Women, in particular, took to the wheels enthusiastically as a leisure activity, a sport, and a form of transportation. Anyone who has ridden bicycles in a skirt that reaches below the knees can appreciate the hassle. And if you haven't tried it, there are good reasons why you shouldn't. Just imagine the difficulty of moving your legs under those skirts and the mortification—and mortal danger—of the skirts getting caught in the pedals and chains.

The bloomer dress came to the rescue, in several variations: the skirt over the pantaloons became shorter, or it was eliminated altogether, turning the dress into baggy knickerbockers (knee pants) or tight knee breeches (figure 3.3).

Despite their obvious convenience and safety on the wheels, these reform garments were ridiculed—hence, the unmistakable tone of derision depicted in figure 3.3. Many women dared not wear them, at least not downtown in broad daylight, and opted instead for frill-free and *slightly* shorter skirts that could clear the pedals and chains. Divided skirts were also invented

FIGURE 3.3 Bloomers for the wheels. "Bloomers in Paris," *Philadelphia Inquirer* (Philadelphia, Pennsylvania), March 15, 1896, 31.

specifically for bicycle riding. These skirts feature two separate leg compartments to allow women to sit on the bicycle. The compartments are fastened together in the front and made full in the back to resemble the look of an ordinary skirt.[41]

The women who did brave the bloomers—whether in the comfort of water-cure resorts, on top of soapboxes, astride bicycles, or in dance halls (bloomer balls were a thing!)—formed a long line of women who used their clothes to express personal and political identities. Every single one of them, though failing to bring widespread dress reform, made the public a little more aware, if not necessarily more tolerant, of the female knees. Every single one of them softened, as it were, the arrival of the flappers.

The earliest recorded use of the word "flapper," according to the Oxford English Dictionary, was in 1570. It means something flat and loose used for flapping or striking, such as a device used to swat flies. Later, the meaning extended to people who flap or strike. In Jonathan Swift's *Gulliver's Travels* (1726), residents on the island of Laputa were so lost in their thoughts that they had to hire servants called flappers, whose job was to strike their masters' mouth to remind them to speak, ears to remind them to listen, and eyes to remind them to watch. Constantly carrying a blown bladder fastened to the end of a stick for their job, the flappers hardly looked fashionable.

In 1747, the word "flapper" took on another meaning: a young wild duck or partridge. Presumably, these young birds were called flappers because they frantically flapped their wings when learning to fly. It is likely that from this meaning then derived, in the late-nineteenth and early twentieth century, the use of "flapper" to refer to a young girl at an awkward age: "the age when pretty little girlhood with its dimples and its curls has been left behind, and slender young womanhood is not yet in sight. The flapper is long-legged, scrawny of neck and forearm. Her hair is not a pretty length. She is getting her second teeth—and very likely having them straightened by some disfiguring dental process."[42]

From these unflattering definitions emerged, in the Roaring Twenties, a far more agreeable one that the flappers gave themselves. This self-definition was recorded in *Flapper* magazine, a publication that was circulated in the early 1920s to celebrate flapperdom: "She's independent, full of grace, a pleasing form, a pretty face; is often saucy, also pert, and doesn't think it wrong to flirt; knows what she wants and gets it, too; receives the homage that's her due; her love is warm, her hate is deep, for she can laugh and she can weep; but she is true as true can be, her will's unchained, her soul is free; she charms the young, she jars the old, within her beats a heart of gold."

How does a woman go about showing that she is independent, saucy, and free in the 1920s? Why, by dancing the Charleston, smoking cigarettes, drinking alcohol, and flirting and partying.[43] These deeds alone, though, were not enough. One's fashion choice, as you will appreciate, is an essential outlet of one's identity—bikers dress very differently from soccer moms, who in turn dress differently from Goth girls. To be a true flapper, one had to *look* the part.

Given centuries of long skirts that pretended women didn't have lower limbs, it made sense that short skirts were flappers' favored symbol of freedom. But how short is short? Ankle length is nineteenth century and clearly old fogie. Calf length is boring as it doesn't reveal any distinctive limb part. Thigh length, like today's miniskirts, was evidently too much even for the flappers. Knees, then, proved to be the sweet spot, as *Flapper* magazine announced:

For the first time since civilization began the world is learning that girls,
   women, females, maidens and damsels have KNEES.
Nevertheless, it's the naked truth.
And it's becoming more evident every day.
'Tain't necessary to roll the sox to disclose them.
The short dresses have revealed them to a gasping world.
And oh, what a shock!
. . .
O, well, I suppose we will be able to worry along, after we get used to
   them. The first hundred knees are the hardest.
After that you get callous. So do they.[44]

In all fairness, the flapper skirt isn't *quite* as short as the above might suggest or as people today tend to imagine. A 1920s skirt that leaves the knees (and more) completely exposed or covered only by tantalizing fringe is again more Hollywood than history. Fringe was not mass producible in the 1920s, and what fringe existed was a dense knit from fibers and looked more like bulky strands than flowy threads.[45] Look closely at the flapper depicted in figure 3.1, who was wearing a fringed dress, and you will see what I mean. As for the knees, most flapper skirts come to just below or

FIGURE 3.4 Rolled stockings. From "Flapper," accessed November 27, 2021. TomandRodna
.com, http://www.tomandrodna.com/Flapper.jpg, contributed by Dave Bumgardner.

at the knee. The knees are meant to be exposed *by accident*, in a "Venus-surprised-at-the-bath sort of way,"[46] when a woman walks in a breeze, sits casually down, or kicks her legs while dancing the Charleston.

Of course, the flappers did count these accidents to happen and rolled their stockings to remove visual hinderance at the knee. Specially designed garters were used to assist with the rolling. Selling between ten and twenty-five cents a pair, roll garters were made from soft pliable rubber in a circular shape.[47] Women would put on stockings in the usual fashion and slide the circular tube up their legs. They would then roll the top edge of the stocking down, wrapping it around the tube and continuing to roll and wrap as they went. The pliable garter would hold the stockings in place wherever the women pleased, as shown in figure 3.4. Women could also fixate the rolled stocking with a flat, ribbon-covered elastic band, which was the method of choice used by the flapper shown in figure 3.1.

Rolled stockings were considered such a naughty and novel thing to do that songs were written about them. Titled "Roll 'em Girls," Billy Murray's 1925 production goes,

Listen girls, listen girls,
I've a word for you,
Just because you're up to date
And do the things you do.
Don't let anyone tell you
That you don't act nice,
For you're as sweet as grandma was,
So take my advice.
Roll 'em girls, roll 'em,
Go ahead and roll 'em,
Roll 'em down and show your pretty knees.
Roll 'em girls, roll 'em, everybody roll 'em,
Roll 'em high or low just as you please.

Don't let people tell you that it's shocking,
Paint your sweetie's picture
On your stocking.
Laugh at ma, laugh at pa,
Give them all the haha.
Roll 'em girlies, roll 'em, roll your own![48]

To further ensure that their knees stood out, and were put forward in the best light, flappers took to another fashion practice: rouging them. As a New York beauty parlor manager explained, women should beautify their knees just as they do their faces, "what with disclosures that might result while sitting . . . or crossing a windy street or boarding a taxi. . . . The idea is to get just the faintest pink effect—a coat of rice powder, the slightest touch of rouge on the knee cap and a film powder over that."[49] The effect was supposed to be quite charming.

This practice soon caught the attention of fashion-forward Parisians. Not to be outdone, Paris claimed that "to powder all the knees in the same color is to leave the new art on a low level."[50] According to Paris experts,

"Pink knees may be very well for pale people—that is, blonds. But for brunettes it is not considered to be good color at all. . . . [T]he ideal for brunettes is peach color."[51]

What about knees of darker colors, like brown and black? How would we rouge those? The New Yorkers didn't say, nor did the Parisians. When we think of the 1920s flapper, we automatically picture a young, white woman—because that's how media and literature portray her. But Black flappers were very much a thing. During the Great Migration of the 1910s and 1920s, more than a million Blacks, including many single young women, moved from the rural South to the urban North and Midwest and found employment that earned them disposable income.[52] For these women, the flapper fashion was not only an aspiration for individual freedom but also for racial equality: to partake in the trending youth culture and defy the stereotype of Blacks as ignorant and ugly.[53]

As their white counterparts, Black flappers donned short skirts, put on makeup, and went dancing. And, as their white counterparts, they invited objections from conservative African Americans. In fact, these women faced the extra charge of letting down their *entire* race: apparently, by ditching "respectable" appearance and behavior, they stalled African Americans' racial progress.[54]

For all we know, Black flappers rouged their knees in what colors they thought most flattering. And the more creative and ambitious among them, like their white counterparts, tried painting their knees, too. No, this was not painting pictures on the stockings, as Billy Murray had sung, or decorating the knee areas of stockings, as some manufacturers did.[55] This was painting one's knees in the flesh, which, according to beauty specialists, was "the latest thing."[56] One may choose from a variety of designs simple and elaborate: "Some girls prefer a flower or a group of blossoms. Others like a portrait or a little landscape."[57]

At Ohio Northern University, girls preferred roses. Evidently, the effect was so tasteful that "some of the professors had not been able to do their best work owing to the profusion of knees in certain classes, that it is difficult for a mere male instructor to think of the Einstein theory, for example, with a tastefully decorated knee—well, staring him in the face, as it were."[58]

In Omaha, Nebraska, seventeen-year-old Mary Bell had a more unique taste and painted on her knees the portraits of Clarence Darrow and William Bryan, who were the opposing attorneys in the Scopes Monkey

Trial, formally known as *The State of Tennessee v. John Thomas Scopes*. The trial was staged to challenge an act recently signed into law that prohibited the teaching of human evolution in Tennessee. Mary was seen taking her decorated knees out for a swim "just to see if the portraits were water tight" before going to a dance, where "[i]t is not expected she will have to sit out many dances."[59]

If a long line of smart and brave women had failed to bring widespread adoption of short skirts, what was it that allowed the young flappers, despite their fair share of social resistance, to succeed in exposing their knees?

If anything, the flappers weren't as organized as the water curists or as politically conscious as the women's rights activists. A National Flappers' Flock did exist, and local chapters were established. But what was it that these flocks did? "In addition to having good times, the members of the [national] flock will hold regular meetings, and conventions when there are a sufficient number of flocks. Town, city and state parades, picnics and parties may be expected when the movement gets under way."[60] Among the events planned by the New York flock were a big Halloween party, hikes, and theater parties.[61]

Another event flappers enjoyed was beauty contests. The announcement for one such contest read: "Girls, here's good news for you! How would you like to get a hundred dollars—just for being a flapper? The money is yours! That is, it's yours provided you are more of a flapper than any of the other entrants in THE FLAPPER beauty contest. Your photograph will tell. . . . ALSO CHANCE IN THE MOVIES."[62]

Not especially political, is it?

Now, of course, the personal *can* be political. Women's choices in how they present and perform their bodies, like their choices to vote or to work, *can* be political. Yet, some of flappers' choices seem to encourage the wrong kind of political message and female power.

The flapper dress, with its relaxed fit, dropped waist, and raised hemline, is supposed to look youthful and boyish. But to pull off this *garçonne* look, one had to be slim and flat-chested. To achieve this body type, women took to strenuous dieting, setting magazine models and movie stars as heroines.[63] More ironically, in tossing the corsets, women picked up another piece of garment to police their bodies. Variously called the brassiere or the bandeaux, these undergarments were flat in the front and were used to flatten one's bust. Some of them were made with the same stiff corset material; others were made of rubber to, supposedly, cause perspiration and shrink the bust.[64]

Women with the right sort of body—and sportiness—did well in the 1920s new dating system: a system in which women would go out with strangers for little more than just a good time:

> On their own, working-class women [of the 1920s] could scarcely afford to indulge themselves with fancy clothes, movie tickets, or a thrilling afternoon at Coney Island. . . . The same could also be true for many middle-class teenagers . . ., who grew up in considerably greater comfort but were still dependent on their parents for money. This was where men came into the picture. A central component of the new dating system came to be known as "treating," whereby men paid cash for dinners, theater tickets, and amusement park admissions and women carefully estimated how much physical and romantic attention they needed to provide in turn.[65]

No, flappers were *not* Girls Gone Wild. But was there not a trace of twenty-first-century raunch culture going on? That is, women were using their sexuality as a commodity, mimicking male "comrades" in sexual sportiness, and mistaking these behaviors for female power?[66] Surely, more profound forces beyond these attitudes were at work that raised the hemline and freed female knees in the 1920s?

Before we get to these forces, let's make one thing clear: the hemline change didn't happen overnight (see figure 3.5). At the turn of the century, the Gibson girls' skirts were floor-length, barely exposing shoes. Hemlines started to rise in the 1910s, were well above the ankle and showing calves in 1920, hovered around the knee by 1925, and then threatened to expose the knee in 1927. This change wasn't linear either.

1900　　　　1908　　　　1915　　　　1920　　　　1925　1927

FIGURE 3.5 The gradual rise of the hemline (based on *Delineator* magazine dress patterns).

In fact, hemlines *dropped* in the first few years of the Roaring Twenties before going up, really up.[67]

Behind this gradual change was the rising public awareness of women's dress reform, thanks to the seventy years of struggle that came before the flappers, as well as several other factors.

The first factor was World War I. The War considerably damaged France's cotton and woolen mills, and French designers, trying to conserve fabric, adopted a straight silhouette and raised the hemline.[68] With Paris being the world's fashion capital then as now, women's clothes on both sides of the Atlantic followed suit.

Meanwhile, shorter and simpler sportswear entered the scene, owing to the increasing participation of women in sports such as tennis and golf. Take tennis, for example. Before World War I, "approved tennis dress for women included a long skirt with many petticoats, a fussy long-sleeved blouse, and a wide-brimmed, flower-trimmed hat . . . After the War, a more or less modern tennis dress was introduced by the continental star Suzanne Lenglen. It consisted of a sleeveless middy blouse, a gored skirt which stopped well short of the ankles, and tennis shoes with short socks."[69] Sportswear was initially sold to a high-end market, to women who had the leisure to play sports, but it soon reached other market segments and became acceptable for informal daytime wear. By 1926, Neiman-Marcus in Dallas was advertising sports dresses with hemlines barely below the knee.[70]

The blessing of French designers and the fashion industry at large led to one crucial difference between the failed bloomer costume and the popular flapper outfit: aesthetic perception. Although nineteenth-century American women conceded that the bloomer costume was suitable for walking, hiking, and cycling, they nonetheless thought it was "ugly," "very ugly," and "supremely ugly."[71]

"They make a woman look short and stocky and give her a baggy aspect," complained one writer to the *San Francisco Examiner*.[72] "They cut her off, as it were, and throw her figure entirely out of proportion. They make her waist look large, and their fullness at the knees only accentuates this," explained another to the *Chicago Tribune*.[73] Tourists like Olive Logan made the most bitter wearer of the bloomers,

> I was informed by one of the few ladies who had been to the [Yosemite] Valley, whom I met in San Francisco, that it was next to an impossibility to accomplish the journey without arraying myself in a Bloomer costume.

Pardon me that I recoiled at this. I feel that my charms are not so numerous that I can afford to lessen them by the adoption of this most ungraceful and unbecoming of dresses; but when she assured me that it was almost a necessary precaution against being thrown from the horse to ride astride, I saw at once that my time had come, and a Bloomer costume I must wear. The dressmaker to whom I applied had made others, and needed no instructions when I told her I was going to the Yo Semite. She carved me out a costume; but pardon me once more if I shrink from the task of describing it. It was simply hideous.[74]

The flapper dress, by contrast, was pretty—if not in absolute terms, then at least advertised as such through Sears catalogs, *Good Housekeeping* magazines, and movie stars like Clara Bow. The design was advertised as Paris-inspired, featuring straight, youthful silhouettes, simple lines, and charming details.

A considerable part of this "prettiness" is tied to youth. In fact, the so-called flapper outfit was originally designed for schoolgirls ages thirteen to seventeen—the young girlhood that bridges childhood and womanhood.[75] Vying for this youthful look, "small woman" would shop the girls' section rather than the matronly women's section, to the "envy of her more weighty sisters."[76] Sensing this untapped market, by 1922, manufacturers were purposefully selling the flapper outfit to the young woman who had "publicly established herself" and "demanded that costumes be designed for her individually in place of the school girl garments from which she was forced to choose."[77]

Aside from being pretty, the flapper outfit differed from the bloomer costume on another crucial front: the flapper outfit was, through and through, a dress; the bloomer costume was dangerously ambiguous by utilizing trousers. Although the trousers were baggy and hid women's legs well, they were still deemed indecent—because by nineteenth-century standards, trousers, or anything bifurcated, belonged to men. Back then, little kids of both sexes wore dresses. But when a boy was toilet-trained and "could be trusted in non-washable wool trousers rather than washable cotton petticoats," he was "breeched": changing from infants' dresses to shorts and then trousers.[78] This change of clothing was symbolic, marking the transition from helpless infanthood to emancipated boyhood and then manhood. Girls never had this rite of passage. They were to remain in dresses, and to be dependent, throughout their lives.

The bloomer costume, by co-opting trousers, was thus an abomination, a usurper of male superiority and power. If women could wear trousers, what would become of men and their place in society? At stake here was not a mere piece of garment, "but the developments that must inevitably follow, with certain women at any rate."[79]

Mary Walker was seen as one such woman. A medical doctor, Walker experimented with various reform garments and wore short dresses with trousers in medical school.[80] She was the first female surgeon employed by the Union Army during the Civil War. To make her job easier while tending to the wounded on the war front and in field hospitals, Walker wore men's clothing.[81] For her service during the war, she was awarded the Congressional Medal of Honor in 1866, the only woman to ever receive the Medal. Yet, in the same year, she was arrested on the street of New York for "wearing 'male attire.'"[82] A woman who dressed like a man and did a man's job—apparently better than men could—was (and often still is) a grave cause for concern.

This concern was made all the more alarming by the rapidly increasing urban female workforce. Between 1860 and 1920, the number of people living in American cities increased from 6.2 million to 54.3 million.[83] Urban women were entering the workforce in droves, doing jobs previously done by men, jobs that were created by America's industrialization and vacated by men killed or incapacitated in World War I. By the year 1930, 10.7 million American women, or 22 percent of all women ten years or older, were employed in factories, offices, stores, and various professions.[84]

These women were the ones who purchased the pretty flapper fashion using their hard-earned money. If there was a single most important contribution that flappers made to raise their hemlines, it was this: shopping. No doubt, simpler and shorter dresses helped women to function in the modern workplace, but, on the whole, the flappers were less motivated by conscious political aspiration than "a steady diet of bright and glitzy department store windows, advertisements and amusements, consumer products, and magazine articles—all urging them to let go, enjoy life, and seek out personal happiness."[85]

Certainly, the ability to consume, which signifies economic independency, *is* power, especially when a consumer could influence the products presented to her for consumption, as the flappers did in the late 1920s. By then, the fashion industry had grown alarmed that the flapper design has remained stable for several seasons. Fashion thrives on change: if there is no change, there is no selling of new clothes. Not to mention that the flapper design, with its simple lines and short hemline, used less fabric and had a small profit margin to begin with. Designers and manufacturers thus tried to bring back long skirts and a fitted silhouette, which would create a different look, sell more fabric, and also boost the sales of corsets and girdles.[86] But women refused to buy into the new design. Some did so for aesthetic reasons; others saw the long skirt as a return of sartorial shackles, an attempt to put women back into trains and eleven-yard gowns and to take away their votes.[87] This battle of the skirt ended with a partial victory on both sides: while evening and formal wear lengthened, day wear remained short, until the market crashed, that is.

Between August 1921 and September 1929, the Dow Jones Industrial Average climbed sixfold.[88] The economy was booming, and consumer confidence was high. People bought stocks with credits and loans, expecting their values to soar.

This economic bubble started to burst as the decade drew to an end. September 1929 saw the stock market peaking and then starting to fall. On October 24, spooked investors sold shares en masse, ending the day with a record trading of 12.9 million shares; on October 28, Black Monday, the Dow Jones fell 13 percent; the next day, Black Tuesday, it fell another 12 percent.[89] If this all sounds random, it was not. Before the market crashed, the house of cards that was the 1920s boom was already unraveling.[90] Consumer demand dwindled as people reached their spending limit, factories cut down production and laid off workers, and farmers struggled to make a profit.

By 1933, fifteen million people, or 30 percent of the workface, were unemployed.[91] Losing confidence in the economy, people stormed the banks to withdraw their money, forcing banks to liquidate assets at a loss. Soon, nearly half of America's banks failed.[92]

As the Great Depression set in, hemlines plummeted, and women's knees disappeared.

The shortening of women's skirts during the 1920s and their subsequent lengthening during the Depression was often cited as evidence for the hemline index theory, or the bare knees, bull market theory. Proposed by economist George Taylor in 1926, the theory suggests that the hemline of women's skirts is somehow indicative of the economy: when the market rises, so too does the hemline, begetting shorter skirts; when the market drops, so too does the hemline, resulting in longer skirts.

One explanation for this theory is that when the economy is down, women can't afford hosiery and have to wear long skirts; when the economy is up, they want to wear short skirts to show off their beautiful stockings. I don't know about this. It seems too flippant of an explanation to me.

A more complicated, but probable, explanation is that when the economy is down, women are more likely to use marriage as an economic strategy. Because men generally value sexual restraint in a marriage partner (for it enhances confidence in the marriage), women will wear long skirts to appear conservative and family-oriented.[93] When the economy is up, women are more likely to be gainfully employed and not reliant upon a husband, so they care less about appearing conservative. In addition, short skirts also made women appear to be more progressive, independent, and thus attractive to employers.[94]

Most economists snub the hemline index theory, considering it an urban legend. Economy, especially today's economy, they say, is much too complex to be tied to fashion. But a few studies (granted, a small number and *not* peer reviewed), after measuring hemlines in fashion magazines from 1921 to 2014, support the gist of the theory: short skirts were popular in prosperity; long skirts, in hardship.[95]

A case in point is the 1960s, when the Dow Jones reached new heights, as did the hemline, ushering in the miniskirt. Popularized by English fashion designer Mary Quant, miniskirts were proclaimed "the most important British import [to America] since the Beatles."[96] A miniskirt is supposed to sit at midthigh, about five inches above the knee; its cousin the microskirt ends at the upper thigh, about ten inches above the knee.[97] Next to the micro, the flapper dress looks positively prudish.

Though flappers invented knee-painting, it is not clear how prevalent that practice really was in the 1920s. Women, unless extremely talented, can't be expected to paint their own knees upside down in a satisfactory manner, and beauty parlors weren't regularly staffed with artists who could paint knees.[98] It was not until the 1960s when miniskirts were the fad that the beauty industry took knee decoration seriously. On the market in 1966 were decals, fake flowers, and jewels with waterproof adhesive that could be applied to the knees; do-it-yourself, knee-painting stencils from Helena Rubinstein that could be filled out with regular makeup; and leg make-up kits from Revlon with special poster-color paints.[99] It was, as *Philadelphia Inquirer* proclaimed, the year of the knee.

And America was ready for it. One did hear the same old tune that the miniskirt was immodest, but on the whole, people were good-natured about them. Women shrugged and said that the skirts were fine "if you've got the legs for them,"[100] while men, young and old, eagerly agreed that the view got better when the skirt got shorter.[101]

Although the 1920s educators seemed easily distracted by the mere sight of knees, their 1960s counterparts were more poised. "I guess I don't pay much attention to what the students wear," said history professor Tom Murphy at the University of Hawaii.[102] English professor Bruce Stillians volunteered that the effect of the miniskirt may be discipline specific: "In the English department we're involved with much more elevating thoughts," he said. "However, I understand that professors in other departments talk a lot about mini-skirts, particularly in the sciences."[103] The joke may be lost on readers who haven't spent considerable time with both the English and science sides of the academia, but I couldn't resist.

Similarly, while the 1920s legislators seriously attempted to ban short skirts, in the 1960s, they seemed to regard the matter as fodder for political entertainment. When Republicans in California attempted to ban miniskirts, the Democrats replied: "We are dangerously close to imposing on people's constitutional rights when we tell them how to dress. . . . More importantly, we're imposing on the constitutional rights of those of us who like to look."[104] In New Hampshire, when Mrs. Caroline Gross, special assistant to Republican Governor Walter Peterson, was criticized for wearing miniskirts, she protested that she was just being an obedient wife: "My husband picks out all my clothes," Mrs. Gross explained. "He's a lawyer and Democratic ward chairman. Maybe it's some kind of sabotage."[105]

On a more serious note, miniskirt wearers lived—and continue to live— the ideological quandary that faced the flappers:[106] Were their short skirts a symbol of female liberation, of women being in control of their bodies and sexuality? Or were they a symbol of consumerism and sexual commodification, of women objectifying their bodies for the male gaze? For women who aspire to the former, the unintended effect of the latter is absolutely mortifying, as British feminist Sheila Rowbotham recalled.

The year was 1968, and Rowbotham was at the founding meeting of the Revolutionary Socialist Students' Federation at the London School of Economics. She was about to make a speech: "I had never spoken to so many people before. It was a warm, sunny day and I was wearing a black and gold summer miniskirt. To my horror, as I walked to the mike, I was greeted by a tumultuous barrage of wolf whistles and laughter. I remained frozen for what seemed an eternity. . . . I had ceased to be an individual and had become an object of derision. It was like a living nightmare."[107]

On a warm, sunny day fifty years later, in a college town in the American Midwest, I too was wolf-whistled walking to the gym in my skort. A skort, by the way, is such a curious garment: you feel liberated and secure by wearing shorts, but all *they* see is a tantalizing little skirt. That seems a perfect summary of the female power struggle.

Like Rowbotham, I was mortified. I hastened my steps and ran away, not knowing *what* to think. A woman can choose her clothes for comfort, efficiency, and power, but it takes only one man's derision to put her in her place. Frustration is a gross understatement.

Hemlines rise and fall, knees come and go, but some things haven't much changed, have they?

4

The Weaker Sex?

Woman's glory rightfully came through the number and quality of children she produced, and that where sports were concerned, her greatest accomplishment was to encourage her sons to excel rather than to seek records for herself.

I personally do not approve of women's participation in public competitions. . . . In the Olympic Games, just as in former tournaments, their primary role should be to crown the victors.

—BARON PIERRE DE COUBERTIN (1863–1937), FOUNDER OF THE MODERN OLYMPIC GAMES

In rare pockets of the human history, female athleticism was extolled as an attraction, a sign of reproductive vigor and mating desirability. So it was that Spartan girls trained for foot races, and Medieval maidens played violent football.[1] But, by and large, athletic pursuits were considered ill-suited for women. Physical exertion leads, naturally, to sweats and muscles, which were supposedly unfeminine and unattractive. Moreover, exercise was suspected to compromise women's number-one sanctioned responsibility in this world, reproduction: "muscular arms and legs," for example, sapped "the strength of the internal organs," making for difficult childbirth,[2] and high-impact jumping and landing were thought liable to hurt women's uteruses.[3]

In the United States, the second half of the nineteenth century finally brought the realization that physical exercise was beneficial for female health—and, yes, her reproductivity. After all, as water curists so eloquently put it, no frail women could give birth to strong sons (see chapter 3). But, make no mistake, all exercises were not created equal. "*Vigorous* exercise, it was believed, could do much to transform boys into 'manly' men. *Mild* exercise was intended primarily to turn inherently weak girls into 'fit' mothers."[4]

Women's initial foray into the world of sports started in "gentlemanly" games, such as golf and tennis; from there, they graduated into sports such as boating that did not require "physical contact, awkward positions, endurance and great strength."[5] Track was decidedly off-limits. No spectator wanted to see women run with brute force, gasp for air, and exhaust their bodies. Indeed, when the women's eight-hundred-meter race was first introduced at the 1928 Olympics, it caused *mass* collapsing among the female contenders—according to widespread, but erroneous, media reports. The women were, as you would expect, pretty tired, but they were just fine. Still, the women's eight-hundred-meter race was banned for thirty-two years before being reintroduced in Rome in 1960.

Ironically, another sixty years on, we are seeing evidence that women are just as good as men, if not actually better, in ultra-long-distance races that last one or two hundred miles. Yes, women are less explosively powerful, so they don't jump as high or sprint as fast, but their muscles fatigue less, their tendons are more flexible, and they are often more resilient when it comes to dealing with pain.[6] All of these attributes make women well equipped for physically and mentally demanding ultra-long-distance races, where they have, on multiple occasions, outrun men.

If women can do *that*, then no one can possibly deem them too tenderly built for sports. Recent U.S. history offers a particularly telling example that systems rather than physiques are the barrier to female athleticism. In 1972, Congress passed Title IX, requiring equal access of males and females to any federally funded programs, including sports programs. Since then, female participation in sports has increased 990 percent in high school and 545 percent in college,[7] and more women than ever are competing at the highest level of their sports. At the 2012 Olympic games in London and the 2016 Olympic games in Rio (with 2020 Tokyo canceled), the U.S. team featured, for the first times in history, more women than men. These women also captured the majority of the U.S. medals and gold medals.

Really, every way you look, women are *not* the weaker sex in sports—except, perhaps, when it comes to the knee.

As more girls and women turn to competitive sports, more female knee injuries are to be expected. Accidents, after all, do happen. The problem is that the increase has been alarmingly disproportionate. Studies from a variety of sports and play levels show that women are far more likely to sustain knee injuries and require surgeries than men who play the same sports.

For example, in high school basketball, girls are 44 percent more likely to suffer knee injuries, 165 percent more likely to have knee surgeries, and 315 percent more likely to have anterior cruciate ligament (ACL) surgeries.[8] High school soccer follows similar patterns.[9] In collegiate soccer, female players are 138 percent more likely to sustain ACL injuries and 79 percent more likely to have meniscal tears.[10] Collegiate basketball sees even worse trends.[11] Among national championship-level volleyball players, women are 300 percent more likely to suffer knee ligament injuries.[12]

As a woman, with a bad knee, who plays sports, I can't help but wonder, what the heck?

Some say it's the anatomy. Evolutionarily speaking, modern men and women have the same knee. But, anatomically speaking, that's not necessarily true.

Women, because of their childbirth duties, have a wider pelvis than men. A wider pelvis creates a wider internal opening (i.e., a wider pelvic canal), so babies can squeeze through during labor. But a wider pelvis also creates a wider angle between the thigh bone and the shinbone as the two connect to form the knee joint (see figure 4.1). This angle is called the *Q angle*.[13]

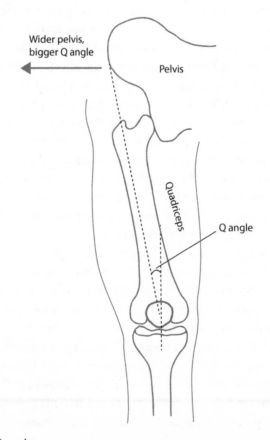

FIGURE 4.1 Q angle

Q stands for quadriceps (or quads), the muscles that run down the front of your thigh. Men, too, have this Q angle,[14] but in men, because the pelvis is narrower, the angle is smaller.

Women's pelvis is often considered an evolutionary compromise between two pressures: the pressure to reproduce and the pressure to move bipedally. A wider pelvis is better suited for delivering the larger-brained babies that early humans have become, but a wider pelvis isn't exactly well suited for bipedal walking and running. With a wider pelvis and a larger Q angle, the quad muscles will exert a strong sideway pull on the knee. This interferes with the smooth movement of the kneecap, causes excessive rotation of the lower leg, and endangers the proper function of the knee.[15] Women, in trying to meet both evolutionary demands, got the short end of the stick. At least, that's the conventional belief.

Fortunately for women, contemporary research is starting to challenge this classic view and shows that a wider pelvis is not necessarily a hinderance to bipedal walking. In fact, it may be a unique evolutionary advantage for women.[16] A wider pelvis allows women to rotate the pelvis through a larger angle and increase the length of their stride, which reduces the *number* of strides women must make to cover a same distance and thus reduces energy cost. Because women tend to be shorter than men, this gave our female ancestors the edge they needed to keep up with men as they traveled in groups. A wider pelvis also begets more bone and muscle tissues lower in the body, which increases stability and reduces energy cost when walking while carrying a newborn baby or foraged food.[17]

In addition, it turns out that not all women have the same wide pelvis— come to think of it, that's hardly shocking. Depending on geographic locations and population groups, the size and shape of the pelvis differs, by quite a bit. For example, Native American women tend to have a wider pelvis; sub-Saharan African women have a narrower yet deeper (from front to back) pelvis; and Asian, European, and North African women fall somewhere in between.[18] Not all women, then, have a categorically large Q angle.

That said, within the same population, women do have a larger Q angle than men. In one study of healthy young adults from Chapel Hill, North Carolina, women's average Q angle was about 15.8 degrees, whereas men's average was only about 11.2 degrees[19]—a difference of 4.6 degrees. In another study of healthy adults from India, women's average Q angle was about 14.5 degrees, and men's average was about 11 degrees—a difference of 3.5 degrees.[20]

The question, then, becomes what we make of these differences. Are they large enough to create a meaningful impact on knee function and health?

Some believe so. Others, however, think that they are little more than natural sex differences: women tend to be shorter than men, and everything else being equal, a shorter person will have a slightly larger Q angle.[21]

Which side is right? Despite decades of research, we haven't been able to decide. Notably, we have intensively studied one knee condition for its possible connection with the Q angle: the runner's knee. In medical jargon, this condition is called the patellofemoral pain syndrome—or pain at the front of the knee around the kneecap. It is the most common running-related injury,[22] and it also affects those who jump, squat, or cycle as part of their sports routines.[23] In some studies, a larger Q angle is associated with an increased risk for the runner's knee,[24] but in other studies, it apparently is not.[25]

Similarly, when comparing female athletes with healthy knees and those with ACL and other knee injuries, some studies found that the injured athletes sport a larger Q angle,[26] whereas other studies found no such difference.[27]

I can't say that that's bad news. No one can change his or her Q angle, so if this anatomic feature were indeed a determining factor, then women would always have weaker knees. Also, given how reproduction has been used to keep women out of sports, the further we can get away from reproduction-related causes for women's weaker knees, the better.

Aside from the pelvis and Q angle, another anatomic feature that may have rendered women's knees inferior is the *intercondylar notch*. Recall that the thigh bone has two round knobs at its end called condyles, which fit into the shinbone plateaus to form the knee. The space between the two condyles is called the intercondylar notch—like a little house with bones for its roof and walls (see figure 4.2). This house is where the ACL and posterior cruciate ligament (PCL) cross each other, where the ligaments live, stretch, and twist about as the knee moves.

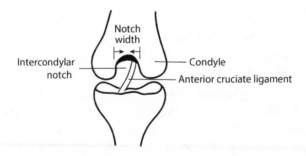

FIGURE 4.2 The intercondylar notch

This notch first gained attention in 1938. That year, the pioneer Swedish knee surgeon Ivar Palmer suggested that the intercondylar notch may be a risk factor for ACL injuries.[28] According to Palmer, if the notch house is too narrow, as the knee moves about, the ACL will be crowded, impinge over the edge of the notch, and then rupture.

Later research supported this prediction.[29] The findings seem particularly convincing as they are repeated around the world. In Norway, for example, handball players with an intercondylar notch width seventeen millimeters or less were found to be six times more likely to suffer ACL injuries.[30] Similarly, from Spain to the United States, people with smaller intercondylar notches are more likely to have ACL tears and insufficiencies.[31]

As you may have guessed: compared with men, women have a narrower intercondylar notch. In a study of 50 cadavers (32 males and 18 females), the male intercondylar notch width averaged about twenty-two millimeters, whereas the female notch width averaged about eighteen millimeters.[32] In an magnetic resonance imaging (MRI) study of one hundred high school basketball players (fifty males and fifty females), the male notch came to about twenty-four millimeters wide, and the female notch was twenty-one millimeters.[33]

If it is true that women's smaller notches made them more likely to have ACL injuries, then something *can* be done about it: notchplasty, a surgical procedure that removes part of the condyles to enlarge the intercondylar notch. This isn't something that one just up and does, however. It is an add-on surgical step during ACL reconstruction to prevent future impingement of the new ACL graft. How much of the condyles to remove depends on the surgeon, with some gung-ho surgeons going as far as 25 percent.[34]

But notchplasty is not without serious complications. It turns out that hacking away the condyles can cause degeneration and arthritic changes in the knee[35]—again, come to think of it, that's hardly shocking. The effort to enlarge the notch house is also somewhat futile as bone spurs and scar tissues will soon grow on the surface of the chopped notch and fill it back up.[36] Moreover, notchplasty can negatively affect the function of an ACL graft by shifting the graft from its originally correct insertion site on the notch to one less anatomically desirable.[37]

Given these complications, it seems wise to take a step back and ask: Was Palmer correct? Does the ACL really need a roomy house for its protection?

From a biomechanical standpoint, that's a difficult question to answer. While we can demonstrate the ACL being impinged in its notch house, we can't confidently say, "so what?" In a study of cadaver human knees, robotic arms were used to rotate, bend, and extend the knee to cause ACL impingement.[38] And sure enough, impingement did happen and did increase the stress put on the ACL. This added stress, however, was minor compared with the stress caused by the robotic movements. In other words, if the ACL were to rupture, it was because of the forceful movements, not because of its subsequent impingement on the notch.

To this finding, some other researchers weighed in: even if impingement doesn't matter, a narrow notch will house a narrow ACL, and a narrow ACL, by nature, will be more likely to rupture.

But there is quite a big assumption being made here: how do we know a narrow notch will necessarily beget a narrow ACL? The size of your house, as you can appreciate, doesn't necessarily determine the size of the family living in it, much less the size of everybody in the house. Indeed, using both cadaver knees and MRI scans, research findings are all over the place. Some assert that a correlation definitely exists between the notch size and the ACL size,[39] others suggest that the correlation exists in men but not in women,[40] and yet others declare that there most certainly is no correlation.[41]

I'm not sure the dust will ever settle on this debate, and it probably doesn't matter—because our belief that a bigger, wider ACL is necessarily stronger is also an unproven assumption.[42] I mean, this makes good sense. When it comes to ACL reconstruction, a bigger and wider ACL graft *is* more likely to succeed.[43] But does size matter in the native ACL that we were born with? That we don't know.

Although the Q angle and intercondylar notch have attracted most research attention, various other sex differences in anatomies were also blamed for women's weaker knees. It really is a rabbit hole if one is inclined to go down it. Anything from the shape of the shinbone, to foot pronation, to various leg and foot angles (front-back, left-right, inside-out) have been studied.[44] In all of these cases, definitive evidence is similarly lacking, and not one factor has been found guilty beyond a reasonable doubt.

Perhaps, then, the clue to women's weaker knees resides not in concrete bones and ligaments, not even in the knee itself, but in something invisible, something pervasive in the female body?

Puberty is exciting (and also weird and sometimes downright difficult). The body is changing, the brain is changing, and emotions seem out of control. Now, I'm not saying that puberty is easier on boys than on girls, but as far as the knee goes, it is. Before the onset of puberty, there is no sex difference in ACL injury, but as soon as girls hit eleven or twelve years of age, it's like a switch is turned on, and they go and rupture their ACLs twice as often as boys.[45]

Another switch, as we all know, turns on in young girls: estrogen, the sex hormonal responsible for maintaining the female characteristics and reproductive system. Doesn't it seem suspicious that knee injury should suddenly increase when estrogen starts to surge? Could it be that the injury switch in women is none other than estrogen?

Many researchers think so—because of something known as estrogen receptors. Receptors are protein molecules that respond to different stimuli to initiate specific cellular responses. Estrogen receptors, as the name suggests, respond to estrogen. As you can imagine, these receptors exist in female reproductive organs, such as the uterus and ovaries. But surprise, surprise, in 1996, estrogen receptors were also found in the human ACL.[46] This seems to provide bona fide evidence that estrogen will travel to the knee and somehow interact with the ACL and change its properties.

But what properties? If we are to break down the ACL, the main protein we will find is collagen, which is the major load bearer in the ACL. To be precise, the ACL has more than one type of collagen. There it Type 1 collagen, and there is Type 3 collagen. Type 1 collagen provides mechanical strength—so the more Type 1 collagen, the stronger the ACL.[47] Type 3 collagen, by contrast, provides elasticity—so the more Type 3 collagen, the more stretchable the ACL.[48]

The production of these collagens, experiments show, is influenced by estrogen. When ACL tissue obtained from a thirty-two-year-old woman was exposed to estrogen, production of Type 1 collagen immediately went down.[49] The more estrogen the tissue was exposed to, the less the collagen production—and presumably, the weaker the ACL. The same happened to ACL tissue obtained from a nineteen-year-old woman.[50] Interesting enough, in both cases, no significant change happened to Type 3 collagen production, so the ACL remained stretchable. The net effect, then, is that the ACL becomes weaker, looser, and presumably more prone to injury.

Now, two people are clearly not a large sample, but it is difficult to obtain ACL tissues from human donors, so we had to make do with animal specimens. Sheep ACL tissues, for example, have been exposed to estrogen, manipulated, and tested. Unlike in humans, however, the exposure caused no change in collagen production, nor were the exposed sheep ACLs more likely to fail under stress.[51] In rat ACLs, estrogen similarly had no effect.[52]

It's difficult to say which side of the evidence is stronger. On one side, we have human studies, but the number of humans is very small, and they are patients with knee injuries, so their ACL tissues may be compromised. On the other side, we have plenty specimens from healthy sheep and rats, but sheep and rats, no matter how healthy, are decidedly not humans.

To resolve the debate, some researchers turned to a different angle: the timing of women's ACL injury. Throughout a woman's menstrual cycle, estrogen ebbs and flows. If estrogen does weaken the ACL, then injury rate may differ depending on where women are in their cycle. Pretty clever, huh?

This research approach is not limited by ACL donations, but it has its own set of problems. As many women will appreciate, it is difficult to standardize the menstrual cycle. Not every woman experiences the classic twenty-eight-day cycle, and even if she does, this perfect cycle doesn't happen all the time. Different studies also classify the cycle differently: some break it down into two phases, and some into three phases. Some determine phases according to hormone tests; others rely on women's memories. Given these variables, research findings are not always comparable.

Still, from multiple studies, a general trend emerged that women are more likely to injure their ACLs during the first half of their cycles, especially toward the end of that first half, so something like day eight to day fourteen.[53] For example, in one study,[54] sixty-nine women who sustained an acute ACL tear were studied within twenty-four hours of injury. Urine samples were collected to assess hormone levels and determine where the women were in their respective cycles—and most of them turned out to be between day nine and day fifteen.

This timing is interesting because, in a classic twenty-eight-day cycle, ovulation happens at the midpoint around day fourteen. The days leading up to and through ovulation are precisely when estrogen levels surge to prepare for ovulation. In other words, in multiple studies, rising estrogen levels seem to accompany rising ACL injuries. More interesting, this is also the time window when a woman is most fertile and most likely to get pregnant. Pardon me for being cynical, but it seems that nature thought ovulating women ought to be doing certain other things than playing sports.

Following ovulation, estrogen drops rapidly. But it is not done. Within days, estrogen rises again to prepare for a possible pregnancy. In multiple studies, the timing of this second estrogen rise is also significant: it is linked to increased knee laxity,[55] which means knee ligaments become looser and the knee becomes less stable, potentially rendering the ACL more likely to stretch and injure.[56]

Well, by now, there aren't many days left in the month where women can confidently schedule sports activities. But I digress. With joint laxity, estrogen may not be the only hormone to blame. Another female sex hormone may be at fault: relaxin.

In 1929, researchers found that when virgin guinea pigs received blood injection from pregnant rabbits, the guinea pigs would *act* as if they were pregnant.[57] Their pelvic ligaments would relax—something that pregnant guinea pigs do—which would open their pubic bones and pelvic canal for childbirth, even though no child was obviously coming. A year later, researchers extracted from pig ovaries the hormone that's responsible for this "ligamentus relaxation" phenomenon.[58] It seemed only proper to call the hormone relaxin.

In human pregnancy, which is much more complex, relaxin wears more hats: it prepares the uterus lining to establish pregnancy, regulates blood vessel development to maintain pregnancy, and orchestrates a variety of cell and biological signals.[59] Amidst all this work, it doesn't forget to live up to its name: to relax, the cervix, that is, loosening it up and dilating it to facilitate labor.[60] The problem is that aside from the cervix, relaxin may be shaking up other soft tissues, too. In pregnant women, significantly higher joint laxity in the limbs, including knees, has been reported.[61]

Of course, most female athletes are *not* pregnant, much less at the time of an injury. But relaxin is not exclusive to pregnancy. It is found in unpregnant women too, and its level rises during the second half of the menstrual cycle—joining hands, it seems, with the second wave of estrogen to afflict the knee.

Several pieces of evidence support this theory. Women in general have more knee laxity than men,[62] which, once again, may start with the onset of puberty. In one study,[63] researchers assessed the joint laxity of 275 female and 142 male middle-school and high-school basketball and soccer players. Before puberty, the players demonstrated similar joint laxity; following the onset of puberty, joint laxity increased in the female players, but no change was found in the male players.

In addition, relaxin, like estrogen, also has its receptors in the ACL. In fact, relaxin receptors exist *only* in women's ACLs, not men's.[64] Presumably, then, relaxin will target a woman's knee, interact with her ACL, and disrupt the ACL's collagen density, loosening it up, so to speak.[65]

We can test this by taking advantage of the fact that relaxin levels differ from woman to woman. The average peak is around fifty picograms per milliliter, but actual levels can range from thirty to one hundred and fifty picograms per milliliter.[66] In some outlier women, the relaxin levels are so low that they are not measurable, and in other outliers, the levels are so high they reach those of early pregnancy.[67] If relaxin is indeed detrimental to the knee, then we can expect to see a disparity in who gets injured. This was demonstrated in a study of 143 Division I female athletes who played a variety of sports, including basketball, volleyball, and soccer.[68] Among these athletes, those with high blood relaxin levels were more than four times more likely to tear ACLs during their collegiate athletic careers.

But now we have a new problem. Did you notice, the dates don't exact line up? If the rise of relaxin, along with the second surge of estrogen, *after* ovulation causes knee laxity, why would ACL injuries primarily happen, as mentioned earlier, *before* ovulation during the first surge of estrogen?

This, some researchers explain, is because there is a "delayed effect" caused by the turnover of ACL tissue.[69] Others, more radically, believe that there is no perfect, one-to-one correspondence between hormone surges and knee injuries. Rather, it is women's chronically high levels of estrogen and relaxin and their rapid fluctuation that do their knees in.[70]

Whether the damage is immediate or in the long term, women, it seems, are once again caught in an evolutionary bind: what's essential for reproduction, for the spread of the human species, is detrimental to their knees, to their mobility, and to that which makes them bipedal humans. Nature, perhaps, did the best compromising it could.

What nature didn't foresee, however, is voluntary birth control and the invention of oral contraceptives. By taking the pill, women can stop estrogen and relaxin surges and prevent ovulation and pregnancy.[71] They don't get to have athletic sons this way, but they may end up with more athletic knees. At least, that's the promise coming from Denmark.

In Denmark, the National Health Service provides tax-financed health care to all residents, providing them free access to hospital care and general practitioner visits.[72] Each patient is registered with a unique ID in different medical registries, making it possible to link their data from across registries. Taking advantage of this set-up, researchers compared the Danish Prescription Registry and the Danish Knee Ligament Reconstruction Registry to look for a connection between contraceptive prescriptions and knee surgeries. The study located some 6,200 women who were on the pill in a five-year window and some 7,000 women who were not on the pill.[73] Between the two groups, the pill users were 20 percent less likely to receive an ACL-related surgery, which suggests that the pill had a protective effect.

Unfortunately, the report coming from Austria shows something different. Austrian researchers didn't dig up public health records; they looked at the national pastime: alpine skiing. Skiing being a popular sport in Austria, ski-related knee accidents are common. When researchers compared ninety-three female recreational skiers who sustained an ACL injury with ninety-three skiers who did not, their contraceptive use was essentially the same: 34–35 percent, suggesting that the pill offered no protection.[74] The same negative results were reached by American researchers who studied National Collegiate Athletic Association (NCAA) basketball and soccer players. Among 3,150 female players across three seasons, they did not find any difference in injury rate between those who used contraceptives and those who did not.[75]

Whether positive or negative, these results come from "observational studies"—that is, studies in which we merely look at women's pill use and knee health, rather than randomly providing some women with the pill and others with a placebo to test the effect. Without such placebo-controlled trials, which are the gold standard in drug development, we can't be sure of the pill's specific effect on the ACL—or the lack thereof. To the best of my knowledge, no human trial data are currently available, and the best we have is animal studies—rat studies, to be precise.

It is not especially flattering, but female rats have surprisingly similar cycles and hormonal changes as female humans, making them good experimental subjects.[76] A rat's typical cycle is much shorter though, lasting only five days. In one experiment,[77] forty sexually mature female rats were divided into two groups. One group was given oral contraceptives, weight-adjusted to mimic the human dose. The other group was given placebos. The rats were studied for sixty days, or twelve cycles, and then were killed (sorry, they had to be) so their ACLs could be harvested. Compared with the placebo group, rats on the pill had significantly tougher ACLs that were less likely to rupture under stress.

This is an intriguing finding, but rats, again, are not humans, and too often in drug development, animal findings do not translate well to humans. If you are already on the pill, revel in the *possible* knee benefits they offer. But until we have more evidence, it is premature to start taking the pill *just* to protect the knee, especially because women may very well want to have it all: sports records for themselves as well as for their sons—and their daughters.

Discussions of sex-based cognitive differences can be touchy, and for good reasons. Compared with the Bloomerites and flappers, contemporary women have come a long way in obtaining civil and political rights, but real inequalities continue to exist in work and life. I mean, who wants to earn ninety-five cents for *every* dollar earned by a co-worker in the same position, at the same company, with a similar background and experience?[78] Apparently, women do.

And, perhaps, "having come a long way" is a blissful overstatement altogether. In some ways, we are going *backward*. In the summer of 2022, as I put finishing touches on this book, the U.S. Supreme Court reversed *Roe v. Wade*, ruling that women's right to choose an abortion, a right that was upheld for almost half a century, no longer exists. If nature hasn't compromised women's bodies for the sole purpose of reproduction, we can always count on societies to help out.

Because of this reality, many (myself included) have a knee-jerk reaction when they hear that men and women are somehow cognitively different, that they, by nature, have different capabilities. Consider the widely spread but sketchy belief that girls are, by nature, poor at math. This is the kind of self-fulfilling prophecy that would keep women out of high-paying jobs in science, technology, engineering, and mathematics. No wonder educators, social scientists, and the public at large are eager to pooh-pooh such ideas as "neurosexism."[79]

It is, therefore, with trepidation that I write this: despite being the *same or similar in many respects*, the male and female brain structure and function do differ. These differences may range from a little to a lot, their implications may be up for debate, and their interactions with "nurture" and "culture" may be difficult to disentangle, but the differences do exist.[80]

And herein lies the third and last theory to explain why women have weaker knees: compared with men, women have poor neuromuscular control.

What is neuromuscular control, exactly? Imagine running merrily down the sidewalk on a brisk winter morning, feeling good about living up to your New Year's resolution. Out of nowhere, a neighbor's dog darts out in front of you. You startle, make a false step, and your knee starts to buckle.

At this fleeting juncture, two things can happen: One, you fall on the hard concrete (with or without the dog) and do a number on your knee, ruining your New Year's resolution before it takes off. Two, your leg muscles react and stiffen up, and you make a few stumbling steps to dodge the dog but remain on your feet. This rapid muscle activation to gain dynamic stability is neuromuscular control. Some of it is conscious, and some of it is an unconscious reflex. Either way, it has something to do with your neurons.

Usually, when we hear the word "neuron," we imagine cells buzzing and synapsing inside the brain, allowing us to solve math problems or read the *New York Times*. Nothing is wrong with that picture, but that's not all there is to neurons. When it comes to neuromuscular control, we are primarily concerned with neurons that live outside of the brain in the so-called peripheral nervous system. The word "peripheral" means "on the edge," so this system contrasts with the *central* nervous system (which consists of the brain and the spinal cord).

Integral to the peripheral nervous system are sensory neurons. The bodies (i.e., the nuclei) of these neurons reside near the spinal cord, but they have lengthy, cable-like projections that extend to our joints, organs, and every inch of our skin. These lengthy projections are called dendrites. At the end of the cable lines are receptors, which can sense the stretching, tension, and pressure of muscles and the position of joints. So, when you make a false step on the sidewalk, the receptors inside and around your knee will pick up the sensation, run it up the cable lines, feed it to sensory neurons, and, from there, take it to the central nervous system.

The central nervous system, through both reflex and conscious decision making, decides that you are in danger of an imminent fall and should activate leg muscles and stiffen up your knees. Without a moment's delay, this command is sent to another type of peripheral neuron: motor neurons. The bodies of motor neurons reside in the central nervous system, but they too have long projections called axons that reach the muscles in our bodies. Through these axons, the movement command is delivered to leg muscles, which react rapidly, saving you from the fall.

Or, it does not. Apparently, if you are a woman, you may fall anyway, because your neuromuscular control isn't keeping up.

To have good neuromuscular control, the first step is to quickly and accurately collect sensory information. In the case of the jogger–canine collision, the jogger must skillfully sense the position of the knee, the angle it is in, and the amount of tension in the surrounding muscles before executing effective motor movement. This ability to collect sensory information is called proprioception (pronounced *pro-pree-o-ception*), or the so-called sixth sense. It is what allows us to be aware of our bodies in space without having to constantly stare at our body parts. It is what allows us to bend our knees or to touch our noses with eyes closed. It is, as you can imagine, crucial to not falling on the sidewalk and to maintaining dynamic stability in just about any activities or sports.

Women, as it turned out, may have poorer proprioception than men. In one study, male and female participants were seated in a testing device that could either extend (straighten) or flex (bend) their knees.[81] The participants were blindfolded, wore a headset, and had their feet in air-cast boots to deprive them of other sensory input. The testing device would then ever so slowly, without warning, move their knees at a random point during a ten-second window. How quickly the participants could sense knee movement is an indication of their proprioceptive acuity. Evidently, men and women were similarly efficient at detecting knee flexion, but women took significantly longer to detect knee extension past fifteen degrees. Because knee extension past this angle increasingly stretches the ACL, the lesser ability for women to detect this movement may contribute to their higher risk for ACL injury.

With a similar setup, researchers also tested the ability for men and women to detect knee rotation.[82] In this case, men and women were equally sensitive to knee external rotation (rotating a knee away from the midline of the body), but women were significantly less sensitive to knee internal rotation (rotating a knee toward the midline of the body). As you may have guessed, yes, internal rotation is what loads and stresses the ACL, so this finding provides additional evidence that proprioceptive disadvantages are putting women's knees at risk.

But why are women worse than men at proprioception in the first place? Some blame female sex hormones. Earlier, we learned that sex hormones could directly impair ligaments. Could they also change the central nervous system's sensitivity to sensory signals?

In one study, thirty-two female participants underwent the knee-extension-flexion proprioception test described earlier.[83] The test was repeated three time, each time during a different phase of the women's menstrual cycle. Test results showed that participants had significantly worse proprioception when they were in the postovulatory phase of their cycles, a phase when both estrogen and progesterone levels were high. That said, conflicting findings do exist, and elsewhere, female collegiate athletes' ability to sense knee flexion was not influenced by their menstrual cycles.[84]

So, maybe it is the hormones, or maybe it isn't. Either way, women just seem to lack proprioceptive acuity compared with men.

Unfortunately, that's not all. There is also that second stage of neuromuscular control: the activation of muscles.

During everyday activities, we generate forces on the knee that are one to two times our body weight; sports activities further increase those forces to the range of five times our body weight.[85] Ligaments in the knee alone are not sufficient to withstand such forces, and muscles must be recruited to help stabilize the knee during dynamic movements.

Two groups of leg muscles are especially important: the quadriceps (or quads), which are the bulk of muscles in the front of your thigh, and the hamstrings, which are the bulk of muscles on the back of your thigh. Although both muscle groups are essential to dynamic movements and stability, a long-established understanding is that the hamstrings protect the ACL, whereas the quads endanger it.

When our legs are straight or only slightly bent (like what happens when we just landed from a jump), the hamstrings and the ACL work together to prevent the forward movement of the lower leg.[86] In other words, the hamstrings can relieve the stress put on the ACL. By contrast, in the same leg position, the quads apply a forward force on the lower leg and increase the stress put on the ACL.[87] So, if the two muscles are not balanced, that is, if the quads are strong while the hamstrings weak, a person is more likely to suffer ACL injuries.

And guess what? Women just happen to be quad-happy. In one study, male and female athletes sat in a testing device with their legs slightly bent at thirty degrees.[88] The device would deliver a thirty-pound force from behind their lower legs. The reactions of the athletes' leg muscles were measured using attached electrodes. The test showed that female athletes were more likely to contract their quads to resist the force, whereas the male athletes were more likely to use their hamstrings. Similar findings were reached in studies in which participants were asked to perform running, jumping, and cutting tasks.[89]

Sure, the activation of the quads is only harmful when the legs are relatively straight. If we bend our knees beyond thirty degrees, the effect is reduced, and beyond forty-five degrees, the quads add no significant stress to the ACL.[90] But, that's just the thing: women do *not* bend their knees, not enough, during dynamic movements.

Two examples of dynamic movements are running and cutting, and studies show that healthy young men and women perform them with notably different amounts of knee bending.[91] In the running task, participants were asked to run forward and hit a marked area with their dominant foot. In the cutting task, they ran and hit the marked area with their dominant foot while turning forty-five degrees to the left or right. Participants wore reflective markers at various landmarks on their legs, and multiple video cameras were used to record their movements. As the results showed, during both running and cutting, women bent their knees a lot less than men did. While men generally bent their knees beyond forty degrees, women were at or below thirty degrees.

Rather than bending their knees, women like to "collapse" their knees, in the posture of so-called valgus collapse, a frequently cited risk for knee injury. In valgus collapse, your two knees move in toward each other rather than staying straight over your two feet. Imagine forming an upside-down V shape with your two lower legs—that's what puts your knees in valgus collapse. In this position, the inside of your knees is locked and the outer side of the knees pushes inward and forward, stressing the ACL.[92]

When researchers analyzed the postures of National Basketball Association (NBA) and Women's National Basketball Association (WNBA) players caught on tape rupturing their ACLs, they found that at the moment of injury, the WNBA players displayed more valgus collapse.[93] The same tendency is witnessed in experiments in which healthy young men and women performed a single-leg drop task, in which they plopped their body down from a height of two feet and balanced on one leg for two seconds. Video analysis showed that the female participants landed in valgus collapse, whereas the male participants actually went in the opposite direction, pushing their knees outward.[94] In a different task, participants dropped from a height of one foot on two legs and immediately jumped upward as high as they could.[95] Between landing and jumping, the women collapsed their knees inward eleven degrees more than the men did, which would increase the load on their ACLs by up to four times.[96]

From proprioception to muscle activation to leg postures, the female knee can't seem to get a break. Fortunately, unlike anatomical features or hormonal cycles, neuromuscular control *can* be trained. Just like we can train ourselves to run backward (something that you see soccer players do), we can train ourselves in the correct knee positions for jumping, running, or cutting. Enough repetition and practice will allow the associated neuron-muscle connections to become preprogrammed so the movements happen subconsciously,[97] or, as the saying goes, through muscle memory.

In multiple studies, neuromuscular training programs were found to reduce knee, and especially ACL, injuries among female athletes in a variety of sports. In one program, female high-school volleyball players were drilled in various jumping tasks: broad jump, vertical jump, single-legged jump, squat jump, and more.[98] The emphasis was placed on maintaining good techniques: keep the spine erect and shoulders back, point the knees forward, jump with the chest over knees, and land softly with bent knees and toe-to-heel rocking. After six weeks of training, the participants were able to reduce valgus collapse stress by about 50 percent, increase hamstring power by up to 44 percent, and reduce landing force by 22 percent.[99]

In another successful program, female soccer players completed, among other things, leg stretching, jumping tasks, and strengthening exercises.[100] Similar emphasis was put on correct landing techniques. Compared with soccer players in the same league who did not enroll in the program, those who were enrolled saw an 88 percent reduction in ACL injury in the first season and 74 percent reduction in the second season.

Granted, not all training programs were this effective: some saw only small benefits, and others saw no benefits at all.[101] Cross-program comparison is difficult because each program is quite distinct. Some programs focus on jumping techniques, others like balance exercises, some include strengthening exercises, and others favor flexibility exercises. Some program routines are two hours a day, three days a week; other programs are ten to fifteen minutes a day, every day of the week. Given these variations, current research hasn't been able to sort out exactly what training works the best, when, and how.

Still, looking at multiple studies, jumping techniques were the single-most-important component for a successful program.[102] Programs that included multiple types of trainings were also more beneficial.[103] In addition, programs with high compliance rates were nearly five times more likely to reduce ACL injury than those with low compliance rates.[104] In other words, the more faithfully you follow through with a program, the more likely you will see an effect.

If thus far, I have failed to impress you with my talk of a weaker female knee, I understand. As mentioned earlier, I get it that some people do not like any talk that hints at innate female "deficits." But here is the thing:

pretending that differences do not exist doesn't bring women "equality"; acknowledging women's unique needs and addressing those needs may stand a chance. As a woman with a bad knee, I don't see my knees as innate deficits. I accept them as what I have, and I want to know what I can do about them. Being told that I'm just as good as a man doesn't help stop the pain. Finding an all-around training program and learning how to jump may stand a chance.

5

To Kneel, or Not to Kneel

The *Ko-teou* [kowtow], or adoration, as the Chinese word expresses it, consists in nine solemn prostrations of the body, the forehead striking the floor each time. It is difficult to imagine an exterior mark of more profound humility and submission, or which implies a more intimate consciousness of the omnipotence of that being towards whom it is made.

—SIR GEORGE L. STAUNTON (1737–1801), FIRST BARONET, BRITAIN

To believe that patriotism will not flourish if patriotic ceremonies are voluntary and spontaneous instead of a compulsory routine is to make an unflattering estimate of the appeal of our institutions to free minds.

—ROBERT H. JACKSON (1892–1954), ASSOCIATE JUSTICE, SUPREME COURT OF THE UNITED STATES

On September 26, 1792, in the Southern England port city of Portsmouth, Lord George Macartney (1737–1806) and an entourage of ninety-five people climbed aboard His Majesty's ships *Lion* and *Hindostan*.[1] The men were an assorted bunch: diplomats, priests, scientists, technicians, artists, and soldiers. They carried onboard an equally sundry collection: portraits of the British royalty, a planetarium that showed the movements of the heavens, telescopes and clocks, guns and cannons, a diving bell used to transport divers underwater, and a state-of-the-art hot-air balloon.[2]

With every last person and trinket accounted for, the captains pointed their ships southward, down the Atlantic, around Africa, up the Indian Ocean, destination China. The voyage would be the first encounter between Great Britain, the world's rising industrial star, and China, the mysterious Celestial Empire of the East.

Weather, alas, had other plans. Sudden and violent gales broke out, and the ships soon lost company of their tender and were forced to take shelter in Torbay. There, they remained stranded until October 1 before finally resuming their departure out of England.[3]

The Chinese would have considered this to be a seriously bad omen and second-guessed the voyage, but not the modern, rational, and scientifically minded Britons. All they could think about were their two missions.[4] One was to investigate China's economy, court, and way of life—to reconnoiter, if you will. The other, and more immediate, mission was to avail Britain a better trade condition with China.

Trade between the two empires had expanded throughout the eighteenth century, but the condition wasn't exactly to the liking of the British. The British couldn't get enough of China's tea, porcelain, and silk, but the Chinese didn't much care for Britain's manufactured goods. In fact, the ancient Eastern Empire seemed to disdain foreign trade and go out of its way to discourage it.[5] Foreign traders could conduct business in only one port in southern China, Canton, and could deal with only a handful of merchants. Foreign traders were allowed in Canton only during the trading season and had to stay in factory buildings that doubled as warehouses and residences. They were not allowed to wander beyond a narrow strip of coastal land, and no foreign women were allowed to set foot onshore. Teaching any foreigners the Chinese language was a crime punishable by death.

Eager to shake things up and increase profit, the East India Company, which had a monopoly on trade with China, sponsored Lord Macartney's voyage, with King George III's blessing. The pretense of the visit was to bring gifts to the Chinese Emperor Qian Long (1711–1799 AD), the sixth emperor of the Qing dynasty (1636–1912 AD) in honor of his eighty-third birthday. But really, the gifts were meant to impress the emperor with Britain's art, science, and technology so as to break into the large Chinese market.[6] Through skillful diplomatic negotiation, it was thought, Lord Macartney would make China see reason and agree to several trade propositions: improve the trading condition in Canton, open up new ports, grant British traders territories for residence and business, reduce tariff, and establish a permanent British embassy in China's capital city Peking.[7]

A reasonable monarch, Emperor Qian Long was flattered by the large embassy that traveled for eleven months from across the world to see him, bearing gifts and, assuredly, much admiration and loyalty toward his empire. He personally saw to the Britons' reception, extending to them the lavish hospitality typical of the Chinese. In a single delivery, the Chinese officials sent the embassy the following supplies: "[t]wenty bullocks, one hundred and twenty sheep, one hundred and twenty hogs, one hundred fowls, one hundred ducks, one hundred and sixty bags of flour, fourteen chests of bread, one hundred and sixty bags of common rice . . . forty baskets of large cucumbers, one thousand squashes, forty bundles of lettuce, twenty measures of peas in pods, one thousand water melons, three thousand musk melons," among other sundry items.[8]

The high point of the visit came on September 14, 1793, when the embassy was received in audience by Qian Long. Amidst an elaborate ceremony, Lord Macartney submitted King George III's letter and Qian Long gave him a jade scepter in return.[9] Qian Long even took off the scented sachet he was wearing—a symbol of the uttermost honor—and gave it to the son of Macartney's second-in-command, a twelve-year-old boy who had learned some Chinese on board the ship.[10]

Yet, for all the hogs and melons and gifts the emperor extended, he did not grant a single one of Macartney's trade requests. The British embassy was, from that perspective, an utter failure, and it is commonly suspected that the failure had much to do with the knee—the bending of it, that is.

Emperor Qian Long, as other emperors of the Qing dynasty, demands a complex kneeling ritual from every soul in his presence. The ritual consists of a person dropping down on both knees, prostrating and touching one's forehead on the ground three times, standing up and repeating the same process two more times, hence the name of the ritual: "three kneelings and nine kowtows."

Chinese officials diligently taught Lord Macartney and his men how to perform the ritual, kindly suggesting that they forego their knee buckles and garters and don Chinese loose garments, which would make bending the knees so much more comfortable.[11] But physical discomfort was the least of Macartney's concerns—personal humiliation and perceived allegiance was. Macartney protested that Qian Long was not *his* sovereign, so he shouldn't need to follow the Chinese ritual to show subservience—indeed, even in front of his *own* king George III, he didn't go down on *both* knees.

Chinese officials reasoned that the ritual did not signal political allegiance and was "a mere exterior and unmeaning ceremony."[12] Macartney responded that he would go through with it on the condition that a Chinese official of equal rank performed the same ceremony in front of a picture he had of King George. The Chinese balked at this request, rather confirming the Britons' suspicion that the ritual was, after all, "of serious and momentous import."[13]

What happened after that is a bit of a historical dispute.[14] According to the British and their writings, the embassy resolutely did not perform the Chinese ritual and merely got down on one knee, in accordance with how they greeted their British sovereign. According to the Chinese and their court records, the embassy most definitely knelt three times and kowtowed nine times, in accordance with how everyone greeted the celestial emperor.

FIGURE 5.1 James Gillray, *The Reception of the Diplomatique and His Suite, at the Court of Pekin*. Published by Hannah Humphrey, September 14, 1792, hand-colored etching, Courtesy of Wikimedia Commons.

Paintings about the event don't help, as they were, on the whole, more inventive than factual.[15] In James Gillray's (1756–1815) depiction, *The Reception of the Diplomatique and His Suite, at the Court of Pekin* (figure 5.1), part of the embassy was standing, others prostrated with heads on the ground, and Lord Macartney was on one knee submitting King George III's letter. At the risk of stating the obvious, I shall add that Gillray's depictions of both the Chinese and the British were rather unflattering. Emperor Qian Long looks like an eighteenth-century Jabba the Hutt from Star Wars: fat, contemptuous, pipe-smoking, and curled up on a dais with claw-like fingernails (for an actual sketch of Qian Long drawn by a member of the British embassy, see figure 5.2). Meanwhile, the Britons, try as they might, couldn't hide their cunning greed, the absurdity of their gifts, and those ridiculously tight breeches.

Some historians suggest that the truth is something in between: both parties compromised.[16] On some occasions, the Britons got on both knees

FIGURE 5.2 Emperor Qian Long, by William Alexander (1767–1816). George Staunton, *An Authentic Account of an Embassy from the King of Great Britain to the Emperor of China*, vol. 1 (London: W. Bulmer for G. Nicol, 1797), frontispiece.

and bowed deeply (with or without their heads touching the ground); on other occasions, they bent only one knee. Each party then selectively recorded what felt more face-saving to them.

Whatever the case, twenty-two years later, when the British dispatched a second embassy to China headed by Lord Amherst, the kowtow ritual was again a point of contention. This time, the embassy wouldn't budge, was denied an audience with Qian Long's successor Emperor Jia Qing (1760–1820), and was expelled from his court.[17] Therein ended Britain's patience for peaceful embassies.

To what extent did the Britons' reluctance to kneel botch their trade mission and plant the seed of war? Probably not much, if we look *just* at the events that transpired, the facts, as they say. When the British sent embassies that bear gifts, they had, in the Chinese's eyes, already consented to a vassal relationship.[18] So, the Chinese emperor had no reason to treat Britain as a sovereign entity for trade. After all, earlier Dutch embassies had committed the kowtow ritual when they visited, but they didn't get what they wanted either.[19] So, whether or not the good lords knelt, it probably wouldn't have made a material difference.

That said, if we look beyond the surface, beyond the mere facts, then the questions of whether or not to kneel and *how* to kneel become profoundly significant. Embedded in these questions is something that the British Empire couldn't or wouldn't accept, something that Imperial China perfected and relished, something that's so entrenched that nothing short of wars and revolutions could change.

But is any of this relevant in a book that's supposed to be about the human knee? Well, once again, if we look just at the surface, at the scientific facts of anatomy or medicine, then I guess not. But, our bodies—and lives— are more than, *so much* more than, static facts or scientific curiosities. It is what we do with our bodies that matters. If walking upright with two bipedal knees makes us human, then bending (or not bending) those knees complicates the meaning of humanity. If it matters how we use our knees to play sports, then it matters *even more* how we use those knees to function in societies.

For a story of kneeling, there is no better place to start than China.

The Chinese people pride themselves on their ancestors' ingenuities, their inventions that changed the course of human civilization: the compass, gun powder, printing techniques, and papermaking techniques. But something that seemed simple and ordinary, dull really, was beyond ancient Chinese's imagination: chairs. For thousands of years, the good people went about their daily life—eating, cooking, and socializing—by way of kneeling rather than sitting. Or, more precisely, kneeling *was* sitting.

This "kneeling-sit" is done by dropping both knees to the ground, with lower legs tucked under the thighs. The tops of the feet are flat on the ground, and the butt sits on the heels. Depending on how straight the upper torso is, one kneeling-sit is different from another. A slouched upper torso constitutes a "comfortable sit" (安坐, pronounced *an zuo*). A straightened upper torso constitutes a "tense sit" (危坐, pronounced *wei zuo*). Generally speaking, the straighter the upper body, the more formality and politeness, with a ninety-degree angle between the upper torso and the ground being the recommended posture for formal occasions. In this posture, one is meant to look simultaneously dignified and modest.

While one may look that, one is also, after a while, quite uncomfortable. Surely, even without the concept of chairs, ancient Chinese could simply lounge about on the floor, stretch out their legs, cross them in the front, or something? They could—and some would—but only in private. Ancient Chinese, under the influence of Confucianism, were dead serious about virtue and propriety, and stretching out one's legs and feet in public wasn't decorous or proper. Precisely *because* the kneeling-sit wasn't comfortable, it was considered good manners and a reflection of good character, so much so that the self-disciplined would insist on kneeling even when no one was looking.[20] In contemporary Japan, the same posture, known as *seiza*, continues to be performed in formal ceremonies and is used by some parents as a way to discipline their children.[21]

Because people were already kneeling, kowtow became a natural next step.[22] Think about it: when sitting or standing, we bow our heads or upper bodies to show respect and gratitude; in a kneeling position, when we attempt to bow, we have no place to go but to prostrate, bringing the head close to the ground. And there you have it: kowtow.

For Lord Macartney, to kowtow, to be "undisciplined in personal postures," was to be weak, pretentious, and even morally deviant.[23] But it wasn't meant to be so for the Chinese, at least not at first, because remember this: at that time, *everyone* was on their knees. So, when a courtier kowtowed to a monarch, the courtier wasn't physically—and thus wasn't symbolically—much lower than the monarch. Indeed, the monarch would, in one way or another, kowtow back. Similar to kneeling, kowtow wasn't just one posture to the ancient Chinese. Depending on how long the head stayed on the ground and whether it touched the hands on the ground or bowed still lower to the ground, one kowtow was different from the next.[24] Generally speaking, the lower and the longer the head bows, the more respect is shown. So, a courtier might greet a monarch with an extended kowtow to the ground (稽首, pronounced *ji shou*), whereas the monarch might return the gesture with a brief kowtow to the hands (空首, pronounced *kong shou*).

These floor-based ritual exchanges are thought to undergird a healthy relationship between the monarch and his courtiers: courtiers revere and are devoted to a benevolent monarch, while the monarch respects and values competent courtiers.[25] These two differ in roles and ranks, but not in human dignity.

All of this changed when chairs came to China.

When did chairs appear, and where did they come from?[26] Although the actual dates are likely to be earlier, murals unearthed in Dunhuang, Gansu Province, depict the first unmistakable chair in China and put the date within the years 535–556. Historians believe that chairs originated in Egypt. From there, they spread to West Asia, Europe, Central Asia, and followed the Silk Road to China.

Content with—and proud of—their self-disciplined kneeling, the Chinese didn't take to chairs for their comfort, at least not initially. Rather, chairs arose as a by-product of Buddhism, which officially entered China in the first century CE with the blessing of emperor Ming (28 –75 CE) of the Han dynasty (202 BCE–220 CE). Buddhism temples were erected, and Buddhists traveled to China to preach their religion, doing so whilst sitting on their stools and chairs. This foreign posture (with feet dangling in public, the nerve!) was met with raised eyebrows from the Chinese, so much so that some Buddhists compromised by kneeling on their chairs. But, as Buddhism proceeded to establish dominance in China, the paradigm shifted: sitting gradually became the mainstream and had replaced kneeling by the mid-Tang dynasty (618–907 CE). The fact that it was a lot more comfortable than kneeling sure didn't hurt.

The rise of the chairs fundamentally changed the relationship between emperors and courtiers, rulers and the ruled, and by extension, the collective psyche of the Chinese for hundreds and thousands of years to come.

How did the lowly chairs do that? Simple, really. Although sitting was now the norm in everyday life, the ritual of dropping the knees to greet the emperors remained. When one party sat tall on his chair and the other groveled on the ground, the two were no longer physically or symbolically equal. Over time, they no longer *felt* that they *should* be equal. Emperors became the sons of the heaven, the demigods, while everyone else was, by nature, infinitely inferior.

Emperors, relishing their superiority and shrewdly sensing the ritual's ability to condition the mass, doubled down on it, bringing the ritual to an all-time high in Qian Long's Qing dynasty.[27] Gone were the days when courtiers only knelt to converse with an emperor; now, they knelt whenever in the presence of the emperor. Gone were the days when they knelt and kowtowed only once; now, they knelt three times and kowtowed nine times.

The ritual also spread beyond the imperial court. Anywhere power and status were different, people bent their knees: inferiors to superiors, commoners to officials, students to masters, children to parents. After hundreds of years of bending their knees, the Chinese, or so the idea goes, developed a slavish attitude that runs deep to their cultural core.[28]

Within this historical context, then, yes, the British embassies' refusal to kneel *is* significant. To refuse to kneel is to doubt the omnipotence of the Chinese imperial majesty, the son of heaven. It is to doubt the supremacy of his Celestial Empire, which includes everything under the sun. And it is to reject the very fabric of the Chinese society. Why on earth would the majesty suffer such insubordination? And if he is *so* benevolent as to tolerate the barbaric behavior from a group of foreigners who know no better, he surely is not going to grant any of their wishes.

With this historical context, it is also clear why the contemporary New China wants nothing to do with kneeling. Kneeling to officials and authorities was banned by the Republic of China (1912–1949) after it overthrew the Qing dynasty. Under the socialist banner, the People's Republic of China (1949–) further ruled all "old relics, traditions, and norms" feudal dregs and vowed to ruthlessly irradicate them during the Cultural Revolution.[29] Confucianism bore the brunt of that revolution. Meanwhile, the people, painfully aware of how vulnerable and humiliated they were in the face of foreign invasions and colonialism in the one hundred years since the Opium War, took to heart Chairman Mao's 1949 declaration that "the Chinese people have now stood up."

And so it is that henceforth in China, kneeling becomes a social aberration, an admission of personal and *national* weakness, a sign of inequality that supposedly no longer exists in the New China. When it does occasionally happen, it is a surefire way to attract public outcry, as Samsung learned, the hard way.

In October 2016, the South Korean electronics giant recalled its Galaxy Note 7 mobile phone from the Chinese market—because its battery would spontaneously catch fire and explode. Later that month, Samsung held a stock-ordering event in the Chinese city Shi-jia-zhuang. At the event, some twenty Samsung executives, including both Chinese and Korean nationals, sank to their knees and kowtowed to thank the local distributors and apologize for the recall. When the kneeling picture surfaced online, Chinese customers went livid. They accused Samsung, a foreign company, of forcing its Chinese employees to kneel, all so that the company could get more product orders. Samsung denied that the act was coercive and insisted that the Chinese employees knelt on their own accord—because they were moved by the local distributors' continued support.[30]

I'm not privy to what happened behind closed doors at Samsung, but I know that it would take a heck of a lot more than "being moved" for today's middle-aged Chinese men to kneel and kowtow in public.

Less condemned but still controversial was the collective kowtowing of two thousand high school students to their parents, a school-organized event that supposedly taught gratitude, a virtue that many thought sorely missing in today's Chinese youth.[31] Then there was the collective kneeling of three thousand middle and high school students to their teachers to express respect and appreciation.[32] Less grand in size are traditional Chinese wedding ceremonies during which the newly wed kowtow to parents and families to express gratitude, or Spring Festival celebrations during which families kowtow to their elders to send best wishes on an ancient Chinese holiday.

Are these kneelings and kowtows feudal dregs, poisons of a bygone exploiting class? Or are they traditions, customs, and culture? Respecting one's parents, elders, and teachers *is* an ancient Chinese virtue. To this day, legends of the filially pious are being recited: for example, Xiang Wang from the Jin dynasty (266–420 AD) undressed in the dead of winter and used body heat to try thawing a frozen river and catching fish, all because his mother had a craving. I'm not sure how factually reliable—or physically possible—this story is, but you get the idea. So, is bending one's knees in the same spirit despicable feudal dregs, or a proud cultural tradition?

The Chinese are torn. At a time when the United Nations Educational, Scientific and Cultural Organization (UNESCO) intangible cultural heritage is all the rage in China, and the once-condemned Confucianism is used as a branding tool to promote the global spread of Chinese language and culture, the line between feudal dregs and cultural heritage seems thin. Complicating the question is the matter of cultural imperialism. When young Chinese don Western-style bridal gowns and suits for weddings but turn their noses up at traditional ceremonies, when they go to town on Christmas celebrations but forgo Spring Festival customs, the old-fashioned Chinese want to hold onto what their ancestors passed on—as does the socialist government, to an extent.

So, more than two hundred years after Lord Macartney's visit, the question of to kneel or not to kneel remains relevant. As someone who spent most of her adult life outside the country and removed from its fervent rhetoric and reality, I feel simultaneously qualified and inept at answering this question.

In case you wonder, I had never knelt to my parents or elders—I suppose my family is fairly "progressive," whatever that means. But doing so, in one's private home, doesn't feel alienatingly foreign to me. After all, culture is, in many ways, performed; without performance—the decorations, the clothes, the food—there can be no culture.

Kneeling en masse in public feels a bit odd to me, not necessarily because it feels humiliating, but rather because it feels fake. Respect and gratitude come from voluntary sentiments, not compulsory rituals. When thousands kneel on cue and in unison, it feels like a line somewhere has been crossed, and cultural performance has morphed into a publicity stunt.

So it is that Lord Macartney wouldn't harbor genuine admiration even if he kowtowed a hundred times to Qian Long, no more than today's Chinese youth can learn gratitude by organized kneeling. This seems to be an obvious conclusion, isn't it, certainly one that can be appreciated by the leader of the free world?

In American football, kneeling, or so I am told, is a useful tactic. Players may take a knee before halftime when there is little chance of advancing the ball; they may also take a knee at the end of the game to run out the clock. Multiple terminologies exist to describe the maneuver: a quarterback kneel, genuflect offense, victory formation.

As someone who doesn't really watch football, I find it interesting that a sport known for its raw strength and rough combat would co-opt a submissive, vulnerable posture into its gameplay. Okay, I don't just find it interesting. I find it confusing, especially as I attempt to follow the detailed rules pertaining to the tactics' execution. Still, conceptually, at least, I get the idea. What I find more incomprehensible, utterly inexplicable really, is what happened to American football in the summer of 2016.

On August 26, 2016, two weeks before the National Football League (NFL) season was to start, the San Francisco 49ers played a preseason game with the Green Bay Packers at Levi's Stadium in Santa Clara, California. As is the custom, the U.S. national anthem was sung before any ball could be played. And, as is the custom, players, staff, and audiences stood up for the anthem.

One person, the quarterback of the 49ers, Colin Kaepernick, didn't. He sat ringside, alone and discrete, obscured by some coolers—but still conspicuous enough to be noticed by the NFL media reporter Steve Wyche. Wisely sensing that something was up, Wyche sought out Kaepernick for an exclusive interview, the content of which was published the next day.

What was Kaepernick doing? He was protesting against police violence toward unarmed Black people. Kaepernick was born to a white mother and Black father. "I am not going to stand up to show pride in a flag for a country that oppresses Black people and people of color," Kaepernick said. "To me, this is bigger than football and it would be selfish on my part to look the other way. There are bodies in the street and people getting paid leave and getting away with murder."[33] Alton Sterling, Philando Castile, Charles Kinsey—they were the people on Kaepernick's mind.[34]

With that statement, the young quarterback was thrust into the media spotlight—and voluminous public outcry. The outcry was not about police brutality, or racial inequality, but rituals, the anthem ritual. The idea goes that, by sitting down during the national anthem, "The Star-Spangled Banner," Kaepernick disrespected the U.S. flag and the service men and women who bravely fought for the freedom of the American people.

Knowing what I know about chairs, I get this reproof: posture-wise, sitting is more about being comfortable and superior than being humble and respectful. Clearly, Kaepernick saw that, too. Shortly after making his statement, at the suggestion of Nate Boyer, a former NFL player and U.S. Army Green Beret, Kaepernick changed his tactic from sitting to kneeling, a position that he, as a quarterback, would be no stranger to on the football field.

"Soldiers take a knee in front of a fallen brother's grave, you know, to show respect," Boyer explained.[35] So not only would kneeling absolve Kaepernick of *dis*respect, it could *demonstrate* his respect to fallen soldiers—as well as his Black brothers fallen on the American streets. Eric Reid liked the idea, too. A fellow 49er, and a Black, Reid was among the

FIGURE 5.3 Am I not a man and a brother? "The Image of the Supplicant Slave: Advert or Advocate?," 1807 Commemorated, Institute for the Public Understanding of the Past, University of York, 2007, https://archives.history.ac.uk/1807commemorated /discussion/supplicant_slave.html.

first NFL players to join Kaepernick in kneeling during the anthem. "We chose to kneel because it's a respectful gesture," Reid said. "I remember thinking our posture was like a flag flown at half-mast to mark a tragedy."[36]

Actually, it is more than just "respect." It is submission—the same as it was in ancient China. The kneeling Black is a visual trope that dates back to the eighteenth-century British abolitionist movement (figure 5.3). This image, designed by Josiah Wedgwood's pottery firm in 1787, was reproduced on a variety of items (medallions, jewelry, pottery) to spread antislavery sentiment in Britain.[37] The Black man, chained and kneeling, raises his pleading hands and implores, "Am I not a man and a brother?" A similar image depicting an enslaved Black woman was later used in the American abolitionist movement. In this visual trope, the Blacks don't demand or rebel; they beg for freedom and beseech their white masters for benevolence.

So, whether it is in the East or the West, feudal empires or slaveholding societies, the kneeling of a bipedal human carries something fundamentally similar. Or so I thought.

On September 1, 2016, Kaepernick and Reid put their new tactic into play. Before the preseason game between the 49ers and the Los Angeles Chargers at San Diego's Qualcomm Stadium, as "The Star-Spangled Banner" played, Kaepernick and Reid each took a knee. Everyone else stood. The audience booed.

On social media, Kaepernick's kneeling was called "a disgrace to those people who have served and currently serve our country."[38] Then-Republican presidential candidate Donald Trump was the de facto leader in the backlash: "I think it's a lack of respect for our country. I think it's a lack of appreciation for our country and it's a very sad thing," he said. Maybe the kneelers "should try another country, see if they like it better," he added.[39]

Knowing what I know about kneeling, I was puzzled. I couldn't see how kneeling could be anything *but* a subdued, respectful posture. Maybe in this land and culture that doesn't have a long history with kneeling, the posture means something different, something that people like me, from "another country," aren't privy to?

But, apparently, that's not the case. "Pious pilgrims kneel and so too do nervous lovers stammering to pop the question," writes American art critic Kelly Grovier. "A species of bowing, kneeling brings the body low in order to demonstrate the kneeler's insignificance in the presence of a more worthy or powerful figure," he continues.[40] "While we can't know for sure," writes American authors and scholars Jeremy Adam Smith and Dacher Keltner, "kneeling probably derives from a core principle in mammalian nonverbal behavior: make the body smaller and look up to show respect, esteem, and deference. . . . Kneeling can also be a posture of mourning and sadness. It makes the one who kneels more vulnerable. In some situations, kneeling can be seen as a request for protection."[41]

Indeed, like me, these writers seem to be scrambling to answer why kneeling was misconstrued by their countrymen and women. Smith and Keltner proposed that the misunderstanding was due to the fact that "[o]ur amygdalae activate as soon as our brains spot deviations from routine, social norms, and in-group tendencies." That feeling, they add, would be compounded in high-power people, such as "the president or members of the numerical majority" (i.e., white people) who, supposedly, tend to misinterpret nonverbal behaviors because they are less able to see other people's perspectives.

To be perfectly honest, I didn't find these answers all that satisfying. Science is powerful in many ways, but neurobiological reasons alone can't explain complex social problems.[42] More things than our amygdalae must be at play. Tentatively, I turned to media reports and online commentaries to try and find more convincing answers. I found something alright. Whether or not it was convincing, I'm less sure.

Shortly after Kaepernick's first kneeling, a poll from the Quinnipiac University found that a majority of American adults (54 percent) disapproved of the kneeling protests, while 38 percent approved.[43] Two years later, in August 2018, the numbers held fairly steady in an NBC News/*Wall Street Journal* poll: 54 percent disapproved; 43 percent approved.[44] As one can expect, these public opinions differed sharply along partisan lines. In the 2018 poll, 88 percent of Republicans found kneeling during the anthem inappropriate, compared with 23 percent of Democrats. Similar divides were found along racial lines. Between the two surveys, 70–74 percent of Blacks approved of the athletes' anthem protests, in comparison to 30–38 percent of whites.[45] More interestingly (if that's the right word), public opinions also differed along theological lines. Christians as a whole frowned upon athletes taking a knee. In 2018, only 14 percent of white Evangelical Protestants and 30 percent of white mainline Protestants approved of it, compared with 63 percent of religiously unaffiliated American adults.[46] Among people who attended church weekly, 36 percent supported players' kneeling protests; among nonregular church goers, a small majority (51 percent) did.[47]

Why might Christians be especially averse to kneeling during the national anthem? Supposedly, it's because they believe kneeling is a respectful, submissive posture reserved only for prayers and the almighty God. To kneel to anything else in a secular context would be sacrilegious. In the words of the San Francisco Giants pitcher Sam Coonrod, "I'm just a Christian. I believe I can't kneel before anything but God, Jesus Christ."[48]

It is interesting that Coonrod emphasized that he was *just* a Christian, as opposed to what? A public figure, a role model, a citizen? Coonrod may not have known of Bishop Richard Allen, a devout Christian and founder of the African Methodist Episcopal Church, who knelt in 1787 in the "white section" of his church to protest against segregation. But surely, Coonrod would know about Martin Luther King Jr., who knelt in prayer in 1965 after a group of social workers and Blacks were arrested in Alabama during a voter registration drive? For these religious leaders, faith in God and faith in social equality don't conflict.

At the risk of sounding cheeky, I'm also curious about Coonrod's thoughts on men taking a knee to propose. Apparently, 76 percent of American men believe that that is the right way to propose, or so says a survey from the *Men's Health* magazine and TheKnot.com[49]—if there ever is an authority on the subject, they ought to be, right? So are 76 percent of American men also out of line?

Indeed, athletes on the other side of the debate evoked religion *to* kneel. Taking a knee, for them, is "responding to God's call to build a better world."[50] As Eric Reid put it, "my faith moved me to take action. I looked to James 2:17, which states, 'Faith by itself, if it does not have works, is dead.' "[51]

Not finding the religion-derived answer all that compelling, I kept reading. Unexpectedly, at some point, I noticed something in plain sight that I hadn't at all thought about. It was simple and obvious—so simple and obvious that I'm afraid any attempt to interrogate it may come across as laughable.

In all the media reports about the kneeling controversy, one other word besides "kneel" frequently appears: "stand." Semantically speaking, this is a far more interesting word than "kneel" in the English language—because its meanings are multiple. Literally, to stand means to assume an upright, bipedal posture, as in the following statement from the NFL: "Players are encouraged but not required to stand during the playing of the national anthem."[52] But in addition, and more frequently, the word is used metaphorically, as in the *New York Times* headline that "Colin Kaepernick takes a stand by kneeling"[53] or in Kaepernick's own words that "I have to stand up for people that are oppressed."[54] In these instances, to "stand" means to commit, defend, and support. Related expressions are abundant in the English language: "stand on one's two feet," "stand one's ground," "know where one stands," "stand by," "stand against," and "stand down."

If something is this extensively codified in the language, it must be significant for the people who speak that language—much like how the Chinese used multiple phrases to denote different kinds of kneeling and kowtowing. If kneeling really is the actual and symbolic cause for Chinese subservience, then standing may be the actual and symbolic cause for Western pride, a notion that goes back to the Enlightenment, to Lord Macartney's belief that an upright physical posture indicated moral integrity.[55] Plus, an American would have more reasons to be proud, because the United States is, as the idea goes, exceptional: a country that is virtuous, divinely blessed, the biblical "city on a hill." It is the Celestial Empire of the twenty-first century.

If it is barbaric not to kneel in Qian Long's court, it is just as barbaric not to stand during "The Star-Spangled Banner." At risk here are not mere rituals, but cultural psyches. When Black athletes Tommie Smith and John Carlos raised their fists at the 1968 Mexico City Olympic Games to protest racism in the United States, they admitted to the world that their country wasn't exceptional. That was embarrassing to the proud America. Kaepernick's kneeling hit the same nerve.

So, it is not that Kaepernick's naysayers can't comprehend the fact that kneeling is a physically submissive posture. It is that they, consciously or otherwise, sense that anything other than a proud, bipedal stand is a slap in the face of American exceptionalism and therefore, in *their* logic, it is unpatriotic. Never mind writer and activist James Baldwin's famous words that "I love America more than any other country in the world and, exactly for this reason, I insist on the right to criticize her perpetually."[56]

Incidentally, people who are quick to dole out the unpatriotic charges may not know—or care to know—that prior to 2009, NFL players usually remained in their locker rooms during the anthem.[57] This only changed when the Pentagon contracted with the NFL (and other leagues) to publicly display patriotism. The enormous flags, the Air Force flyovers, the soldier reunions, they cost millions of taxpayer dollars—but they also make darn good military recruitment campaigns. Chinese monarchs learned a long time ago that normalized rituals can condition the masses. The U.S. government and corporations are catching on.

As I reach the answer of the proud American, I'm more or less intellectually satisfied, but I can't shake the feeling that I'm being naively academic in my search for rational answers. Maybe there were no rational answers to begin with. Maybe kneeling per se had nothing to do with anything and was merely being used to hide something less palatable to the American people: race.

From day one, Kaepernick had made it clear that his action was to protest the oppression of Black people and people of color in America. His co-kneeler Eric Reid elaborated: "In early 2016, I began paying attention to reports about the incredible number of unarmed black people being killed by the police. The posts on social media deeply disturbed me, but one in particular brought me to tears: the killing of Alton Sterling in my hometown Baton Rouge, La. This could have happened to any of my family members who still live in the area. I felt furious, hurt and hopeless. I wanted to do something."[58]

Race is an uncomfortable topic, even under the best circumstances. I am a person of color, and *I* find it uncomfortable, on oh so many levels. Deep down, I don't want to believe that my fellow humans can be prejudiced against me, based on no fault of my own. I don't want to be reminded of that possibility and reality, to feel like a victim, to be pegged as a "diversity hire," or to come across as "playing the race card." I imagine that for white individuals, whose race is accused of oppression, even when they agree with that charge, the discomfort is compounded. And for those who deny the existence of racism in this country, discussions of race are irrelevant, if not absurd.

Something else, then, must be invoked to explain why colored athletes are taking a knee and making a scene, something that can circumvent the uncomfortable topic of race. And what better than patriotism?

In the United States, patriotism has a color, and that color ain't Black. Black soldiers who fought in World War I were resented (and feared) because they were seen as a threat to the status quo, as "sons of bitches" who "use military service and patriotism as a pathway to equality of rights and opportunities."[59] In June 1995, an NBC News survey asked, "When you hear about someone being 'patriotic,' do you think of a white man, a white woman, a black man, or a black woman?" Only 2 percent of the white respondents pictured a Black person, while 50 percent pictured a white person.[60] In fact, 66 percent of *Black* respondents also pictured a white person, while only 12 percent pictured a Black person. In a February 2012 American National Election Studies survey, 28 percent of white respondents thought that the word "patriot" described Blacks well, while 51 percent thought that the word was a fitting descriptor of whites.[61]

Before Kaepernick, NFL players have used other tactics that have nothing to do with kneeling to protest against police brutality and racial injustice: a "don't shoot" pose or a justice T-shirt, for example. But these actions, too, invited backlashes.[62] So, really, any demonstration by Black men, it seems, begets outrage, and the best way to rally that outrage is to question their loyalty to the country.

Then-president Donald Trump knows this well. "The issue of kneeling has nothing to do with race," he tweeted. "It is about respect for our Country, Flag and National Anthem. NFL must respect this!"[63] Using one of the most juvenile (and hurtful) tactics, Trump suggested, on more than one occasion, that the protestors find another country. What Trump didn't broadcast was Kaepernick's donations of more than $1 million to organizations that focused on anti–police brutality, youth initiatives, community reform, minority empowerment, and health care.[64] That seems like a lot of unnecessary work, if all one wants to do is to disrespect one's country.

Fortunately, not all American people are naïve. In a 2017 *Economist/YouGov* survey, the majority of American adults (66 percent) said that NFL players' kneeling protest was "a matter of race," while 34 percent thought it was "a matter of patriotism."[65] Many people also took to Twitter to counter the president's message: "Thinking NFL players are 'protesting the flag' is like thinking Rosa Parks was protesting public transportation," tweeted JEFF. "Don't allow racists to reframe #TakeAKnee as being a debate about anthem & flag. It's a protest of police brutality & racism," tweeted Bree Newsome.[66]

What about those people who were, supposedly, deeply wronged by the kneeling: the military, the veterans? The American Legion, the nation's largest veterans service organization, didn't approve and called the players "misguided and ungrateful."[67] The Veterans of Foreign Wars concurred: "[W]earing team jerseys and using sporting events to disrespect our country doesn't wash with millions of military veterans who have and continue to wear real uniforms on real battlefields around the globe."[68]

Curiously, not all veterans agree. What Marine Corps veteran Matt Eidson said is worth quoting at length:

> I served in the Marine Corps from February 2008 until August 2015. In that time, I deployed to Iraq and Afghanistan, where I lost a few close friends. The names of those friends are etched on two small black bands that I wear on my wrists every day. While I certainly won't speak for my buddies, I will speak for myself: if the roles were reversed, and my name was the one etched on a small black band, I would not have been concerned with whether Colin Kaepernick was respecting my sacrifice in an appropriate manner. . . . When I look at my fellow Americans protesting and speaking out against injustice, I smile. I smile because that's why I served in the military: to make the country a better place.[69]

This side of the military voice gave rise to the hashtag #VeteransFor-Kaepernick, which is widely circulated on Twitter. "As veterans, we swore an oath to support and defend the Constitutional rights of all citizens to speak freely and protest. #TakeAKnee #ImWithKaep #VeteransForKaepernick," tweeted VoteVets. "#VeteransForKaepernick I'm sick of these people who believe the military is disrespected by Kaepernick & others kneeling. I'm a veteran and I support the right to kneel," tweeted Jo Wright.[70]

If "patriotism" is more palatable than race, then so, too, is "unity." On September 22, 2017, during an address in Huntsville, Alabama, Trump called the kneeling NFL players "sons of bitches" and urged NFL owners to have them fired. Fired! Facing this personal attack, the NFL rushed to celebrate "unity" in its commercial.[71] People who were uncomfortable to rally under antiracism rallied under the vanilla slogan of "unity." Players, owners and coaches, too, who were uncomfortable bending their knees for racial injustice stood with locked arms for comradery and moral support (figure 5.4).

While the NFL united, Kaepernick, as Trump wished, was fired. He has not played professional football since his kneeling season of 2016 ended. In 2017, Kaepernick filed a grievance against the NFL for colluding to keep him off the field because of his racial protests. Reportedly, he, along with fellow kneeler Eric Reid, who also filed an aggrievance,[72] received a settlement in the neighborhood of $10 million. For context, in the year 2016 alone, Kaepernick earned $14 million playing for the San Francisco 49ers.[73] If only he had stood.

Three years and nine months after Kaepernick first knelt inside Qualcomm Stadium in San Diego, police officer Derek Chauvin knelt outside a Cup Foods store in South Minneapolis. And America woke up to a violent new chapter on kneeling.

Kaepernick was on his knees for about two minutes, the average length of a performance of "The Star-Spangled Banner." Chauvin knelt for nine minutes and twenty-nine seconds, more than enough time to kill George Floyd, a Black man whose neck was under Chauvin's white knee.

It happened on May 25, 2020, Memorial Day, actually—chillingly ironic given everything that has been said about freedom and rights.

At 7:57 p.m.,[74] employees from the Cup Foods store confronted Floyd for using a counterfeit bill to purchase cigarettes. They demanded the cigarettes

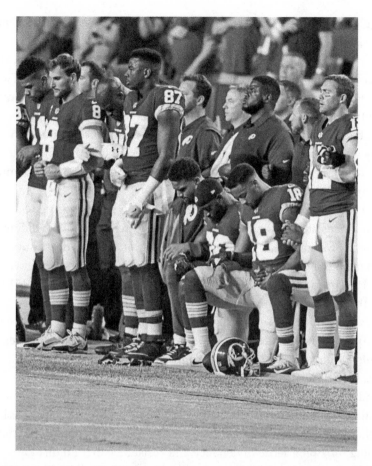

FIGURE 5.4 The then–Washington Redskins during the national anthem before a game with the Oakland Raiders on September 24, 2017, at FedExField in Landover, Maryland. Keith Allison, *Washington Redskins National Anthem Kneeling*, September 24, 2017. Courtesy of Wikimedia Commons.

back, Floyd refused, and an employee called the police. At 8:08 p.m., two officers arrived at the scene. At this time, Floyd was sitting with two people in a parked vehicle around the corner. There was a brief struggle as the officers pulled Floyd out of the vehicle and handcuffed him. The officers then walked Floyd toward their squad car and attempted to put him into the car. Floyd refused, stating that he was claustrophobic and couldn't breathe. At 8:17 p.m., two more officers arrived, including Derek Chauvin, who became involved in the struggle to get Floyd into the police car. During the struggle, at 8:19 p.m., Floyd fell to the ground. The officers called for medical assistance.

Floyd was now lying face down on the ground. To restrain him, three officers pinned him down, with Chauvin pushing his left knee into Floyd's neck. Floyd gasped, pleaded, and told the officers that he couldn't breathe some twenty times, but Chauvin kept his knee on Floyd's neck, even after Floyd became unresponsive. At 8:27 p.m., an ambulance arrived, Chauvin finally let go of his knee, and Floyd was transferred into the ambulance. En route to the hospital, Floyd went into cardiac arrest and was pronounced dead around 9:25 p.m. at a nearby hospital.

The City promptly fired all four officers. Chauvin was charged with murder and manslaughter, and the other officers were charged with aiding and abetting.[75] Meanwhile, bystander video footage of Floyd's death spread on social media, and longstanding crises over police brutality and racism boiled over. Hundreds of protesters took to the streets of Minneapolis.[76] Within days, the protest had spread to more than 140 cities across the United States and, eventually, to all fifty states and Washington, D.C., as well as more than seventy countries worldwide.[77] If ever there was a slap on American exceptionalism, this was it, as people from Ireland to Japan, Australia to South Africa held up signs that read Black Lives Matter and I Can't Breathe.[78]

And people took a knee, to emulate Kaepernick's protest and to echo, intentionally or otherwise, the manner of Floyd's death.[79] As hundreds knelt together, holding their fists and signs, with rage and desperation in the air, kneeling took on a sense of defiance and resistance that I didn't know was possible. It wasn't possible when a handful of athletes knelt inside sports stadiums in front of crowds waiting to have a good time. It surely wasn't possible when enslaved Blacks knelt and pleaded for their freedom.

Some of those in authority also took a knee. From Santa Cruz to New York City to Ferguson, police knelt in solidarity with demonstrators.[80] Politicians, too. Los Angeles Mayor Eric Garcetti knelt outside police headquarters in downtown Los Angeles; the presumptive Democratic presidential nominee Joseph Biden, Jr. knelt during a campaign visit to a Delaware Black church.[81]

In the world of sports, leagues started to backpedal. NFL Commissioner Roger Goodell admitted that the NFL was "wrong for not listening to NFL players earlier" and encouraged "all to speak out and peacefully protest."[82] The U.S. Soccer Board followed suit and rescinded its 2017 policy that all players must stand for the national anthem, stating that "this policy was wrong and detracted from the important message of Black Lives Matter."[83]

The American public also warmed up to the idea of kneeling. In September 2020, a poll conducted by the *Washington Post* showed that the majority of Americans, 56 percent, now think it is appropriate for athletes to kneel during the national anthem to protest racial inequality.[84] Two years ago, that number was 43 percent.[85] In the 2020 poll, the kneeling-approval rates were especially high among Democrats (73 percent), Black Americans (72 percent), and people under the age of fifty (about 66 percent).

Of course, the survey also showed that, post–George Floyd, 42 percent of Americans, including a majority of Republicans (63 percent), still thought it was inappropriate for athletes to kneel during the national anthem, and many continued to express their disgust, not the least President Donald Trump. On July 21, 2020, Trump tweeted, "Looking forward to live sports, but any time I witness a player kneeling during the National Anthem, a sign of great disrespect for our Country and our Flag, the game is over for me!"

None of these developments is surprising for a country reeling from a violent murder but holding onto its deep-seated belief and pride. What *is* interesting (again, if that's the right word) is that post–George Floyd, athletes were compelled to explain why they opted to *stand*, rather than kneel, during the anthem. The tide, you may say, had turned.

On June 27, at a National Women's Soccer match between Chicago Red Stars and Washington Spirit, many players took a knee during the pregame national anthem. Rachel Hill, the twenty-five-year-old forward for the Red Stars, wearing a Black Lives Matter T-shirt, stood. As she stood, Rachel put one hand on the shoulder of a black teammate, Casey Short, who was kneeling (and sobbing) beside her. After pictures of Hill's posture surfaced, she was barraged by criticism for not kneeling and was compelled to offer a lengthy explanation on social media:

> What the black community goes through on a daily basis in this country is unacceptable. The pain. The hurt. Facing racism, prejudice, and social injustice—it must change.
>
> When I stood for the national anthem before the Chicago Red Stars' most recent game, this was a decision that did not come easily or without profound thought. . . . I chose to stand because of what the flag inherently means to my military family members and me, but I 100 percent support my peers. Symbolically, I tried to show this with the placement

of my hand on Casey's shoulder and bowing my head. I struggled, but felt
that these actions showed my truth, and in the end I wanted to remain
true to myself. . . . I support the black lives matter movement whole-
heartedly. I also support and will do my part in fighting against the current inequality.[86]

The widely circulated picture of Hill is cropped. The whole picture
shows other players standing as well. Ironically, by standing beside a kneeling Black teammate and touching her shoulder to try to show support,
Hill unwittingly singled herself out as someone who is callous, unsupport-
ive, and un-just. "Hill tried to have it both ways," wrote *New York Times*
reporter John Branch, but "[t]here is little room for such posturing."[87]

That, I think, is sad.

Why *can't* we have it both ways? If it is irrational (which it is) to equate
kneeling with being unpatriotic, it is also irrational to equate standing with
racial oppression. Standing for the anthem is a ritual, but kneeling to protest can also be turned into a ritual. If one can be co-opted into military's
paid advertising, the other can become politicians' and media's empty rhetoric, as the Chinese have done with their mass kneeling.

At the end of the day, rituals, any rituals, can't bring about equality and
justice. Meaningful actions stand a chance.

6

Treatment, or Placebo

Ice has been a standard treatment for injuries and sore muscles because it helps to relieve pain caused by injured tissue. Coaches have used my "RICE" guideline for decades, but now it appears that [ice] may delay healing, instead of helping.

—GABE MIRKIN, AMERICAN SPORTS MEDICINE DOCTOR AND
INVENTOR OF THE RICE TREATMENT

In an effort to prevent or to minimize these serious [knee] injuries, or to prevent repeated injury, orthotic braces for the knee have been developed and marketed. These braces have gained widespread use . . . despite the paucity of scientific documentation on their efficacy.

—YELVERTON TEGNER AND RONNY LORENTZON,
SWEDISH ORTHOPEDIC DOCTORS

Our knees, as should be abundantly clear by now, are easily injured. And everyone knows that when that happens, our first line of defense is icing. Icing reduces pain, swelling, and inflammation. From peewee football to professional games, whenever a bruised knee pops up, someone rushes for a bag of ice or cold spray.

In the comfort of one's own home, icing can be delivered in more, and more creative, ways. Equipped with a tub or sizable bucket, one can immerse knees in icy water. Unequipped or unwilling to fill a tub, one can wet towels in icy water and apply the towels over the knees. The meticulous among us massage using ice cubes. Those who plan ahead rely on frozen gel packs. Those who don't, reach for a bag of frozen peas.

The latest fashion among athletes is whole body cryotherapy.[1] *Cryo* means involving or producing cold, so cryotherapy is a fancy word for cold therapy. Whole-body cryotherapy is accomplished in a special chamber where the temperature is set to −110°C (−166°F) to −140°C (−220°F)! A person stays in the chamber for two to five minutes while wearing a bathing suit to take in the cold.

Because the knee is (thankfully) about the only part that hurts in this body, I have not sought whole-body cryotherapy. Still, I was an icing enthusiast. I own an assorted array of ice bags and ice packs and routinely carried them. I iced my knee at the first inkling of pain. I iced after exercise, whether or not it hurt, you know, just in case, even in the dead of winter. When the husband suggested that maybe this was excessive behavior, I immediately recognized that he was wrong, because he hadn't caught up with common-sense principles like RICE (rest, ice, compression, and elevation).

Indeed, using ice to reduce pain has a long, long history, dating back to ancient Egyptians and the father of modern medicine, Hippocrates.[2] In *Aphorisms*, Hippocrates wrote that "[s]wellings and pains in the joints, ulceration, those of a gouty nature, and sprains, are generally improved by a copious affusion of cold water, which reduces the swelling, and removes the pain; for a moderate degree of numbness removes pain."[3]

More ambitiously, in 1050, an unnamed Anglo-Saxon monk made the first written record of using cold as a surgical anesthesia. The record goes, "Again, for eruptive rash. Let him sit in cold water until it be deadened; then draw him up. Then cut four scarifications around the pocks and let drip as long as he will."[4]

This anesthetic effect was witnessed by Napoleon's surgeon-general Dominique Jean Larrey after the 1807 Battle of Eylau between Napoleon and the Imperial Russian Army. Larrey remarked, rather coolly, that "amputation could be painlessly performed on soldiers who had lain for some time in the snow."[5]

Finally, in 1938, modern medicine took notice. Dr. Frederick Allen of New York proposed that limbs packed in ice for three hours could be amputated without any anesthetic agent—although whiskey might be offered to patients for emotional distress.[6] Nicknamed refrigeration anesthesia, Allen's method was theorized to have many benefits.[7] By reducing tissue temperature, icing was thought to reduce cells' needs for nutrients, essentially putting them into hibernation. This allowed cells to survive the reduced blood and nutrient supply following an injury. Low temperature also inhibited bacterial growth, which was thought to reduce the risk of infection. Refrigeration anesthesia was deemed especially superior for aged patients with diabetes or vascular diseases who were poor candidates for deep general anesthesia or spinal anesthesia, which might cause post-operative shock or sudden changes in blood pressure.[8]

The New York City Hospital, the "dumping ground" of the old and sick, pioneered refrigeration anesthesia.[9] Ice bags, ice buckets, and ice wrapped in rubber sheets were applied on limbs for anywhere between one hour (for toe amputations) to two-and-a-half hours (for above-the-knee amputations). Meanwhile, a complete surgical team had to be assembled and ready to go because the effect of refrigeration lasted only about an hour. After the surgery was complete, ice continued to be applied at the amputation site for forty-eight to seventy-two hours. According to the hospital, among a series of fifty-eight high-risk amputations conducted with refrigeration anesthesia, the mortality rate was a favorable 13 percent. Still lower mortality rates using refrigeration anesthesia were reported elsewhere.[10]

Popular media were enamored of the idea and declared it a positive miracle: "[A] patient can eat breakfast while his wounded or gangrenous leg is being painlessly cooled to a few degrees above freezing, then can go into the operating room without further anesthetic or any drug whatsoever, chat with the nurse while the surgeon amputates behind a screen, and can return at once to the ward and eat a full lunch, as though nothing had happened."[11]

Such depictions were especially appealing in the early 1940s. As America entered World War II, war casualties and emergent amputations were a looming reality. It was hoped that we could painlessly refrigerate solders' injured limbs for hours and days while transporting them out of the war zone or until proper surgeries could be undertaken.

Unfortunately, the military promptly declared the idea "completely impractical": refrigeration was available only occasionally in a combat zone, and repeated replacement of ice during evacuation was something "no experienced military surgeon could possibly contemplate."[12]

Moreover, the medical benefits of refrigeration anesthesia also came under attack. Although multiple surgeons declared that icing reduced postsurgery infection,[13] animal studies showed that, yes, infection was kept at bay by icing, but once ice was removed, infection came back with a vengeance.[14] Prolonged icing also caused tissue damage, decreased blood flow, and delayed wound healing.[15] In some cases, icing failed to achieve anesthesia, and the mortality rate was no better than that with conventional anesthesia.[16]

With these disputes—plus the inconvenience of constant ice application, patient discomfort, and the improvement of modern anesthesia—refrigeration anesthesia faded from medical history after the 1940s.

However, much the same beliefs in icing's benefits persisted.

In contemporary belief, the effect of icing varies depending on the stage of an injury. At the acute stage, icing is thought to reduce blood flow, swelling, and inflammation. This is essentially the same idea proposed in the 1940s.[17] In a slightly updated—but still quite 1940s—version, icing is said to also reduce secondary damage.[18] The idea is that primary damage occurs as a result of direct trauma. After that, secondary damage occurs because of harmful chemicals released from the dead and dying cells as well as a lack of oxygen and nutrient supplies following trauma. Icing, by reducing cellular metabolism, can slow down chemical reactions and allow cells to survive in a hostile environment.

After the acute stage, icing's primary function is to reduce pain, which is not only comforting on its own but allows earlier and more aggressive exercises, which improve recovery.[19] Doctors in the 1940s already suspected that icing reduces pain by eliminating nerve impulses.[20] Contemporary research inherited the idea but packaged it in different jargon: icing reduces nerve conduction velocity, which is to say that icing slows down nerve impulses.

Interestingly, according to this theory, a "lower" part of your body can be effectively numbed by icing an "upper" part, as long as the two parts are served by the same nerve. Researchers demonstrated this effect by icing participants' ankles and showing that participants not only felt less pain at the ankle but also around the base of their fourth toe, because the two areas are served by the same tibial nerve.[21] Intrigued, I tried this on myself. After a good twenty minutes of ankle icing, I was disappointed to find that I could still feel a good pinch below my fourth toe. Of course, I don't have equipment to quantify my pain perception, nor do I necessarily claim to have pinpointed "the lateral aspect of the shaft of the fourth metatarsal bone in close proximity to its head."[22] At any event, whether or not the effect of icing spreads is of little concern. We don't need to ice our butt to reduce knee pain; if icing works, we can always just ice the knees.

But that's just the thing: how well *does* icing work, for pain, for anti-inflammation, for cell survival, for general recovery? These benefits have been claimed for so long that they are effectively part of our "medical folklore,"[23] adored by the public and medical professionals alike without necessarily understanding its medical basis. In a survey of emergency physician consultants,[24] their most frequent reasons for prescribing ice included experience (47 percent) and common sense (27 percent); only 17 percent relied on scientific reasoning. Moreover, only 23 percent of the consultants had read any literature supporting the clinical effect of ice, 17 percent never had, and 60 percent weren't sure whether they had.

Let's look, then, at some of that literature, focusing especially on icing's effect on the knees.

In one study, forty-five adults who had minor knee surgeries, such as meniscus removal, were put into two rehabilitation groups.[25] One group iced their knees twenty minutes a time, four times a day, followed by rehabilitation exercises. The other group performed the same exercises without icing. One week later, there was no significant difference between the two groups in pain level, knee swelling, or knee range of motion.

Icing performed no better after major knee surgeries. Immediately after anterior cruciate ligament (ACL) reconstruction, 131 patients were randomly assigned to five groups.[26] Four of the groups received cooling pads with temperatures set at 40°F (4°C), 45°F (7°C), 55°F (13°C), and 70°F (21°C). The fifth group received no cooling pads. Groups 1–4 used the cooling pads whenever they were in bed, for an average of fifty-seven hours. Their skin temperatures dropped, and many reported liking the cool feeling. But, there was no difference among the five groups in days of hospitalization, reported level of pain, use of pain medication, knee swelling, or knee range of motion.

On chronic conditions, the effect of ice is also underwhelming. When it comes to the knee, the most common chronic condition is osteoarthritis (OA), which is caused by the loss of articular cartilage, the protective tissue that covers the ends of the thigh bone and shinbone where they meet to make the knee (and under the kneecap, too; see chapter 2). With this cartilage gone, knee pain, stiffness, and loss of function set in.

With no cure to stop the process of OA, pain relief is an essential component in treatment. In one study, fifty-eight patients with knee OA were divided into two groups.[27] One group had their knees iced once daily for twenty minutes, using two plastic bags filled with crushed ice packed around the knee. The other group was treated with the same bags, but the bags were filled with sand instead. Four days later, patients were asked to rate their knee pain on a scale of one to ten, and it turned out that the ability of icing to reduce knee OA pain was "meagre."[28]

Now, don't get me wrong—opposite findings do exist. For example, some studies have reported that ice massage can reduce pain, stiffness, and disability caused by knee OA or that "a novel cold gel" can reduce pain and disability caused by soft tissue injuries, including knee injuries.[29] On the whole, however, I was surprised by how little positive evidence there is to support icing given how much everyone *loves* to ice.

The positive evidence is also tainted with methodology issues.[30] For example, in the study of the cold gel, the gel created a cooling sensation using menthol; whether or not the gel actually lowered tissue temperature wasn't even studied.[31]

More important, because people *expect* icing to work, a placebo effect can throw off research findings. In one study, one hundred patients who had undergone ACL reconstruction were randomly assigned to different recovery groups.[32] Two of the groups received cold therapies at the knee, delivered using bags of crushed ice or a cooling device that ran icy water into a rubber pad. A third group, the placebo group, was outfitted with the same cooling device filled with lukewarm water. As you can expect, the temperatures of the knees in the first two groups dropped significantly. No difference, however, was observed among the three groups in length of hospital stay, knee range of motion at discharge, or use of pain medicine. So it seems that being hooked up to a cooling device that *didn't* cool was enough to make patients feel better.

The placebo effect may also explain why some people, and that includes myself, ice enthusiastically after exercise to recover. In one study, thirty men completed a session of high-intensity cycling.[33] They were then assigned to three recovery groups. In the first group, the men sat in an inflatable bath filled with cold water. The second group sat in a bath with lukewarm water. In the third group, the water was again lukewarm, but as participants were getting in, researchers pored some liquid into the water. Participants were told that the liquid was a newly developed recovery oil.

After fifteen minutes, participants who sat in the cold water reported feeling less pain, less fatigue, and more prepared for exercise, and, indeed, their leg strength recovered. Those who sat in lukewarm water, understandably, experienced no benefits. Those who were treated with lukewarm water plus recovery oil experienced the same recovery as the first group—except, the oil was a ruse. It was simple bath soap. The morale of the story? If we expect something to work, whether it's icing or special oil, we often make it work.

But so what, you ask? If it works, does it matter if it is a placebo effect? Well, it does if icing may have adverse effects. Compared with using a stationary bike to cool down, immersing legs in cold water after strength training reduced muscle mass and strength.[34] Researchers have speculated that this is because muscle protein synthesis depends on blood supply, and icing, by reducing blood supply, suppresses protein synthesis.

In other words, icing can negate the benefits of exercise and reduce long-term muscle development, quite the opposite outcome for people who ice for sports recovery.

More fundamentally, icing may be harmful because it suppresses inflammation. In traditional wisdom, inflammation is pure evil and should be avoided at all cost. But, increasingly, we recognize that inflammation is a necessary, even essential, step toward healing. During the course of inflammation, immune cells work to clear out damaged tissues and then recruit stem cells for tissue regeneration.[35]

Under the microscope, this process can be vividly seen in mice that were subjected to muscle injury.[36] At day three post injury, a robust inflammatory response, along with damaged tissues, was everywhere. By day five, muscle regeneration started to happen. By day seven, damaged tissues were largely gone, inflammation started to subside, and regeneration was well underway. By day fourteen, inflammation was resolved, and regenerated tissues had taken over.

By contrast, in mice that were genetically altered to suppress inflammation, damaged tissues remained abundant at day seven post injury, cell regeneration was sluggish, and regenerated tissue fibers were small even at day forty-two post injury.

At this point, if you remain steadfast in your belief that icing works, you are probably in good company. Old habits die hard. I, for one, am not ready to simply ditch my assorted icing apparatuses. And, just to be clear, I am not saying that icing has zero effect and that you should stop doing it, period. If nothing else, icing does numb pain. After all, people used to have their limbs amputated with nothing but ice for anesthesia and whisky for comfort!

What I *am* saying is that we don't have the kind of robust scientific evidence we imagine there must be for something that's practiced so religiously. And I haven't even gotten to the practical questions yet. Say, you remain committed to icing, do you know what is the right way to do it? Namely, for how long, how often, and with what apparatuses?

Yes, your doctor may have given you certain icing protocols, but whatever the protocol, that would be a personal opinion rather than a medical consensus, because, guess what, no consensus exists. Although plenty of studies have examined icing, it is difficult to compare them to reach "best practices," because different studies use widely different icing protocols: from five minutes to eighty-five minutes, from commercial cooling devices to frozen peas, from twenty-minute intermittent icing to continuous twelve-hour icing.[37] Little can be synthesized from such diversity, although I did glean satisfying tidbits such as frozen peas are better than frozen gel packs at lowering skin temperature.[38] So much for planning ahead!

Multiple other factors also complicate the picture. Some studies have suggested that for icing to reduce pain, local skin temperature needs to stay below 13.6°C (56°F).[39] If we then hope to "hibernate" cells and theoretically avoid secondary tissue damage, temperatures in the neighborhood of 10°C–11°C (50°F–52°F) are probably needed.[40] These kinds of temperatures aren't easy to achieve or maintain,[41] especially if using some kind of protective barrier to avoid frostbite. Thirty minutes of direct ice application can drop skin temperature by 19°C (34°F); the same application over a dry washcloth drops skin temperature by 11°C (20°F); and the same application over padded bandage, alas, doesn't drop skin temperature by one bit.[42]

Then there is the matter of fat. Although icing is applied over the skin, muscles and ligaments under the skin are usually the target tissues. Depending on how much subcutaneous fat one is endowed with and how deep a target tissue lies, the effect of icing changes.[43] For a lean calf, twenty minutes of icing can drop the temperature by 14°C (25°F) at one centimeter under the skin; for a chubby calf, the temperature drops only 5°C (9°F) at the same depth.

As you can imagine, given these complications, icing may not work the same for any two people. *That*, ultimately, may be why icing will continue to occupy a place in both medical science and anecdotal experience. Me personally? Since doing research for this chapter, I have dialed back my icing routine and reach for the ice only when I experience actual, acute pain in the knee and then only ice for short periods of time. I have stopped, you might say, using ice as a crutch.

Worlds apart, people's reaction to icing is also worlds apart. My mother, whose sense of health and well-being is deeply influenced by traditional Chinese medicine, doesn't at all get knee-icing.

Traditional Chinese medicine adores the concept of *Chi*, an invisible sort of energy that supposedly circulates in nature and in our bodies. Multiple kinds of Chi exist. One of them is cold Chi.

Cold Chi emanates from items of low temperature, like ice, but also from food items that are said to be cold by nature. Determining which food is inherently cold is a complicated quasi-science. Things that taste bitter (like bitter melon), for reasons beyond me, are considered cold. Things that grow in winter (like daikon radish), because of their exposure to cold elements, are cold. Then again, oddly enough, things that grow in summer (like watermelon), because of receiving excessive rainwater, are also cold.

It is believed that cold Chi, once it accumulates in the body, can cause various ailments: stiff joints, achy muscles, indigestion, and menstrual cramps, among others too numerous to list.[44] This is why my mother, along with many other old-fashioned Chinese, finds the American custom of drinking ice water positively self-destructive.

The idea of applying ice to knee injuries is similarly unorthodox to traditional Chinese medicine: icing leads to the stagnation of Chi, which will impede, not facilitate, mobility and healing.[45] What Chinese medicine prefers is the opposite of cold: heat. Heat is thought to remove blood stagnation, restore circulation, and encourage healing.

Certainly, heat therapy isn't limited to traditional Chinese medicine. In the West, heat is also a popular remedy to soft tissue ailments—not *as* popular as ice, but still popular. Influenced by both medicinal traditions, I possess an even more diverse collection of heating apparatuses for my knee: ointments that you massage until it feels warm, patches that create a warming sensation, gel packs that can be preheated, battery-powered heating wraps, an infrared lamp, an extra deep bucket for hot water immersion. The bucket, by the way, is the hardest among my collection to come by.

All of these home-remedies are considered superficial heating sources, because they apply heat to the skin. Medical clinics are equipped with deep-heating technologies that can penetrate into the knee. For example, thermal ultrasound transforms continuous sound waves into thermal energy and can raise temperatures by 4°C (7°F) at five centimeters under the skin.[46] Diathermy, which means "through (Greek, *dia-*) heat (Greek, *-therm*)," uses high-frequency electromagnetic waves to induce heat. Depending on the diathermy machines, temperature increases of 11°C (20°F) at two centimeters under the skin are possible.[47]

Since cold and heat are thermal opposites, why should both be therapeutic? More pressingly, if evidence for icing is weak, what actual benefits can we expect from heating?

A fundamental assumption of why heating helps is that it opens blood vessels and increases blood flow,[48] which removes cellular debris, increases nutrient delivery, and hastens tissue repair.[49] Contrary to icing, heating increases metabolism and thus, presumably, healing. For every 3°C (5°F) increase in temperature, tissue metabolism doubles.[50] This effect can be profound at the knee because its local temperature is only about 30°C (86°F), which is 7°C (13°F) lower than the body's core temperature. This means that heating has the potential to quadruple metabolism at the knee.[51]

Like icing, heating is thought to reduce pain, but it does so through different mechanisms. Cold, as mentioned earlier, relieves pain by slowing down nerve impulses. Heat relieves pain by reducing muscle spasms and tension; it also reduces the stickiness of the fluid that surrounds our joints (the synovial fluid) so there is less painful stiffness during movement.[52] More fundamentally, heat may be able to close the "pain gate" in the spinal cord so that pain sensations don't get through to the brain to be registered as pain.[53]

The idea of shutting pain behind a gate may sound a little wacky, but the gate control theory is an influential model for how we feel pain. The gate in question is of course not a physical gate but a "gating" mechanism. According to the theory, nerve fibers that carry pain signals open the little gate so that the sensation of pain can travel to the brain; nerve fibers that carry heat signals close the gate, thereby inhibiting pain signals. A simpler way to think about this is that the feeling of heat can distract us from the feeling of pain.

This theory seems to bear out in studies that compare heat therapy with pain meds, such as acetaminophen (brands like Tylenol) and ibuprofen (brands like Advil), which are commonly used to treat OA pain. When 110 patients with knee OA were randomly assigned to wear heat wraps, to take acetaminophen, or to take ibuprofen for three days, the heat wraps had the best effect at relieving pain and improving knee mobility.[54] An added benefit is that heat has minimal side effects, whereas pain meds like ibuprofen are prone to cause gastrointestinal issues.

That said, because heat *is* a common treatment for soft tissue ailments, we need to, as with icing, question the presence of the placebo effect: are patients feeling better because they expect to feel better? To test this, researchers examined the effect of deep-heating diathermy on fifty-five patients with knee OA.[55] About half of the patients received actual diathermy treatments; the other half was hooked up to a diathermy machine, but the machine wasn't turned on. All participants were informed that with deep heating, they won't necessarily feel superficial warmth. After thirty minutes of treatment three times a week for four weeks, the group that received actual diathermy reported significantly less pain and stiffness and better knee function than the group that received fake diathermy.

In a similarly designed study, sixty-seven patients with knee OA randomly received thermal ultrasound treatment or fake treatment with a disconnected ultrasound machine.[56] After ten treatment sessions spanning over two weeks, real ultrasound proved superior at reducing pain and stiffness and improving knee function. It would seem, then, that heat was benefiting knee OA on top of any possible placebo effect.

For Japanese researchers, however, heat alone is not quite enough; there must also be steam, which creates better heat transfer by condensing on the skin. At the Juntendo University School of Medicine in Tokyo,[57] researchers outfitted thirty-seven patients with knee OA with dry-heat sheets or heat/steam sheets. Both kinds of sheets contained iron powder that, once exposed to the air, reacted and released heat. The difference was that the heat/steam sheets had a permeable side that allowed water in the heating component to escape as steam. Patients wore the sheets for up to six hours a day for four weeks, and both groups had less knee pain and improved knee function. But, the heat/steam group outperformed the dry-heat group on all accounts, and knee stiffness improved only with heat/steam sheets.

Upon reading this, I immediately went online to see where I could acquire this heat/steam sheet and add it to my collection. That search, disappointingly, came up empty. All I could find were "moist heat" pads that supposedly draw moisture in from the air, which I doubt would produce quite the amount of steam Japanese researchers were so enthusiastic about.

Steam or not, evidence shows that heat may enhance the knee's flexibility and range of motion by increasing the extensibility of collagen fibers, which in turn makes ligaments, tendons, and soft tissues more pliable.[58] Adding regular stretching and exercising, we may be able to remodel the connective tissues in our knees, making the joint more supple and springy. This outcome can be demonstrated in rats whose knees had been immobilized.[59] This is accomplished by bending the rat's knee and then surgically implanting a wire around it. The wire loops around the thigh bone and shinbone, essentially tying the two bones together so the animal can't fully extend its knees. The wire is left in place for forty days, allowing the rat to lose knee function without suffering permanent damage.

A total of fifty-five rats were thus operated. After the wire was removed, all of them underwent stretching to regain knee range of motion. In case you wonder, no, the rats can't be persuaded to stretch on their own. Instead, their ankles were tied to a spring, and a force of about 250 to 300 grams was applied to passively stretch their legs. On top of stretching, some rats also received heat treatments. After two weeks, all rats improved their knee range of motion, but those receiving heat treatments had significantly better results.

The same effects have been seen in humans.[60] For example, when healthy young adults wore a heat wrap or apply a heat pack on their knees, the flexibility of both the ACL and posterior cruciate ligament (PCL) increased, making it easier for participants to flex their knees.[61] Ice packs had the opposite effect, making it 25 percent harder to flex the knee.

Certainly, flexible ligaments alone won't be enough to protect the knee. Anyone who has done physical therapy for the knee will know that those exercises are often designed to strengthen leg muscles. The idea is that stronger muscles can take stress off and thereby protect the knee. And heat, evidence suggests, causes physical changes in the muscles that promote regeneration and mitigate injury. For example, heat increases the expression of heat shock proteins, which facilitate protein formation and prevent muscle damage.[62] Heat also activates multiple enzymes that contribute to muscle regeneration and increase muscle mass.[63]

When rats were subjected to crush injury in the leg, the immediate application of hot packs on injured muscles made a noticeable difference.[64] Day two after injury, rats that received heat treatments started to generate new muscle cells. For rats that did not receive heat treatments, regeneration took an extra day to kick in. Days three and four post injury, the new cells grew larger and more abundant in rats who received heat treatments. Days fourteen and twenty-eight post injury, their newly formed muscle fibers were also larger and more numerous. Similar benefits have been shown in other animal studies.[65]

What this muscle regeneration means for humans is less straightforward. Among junior track-and-field athletes, warm water immersion with water-jet massage reduced the damaging effect of intensive training on leg muscles and helped the athletes regain jumping power.[66] But, among strength-trained men, hot water immersion after intensive training did not help restore explosive power, although it did help restore muscle strength.[67] Among healthy, untrained men, hot-water immersion before a series of jumping exercises did not improve jumping performance but helped with muscle soreness and strength recovery afterward.[68]

So, the results are not perfect, but they are generally positive. And, I have to say, after reading study upon study, I'm more impressed by the amount of evidence we have for heat therapy than cold therapy, whether used to relieve knee OA symptoms, promote exercise recoveries, or improve over-all knee health.[69]

Now, to be clear, just as icing studies, heating studies have been criti-cized for their methodology issues, from small numbers of participants to the possibility of bias in data analyses.[70] There, too, persists the practical question of *how* to heat. As with icing studies, heating studies use vastly different protocols: different apparatuses, temperatures, durations, and frequencies, not to mention participants with different physical conditions and activity needs, thus making it impossible to conclude on best practices.

Some researchers suggest that rather than following strict guidelines, people should just experiment with different heating protocols (or icing protocols for that matter) to see what gives them the most relief.[71] Perhaps, individual differences will make one method more preferable than others.

I came across this suggestion earlier on in my research, and, being indoc-trinated in the standard scientific methods in which "personal preference" is largely irrelevant, I balked at it. But, by now, I'm ambivalent. When it comes to the effect of heat, there are many individual differences and exter-nal variations. Scientifically capturing and analyzing *all* of them, even if that were possible, would require many large, rigorous, and expensive clinical trials. Pharmaceutical companies would have little incentive to invest in such trials because heat is not easily patented for profit.

In the lack of a medical consensus, what else can individual patients do but try different products and protocols, compare them, modify some, and aban-don others? To muddle through, in other words? That, I guess, is pretty much what I have been doing all along with my heat collection. Thank goodness there are no side effects to heating—well, as long as one acts within reason.

"They are itchy. They are awkward. They are cumbersome and largely unattractive. . . . Also, they frequently smell as bad as an unventilated horse stall, and it is debatable whether they work as intended," wrote Sam Borden, a reporter for the *New York Times*.[72] He was paraphrasing Ross Pierschbacher, an offensive lineman for the University of Alabama football team. And the "they" Pierschbacher was griping about are knee braces, in particular, prophylactic braces.

"Prophylactic" means "intended to prevent diseases," so these braces, no matter how ugly or smelly, hold the sacred promise to prevent or reduce knee injuries, which are common in American football. Especially common are medial collateral ligament (MCL) injuries.[73] As described in chapter 2, the MCL connects the thigh bone and the shinbone on the inner side of the knee. If the outer side of the knee takes a blow when a cleated foot is planted, the MCL easily ruptures. Because linemen are often hit on the side of their knees near the line of scrimmage, just about every Division I football team requires linemen to wear prophylactic braces during practice, if not in actual games.[74]

Yet, as Pierschbacher let on, players rather dislike these braces. Moreover, no solid scientific evidence supports the claim that these braces actually prevent injuries.[75] Of course, brace manufacturers say otherwise and can and do find isolated studies that support the efficacy of their products, which can run upward of $1,000.

Probably more troubling than the debate itself is how long we've been having it. Brian Moore, a long-time executive at DonJoy, a big-name knee brace manufacturer, believed that prophylactic bracing began in the early 1990s and gained traction in college football later that decade,[76] but that's shortchanging the history, by more than two decades.

In the late 1960s, Dr. Robert McDavid, a former football coach and later professor of exercise physiology, had developed one of the first, if not *the* first, prophylactic knee braces. McDavid's idea took about a decade to catch on. By then, another brace, the Anderson Stabler, had come along and stole the show. The Anderson Stabler was designed by George Anderson, the head trainer for the Oakland Raiders. Anderson came up with the product after his quarterback, Ken Stabler, suffered a knee injury—hence, the name of the product: Anderson Stabler. The brace, as its name fortuitously suggests, is meant to support the knee. It features a rigid support bar; each end of the bar is hinged to a foam-rubber pad.[77] The pads are secured to the leg by tape so that the support bar lines up on the side of the knee. The bar can be worn on either side of the knee, but it is often worn on the outer side to defuse a blow that would strain the MCL.

Anderson put the Stabler on nine players who had previously suffered knee injuries and reported that, after two years of use, no reinjury occurred.[78] Despite a lack of more specific evidence, the idea caught on. Football players started using braces, and manufacturers started churning out similar products, some with a single bar like the Stabler and others with bars on both sides of the knee. Stories of "near misses" or "saves" also circulated, stories in which, according to players, a blow to the knee *would* have caused an MCL injury had it not been for the brace.[79] Several studies fanned this enthusiasm, showing that prophylactic braces reduced ligament and meniscus injuries and decreased the need for rehabilitation among college football players.[80]

However, it soon became clear that things weren't this simple. Conflicting findings emerged in the late 1980s that prophylactic braces were not able to reduce knee injuries; in fact, they seemed to make players *more* likely to hurt themselves.[81] Alarmed, the American Academy of Pediatrics Committee on Sports Medicine issued a statement in 1990 recommending that these braces "not be considered standard equipment for football players."[82]

If you routinely brace your knees during sports activities to prevent injury, don't worry. Chances are you haven't exposed your knees to undue risks. More recent studies have shown that prophylactic braces don't really cause more or more severe injuries—they just may not have helped you is all. Any benefit they possibly have seems to depend on players and the brace used.[83] For example, when playing defense, football players who wore prophylactic braces had fewer knee injuries, but that was not the case for offensive players.[84] As for which brand or kind of brace performed the best, there is no consensus. Our best guess is that longer and stiffer braces are more likely to help distribute impact away from the knee to the leg, thereby preventing injuries.[85]

Prophylactic braces aren't the only type of knee braces, nor are they the only controversial ones.

Functional braces represent another large market. These braces are meant to provide joint stability and help protect a knee that has seen injury. Most functional braces use hinged posts on both sides of the knee; the posts are connected to each other by straps, rigid shells, or both.[86] Based on which ligaments they are designed to support, functional braces can be further differentiated. That said, most of them target the ACL and are worn after ACL injuries or ACL reconstructive surgeries.

Multiple studies show that these braces *do* limit the front-and-back shift of the thigh bone and shinbone, essentially taking stress away from the ACL, but this benefit only exists when a small amount of force is put on the knee.[87] During daily activities, let alone rigorous sports activities, the knee is under much, much greater stress.

Indeed, when we put functional braces to the test in real life, the effect they showed in the testing lab diminished. In multiple studies, the performance of patients with deficient ACLs was no better when wearing a functional brace: there was no improvement, for example, in the distance patients could hop on their bad knees or their ability to sprint or run a figure eight course.[88] Similarly, wearing a brace after ACL reconstruction provides no benefit for pain relief or knee function.[89] In a large study of 969 athletes who underwent ACL reconstruction, those who wore a brace for three weeks after surgery had the same level of pain, knee range of motion, activity level, and surgery complications as those who didn't wear a brace.[90]

Moreover, if you are at all competitive in sports, you may want to know that bulky functional braces can hurt your athletic performance. They compress on muscles and decrease blood flow, causing early fatigue; they slow down muscle reaction time, impairing your ability to start and stop quickly; they increase oxygen consumption and heart rate, costing you 3–8 percent more energy to run.[91] Long-term bracing also causes muscle decline. Patients who chose to wear a knee brace for more than one year after ACL reconstruction showed a significant decrease in quadricep strength.[92]

If you *have* been wearing functional braces to sports but *haven't* felt any negative effect or actually felt that your physical performance was enhanced, I don't know how to break this news to you gently—because there is a good chance that you felt wrong. In a study with thirty-one participants who had undergone ACL reconstruction five months prior, seventeen participants felt that their running and turning performance was improved by wearing a brace.[93] But, in reality, their running and turning speeds were significantly reduced by bracing.

Doesn't this mean that the braces at least provide some psychological benefits? Yeah, kinda. When wearing these braces, patients can feel more stable in their knees, which increases their confidence and motivation to rehabilitate and exercise.[94] While that is all good, by the same token, there can be negative psychological influences too, such as a false sense of security and risk-taking behaviors that increase injury.[95]

Aside from athletes and people with traumatic injuries, braces are also marketed to people with knee OA. As mentioned earlier, knee OA is caused by the degeneration of cartilage that covers the ends of bones inside the knee. During normal walking, our center of mass falls more on the inner (i.e., medial) side of the knee, so that side supports the majority—about 70 percent—of the total knee load.[96] Given this heavier load, cartilage tends to degenerate faster on the inner side, causing it to collapse over time and

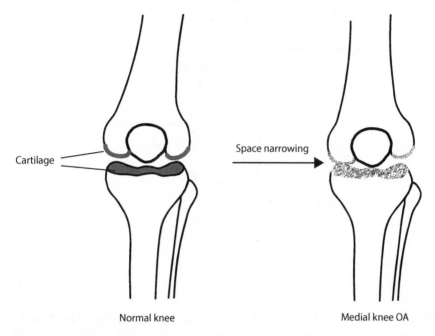

FIGURE 6.1 Medial joint space narrowing

creating pain and loss of function. This is the so-called medial joint space narrowing (figure 6.1).

A specific type of brace is designed to remedy this problem. Known as the unloader brace, it is supposed to unload weight from the collapsed side to relieve pain and symptoms. The brace does so by putting pressure on three points around the knee to bend it away from the collapsed side and to create a more even weight distribution (figure 6.2).

Gait analysis shows that when patients wore unloader braces, their legs became straighter, which presumably shifted loads from the collapsed, painful side of the knee to the open, healthier side.[97] Supporting this theory, considerable evidence suggests that unloader braces relieve pain associated with knee OA.[98] There is also some evidence that they improve knee function for patients with OA.[99]

But, curiously, X-ray results are less convincing. Using a continuous stream of X-ray images, we can monitor joint space in real time as a person walks. Using such techniques, multiple studies tried to determine whether

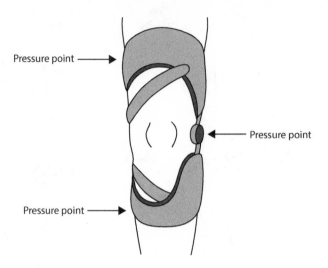

Pressure point ⟶

Pressure point ⟵

Pressure point ⟶

FIGURE 6.2 Unloader brace

an unloader brace actually opened up a collapsed knee joint. The results were yes, no, and maybe, depending on whom you ask.[100]

More curiously, wearing a soft, elastic knee sleeve or even taping the knees also seemed to help with OA: patients straightened up their legs, experienced less pain during movement, and felt more stability in their knees.[101] But these devices do not have mechanical stiffness, so they have little chance to physically change joint space. If they do work, a more likely reason is that the feeling of something tight clinging to the knee distracts us from the feeling of pain.[102] This, then, is the same gating mechanism through which heat therapy is theorized to work.

Either that, or wearing something tight improves our sixth sense. As explained in chapter 4, the sixth sense, or proprioception, is our ability to sense body position and movements and is vital for maintaining dynamic stability. Proprioception is enabled by sensory receptors in our skin, muscles, and joint tissues, and it deteriorates with age, OA, and muscle fatigue.[103] Having something tight around the knee may stimulate receptors and help our knees regain the sense.

In multiple studies, participants became better at balancing on unstable platforms or gauging the movements of their knees when their knees were outfitted with tapes, elastic bandages, or sleeves.[104] These devices, however, seem to benefit only people with impaired sixth sense.[105] For people who already have good proprioception, wearing these devices actually hurt their sixth sense, possibly because their innate good proprioception was now confused by the tight sensation.[106]

Another complicating issue is that we don't know how tight these devices should be for them to work. While some recommend applying them as tightly as is comfortable,[107] others believe that a looser fit is more beneficial. Supposedly, if a sleeve is quite tight and constantly hugs the knee, the joint would get used to the sensation and become insensitive to it, whereas a looser sleeve may continuously provide "new" stimuli.[108]

With all these uncertainties, I don't know what kind of generalized conclusion I can possibly form about knee braces. Personally, I have no love for the $900 prophylactic brace prescribed and fitted for me by a sports physician. It is bulky, clunky, worn a total of five times, and, by the way, I can't get insurance to pay for. The $20 elastic sleeve from Amazon with thirty thousand positive ratings had my hopes up, but, after trying to like it, I had to admit that it gave me no real relief either. Now, I wouldn't be surprised if other people find these products helpful. In fact, I *know* they do. Many of my fellow weekend warriors wear braces or sleeves; some kindly recommend them to me, swearing that theirs are the best designs or brands.

Is the effect of bracing, then, hopelessly individual-specific? Possibly. Everyone's situation is more or less unique: different activities, physical strength, fitness levels, athletic skills, knee ailments, and risk-taking behaviors; every brace is also different: different mechanisms, designs, materials, and fit.[109] It is not a simple math puzzle of counting how many people braced and got hurt versus how many people didn't brace and got hurt to determine the effect of bracing.

Are the benefits, if any, in our heads? Also possible. Wearing a brace *feels* like added support, which is invaluable when one has a bad knee. But is feeling good always a good thing? If a knee is compromised, wouldn't we want to be aware and avoid putting it in positions that might endanger it?

Or, maybe braces do change the mechanical functions of the knee, but modern science hasn't quite figured out how to demonstrate it? Real-life physical activities involve prolonged and erratic movements: from running, jumping, and climbing, to dodging, cutting, and turning. We can't be doing all these things while being hooked up to complicated equipment that measures if a brace has changed the force, torque, rotation, or whatever in our knees. In other words, any mechanical findings researchers come up with about bracing are always already removed from the reality of how we need our knees to work.

On that note, I have no better advice to you than to bring a healthy suspicion to what a manufacturer, a doctor, even a trusted friend says about a brace. Be willing to try one, but, equally importantly, be willing not to wear one. If you care to know, I stopped wearing braces after doing the work I did for this chapter. You probably can't stand hearing yet another extensive collection of mine, so, suffice it to say, I used to brace religiously. At the very least, going brace-free hasn't made my knee any worse. To the contrary, I have gained a sharper awareness of what it is doing, good or bad, at all times.

7

The Hurtful Knee

People with money send their kids to play golf, tennis, swimming or shooting. But poor people can't do that. We can only do boxing.
—SUTHEP SAENGNGERN, UNCLE OF A TEN-YEAR-OLD THAI BOXER

[T]here's no medical literature of what happens to a person's body when someone kneels on their neck for more than eight minutes.
—KARL WILLIAMS, CHIEF MEDICAL EXAMINER,
ALLEGHENY COUNTY, PENNSYLVANIA

Evolutionarily speaking, the job description for our knees is to support our torsos so we can walk on two feet. Our feet, evolutionarily speaking, are "the only part of our body that directly touches the ground."[1] And our feet are well developed for that purpose.[2] They feature large heels that can strike the ground with force and propel us forward. They are arched for shock absorption and energy release. They have big toes in line with the other toes, all short and bendy, to push us off the ground.

Needless to say, our knees have none of these features. Given its anatomy, the knee has no business touching the ground during normal bipedal walking. It is as qualified for that job as our hands or behinds. But, humans are amazing animals. Not infrequently, we do stuff that defy biology, like *shikko*, Japanese for knee walking.

Founded by Morihei Ueshiba (1883–1969), *aikido* is a modern Japanese martial art that emphasizes self-defense. Its specialties are throwing and grappling techniques, all of which can be performed either in a standing or kneeling position. If an attacker is standing and a defender is kneeling, the techniques are known as *hanmi handachi*. If both are kneeling, the techniques are called *suwari-waza*.

Of course, by kneeling, I mean the kneeling-sit. In feudal Japan, as in ancient China, life revolves around the floor. People cook, eat, and socialize in the kneeling-sit position. As described in chapter 5, kneeling-sit is accomplished by dropping both knees to the ground, with lower legs tucked under thighs. The tops of the feet are flat on the ground, and the butt sits on the heels. Japanese people also like to turn their toes inward and their ankles outward to form a V shape with their feet. They call this kneeling-sit position *seiza*.

From this seiza position, if you lift your heels, rest the balls of the feet on the ground, and point your toes forward, you transition to the posture called *kiza*. From this kiza position, you can then proceed to knee walk.

But . . . pray how? Well, you part the two knees, lift one (say the left one) up, swing forward your left hip to make a step, land on the ball of your left foot, and drop that knee back down. Meanwhile, your right foot, which stays tucked underneath your body this whole time, swerves to join the left foot and moves your body along with it. Alternating the two knees allows you to move forward. Swinging the hips backward allows you to walk back. Simple as that.

Not *quite* following? Well, how about figure 7.1?

Still not following? It's not your fault.

For bipedal humans who are used to moving about on two feet, knee walking is difficult to imagine. Like a complicated dance move, it is impossible to capture in words or still images. One way to think about it is to imagine that your two ankles are tied together by a bungee cord. Your two knees and the tied-up feet thus form an elastic triangle. By raising one knee, you lift one anchor of that triangle off the ground. As you use your hip to swerve that anchor forward or backward or in a particular direction, the rest of the triangle rotates, bounces, and follows along. That, my friend, is the best I can do. Now, look up a video online to see it in action.

For people untrained in shikko, it is a *very* awkward way to move around. But, with practice, accomplished aikido martial artists can walk on their knees as swiftly as you and I do standing up. Video sensor analysis shows

FIGURE 7.1 *Shikko*, or knee walking

that when knee walking, expert practitioners have a smooth acceleration in their core,[3] making it look like they simply glide on the ground, footless.

That's all well and good, but, why bother? Why would anyone want to walk on their knees if and when their feet function perfectly well?

Culture and history, that's why. Knee walking is often traced to the samurai, the Japanese military cast that rose to power in the twelfth century. During four hundred years of war-torn Japan, samurai served as military retainers of feudal lords. They were trained in war techniques and various types of martial arts and weaponry, wielding much prestige and power.

As retainers, samurai were expected to show unreserved loyalty toward their masters, so much so that to die for them was an enviable honor.[4] When in the company of their masters, samurai knelt-sit to show humility and respect. Despite being on their knees, they were expected to protect their masters against sudden threats and concealed assassins. To fulfil both duties, the samurai developed the sophisticated techniques of moving and fighting on their knees. Hence, shikko is also known as the samurai walk.

This is a charming explanation, romantic really—except that some aikido martial artists thought it was wrong-headed. On occasion, a kneeling samurai might have needed to defend himself against a standing attacker, so the hanmi handachi techniques, in which the attacker stands and the defender kneels, make historical sense. But why on earth would two samurai opt to both kneel on the ground to fight, suwari-waza style? Shouldn't the devoted samurai value their feudal lords' lives over ritual propriety? Think about it: if a group of samurai were kneeling besides their feudal lord and an assassin suddenly came upon them, drew a concealed weapon, and lunged toward the lord, wouldn't the samurai jump to their feet to protect their master, whose life is supposedly more important than anything else?[5] Plus, in no historical records did the samurai ever move around on their knees the way contemporary aikido martial artists do. When samurai in a formal meeting had to get up for some reason, they bowed, stood up, and walked while remaining low.[6]

The aikido knee techniques, then, may have no romantic historical origin at all and instead were used as modern training exercises.[7] Kneeling on the ground with limited mobility, students are forced to learn hip movement, which is essential for executing aikido grabs and throws. Being on the knees also increases leg strength and flexibility and heightens students' awareness of their center of mass. All of these benefits carry over into learning how to fight while standing upright.

In addition, kneeling techniques can be the saving grace when one is thrown to the ground.[8] Finding themselves in that unfortunate position, aikido practitioners can swiftly rise to their knees, knee walk away from the attacker while defending themselves, and eventually get back up to their feet. It is nothing short of astonishing to see skilled practitioners move smoothly and precisely on their knees, take advantage of their lower stance to dodge attacks, and gracefully grab, pin, and throw an attacker.

To try to *get* that level of skill, however, is not nearly as glamorous. Hijacking the evolutionary intent of our body, which has been in the works for millions of years, is no walk in the park. A quick browse of martial arts online forums, and you will find Western students bitterly complain how difficult and painful it is to practice walking and fighting on their knees. Indeed, many have difficulty sitting on those knees, let alone locomoting on them.

And just as well, the Westerners may have a point. Evolutionarily speaking, all modern human knees, no matter colors or nationalities, are the same. But historically and culturally, they aren't exact replicas. When miniskirts became the fashion statement of the 1960s, young girls in the United States and Europe seized the opportunity to showcase their tender knees and long legs (see chapter 3). But when Japanese teens joined the fad, they revealed, on average, calloused knees and short, sturdy legs.[9]

Now, leg length and stature are complicated matters beyond the topic of this book. But, on the whole, the Japanese, despite experiencing an increase in "long-leggedness" post–World War II,[10] are shorter than their European counterparts. Academic scholars blame this on diet: a high consumption of rice and low consumption of animal protein.[11] Gaku Homma, instead, believes that something else made the miniskirt less flattering on his fellow Japanese: kneeling on the floor from a young age.[12]

Homma is a veteran instructor and scholar of aikido, a direct student of the founder Morihei Ueshiba. Based in Denver, Colorado, Homma is an outspoken voice *against* training his Western students knee walking and knee fighting—because of what he himself had to endure.

When he was a teenager living in Japan, Homma had one instructor who drilled him in knee techniques.[13] Hours of practice made his pants threadbare at the knee, and he had to sew layers of patches over them. The knees cracked and bled, blood soaking through the layers and gluing them to his knees. Just when the knees grew a layer of new pink skin, another lesson would scrape it all away. The only reason he didn't suffer permanent damage, Homma is convinced, is because he had developed strong knees by sitting on them the traditional seiza style and by strengthening them through physical labor.

People outside Japan do not share the tradition of kneeling-sit from a young age, so they do not, in Homma's view, possess the prerequisite knee and leg strength for training what he did. Indeed, today's Japan has widely adopted Western-style furniture, so their youth are losing the advantage their elders once had and are ill prepared for knee techniques.

More scandalously, Homma let on that the teaching of knee techniques outside Japan was an underhanded move. When the first generation of Japanese aikido instructors traveled to the United States during the later 1960s and early 1970s, they were advised by their superiors to teach knee techniques to level the playing field: "Remember that many of the new students you will be encountering will be bigger in stature than you. Suwariwaza techniques will be difficult for them, so practicing suwariwaza will put you at an advantage despite your size difference. To gain control over your students, practice suwariwaza. And during examinations, if there is some individual testing that you are not fond of, have them test last, and make them wait in seiza until it is their turn."[14]

This advice, according to Homma, had far-reaching consequences and caused frequent knee problems among Western students. Hoping to corroborate this, I dug into medical literature but, unfortunately, failed to find formal studies that link shikko or suwari-waza to more, or more severe, knee injuries. This doesn't mean that they don't. Medical studies like to focus on traumatic martial art injuries, injuries that require taking time off from training, emergency room visits, or hospitalization.[15] Traumatic injuries, generally speaking, are caused by explosive actions, such as throwing, falling, or kicking, and manifest in fractures, lacerations, or sprains.[16] Knee walking and fighting, if they do cause injuries, would cause overuse and degeneration that hurt gradually and insidiously.

In my carpeted basement, I gingerly tried shikko, just to see how it feels on my good versus bad knee. Suffice it to say, it was bad enough—or I was bad enough—that I couldn't tell I had a good knee.

With its focus on self-defense, aikido aims to control an attacker without causing injury. Whether knee walking or knee fighting, the knees (and legs and feet) are more a leverage, less a weapon. Rarely, if ever, do you see an aikido practitioner using the lower extremities to forcefully kick an opponent.

If you want to see kicking, you watch taekwondo, the Korea-born martial art that has a particular penchant for kicking. It has, in fact, more than a dozen different kicking moves, carefully categorized based on footwork, body movement, and angles of attack. For example, in the roundhouse kick, you raise the kicking leg to aim an opponent and rotate on the other foot to create momentum. In the back kick, you turn away from an opponent and kick straight back. In the axe kick, you raise your kicking leg up high and bring it straight down onto a target.

Depending on the type of kick, how well it is executed, and a person's body mass, the amount of speed and force generated varies. In different kinematics studies, speeds up to sixteen meters per second[17] and forces ranging from 150 to 230 pounds[18] have been reported, which are sufficient to cause soft tissue and internal damages. More spectacularly—indeed, a little unbelievably—in an episode of the National Geographic TV show *Fight Science*,[19] taekwondo black belt Bren Foster was said to have kicked at a speed of 136 miles per hour (61 meters per second), faster than the record speeds on eight different National Association for Stock Car Auto Racing (NASCAR) racetracks. His kick reportedly generated 2,300 pounds of force, enough to stop the heart.

Naturally, all these kicks involve the knee. It is by flexing (or bending) the knee and then extending (or straightening) it that we kick. Without rapid and forceful knee movements, a kick will be wimpy. That said, in kicking, multiple other body parts join in and are often more important: for example, the pelvis, the hip, the ankle, the quadricep, the hamstring.[20] Notably, Chinese martial arts emphasize the pelvis and the internal energy called Chi (see chapter 6) that supposedly resides in the pelvic region. This ancient teaching is borne out in modern kinematics studies of taekwondo, in which pelvic position and movement were found essential to create powerful kicks.[21]

Aside from kicking, the knee can inflict more direct damage in another martial move: knee strikes. In knee strikes, you make contact with an opponent's body not through the feet or shin as in kicking, but through the knee itself. Various martial arts from Karate to taekwondo incorporate knee strikes, although none is as dedicated as Muay Thai, the national sport of Thailand. In Muay Thai, or Thai boxing, the knees are treated as two extra limbs. The other limbs are the two fists, two feet/shins, and two elbows, making Muay Thai the Art of Eight Limbs.

Living up to this title, Muay Thai has a complicated system of knee strikes.[22] The most basic but plenty powerful is the straight knee, which is executed by raising the knee up, rocking the body back, and then thrusting the knee forward to hit an opponent straight in the abdomen. Ouch. If the opponent's head is forced down with a grip, the knee can also land on the head. Double ouch. Kinematics studies have shown that when performing the straight knee, professional fighters achieve a speed of fifteen meters per second.[23] More spectacularly and, again, a little unbelievably, according to the History channel's TV show *Human Weapon*, a straight knee can deliver up to five thousand pounds of force, enough to lift a car.[24]

An adaptation of the straight knee is the diagonal knee. Instead of striking straight up, the knee is thrust up and inward, landing on the opponent's sides. Although slower than the straight knee at only six meters per second,[25] when hit just right at a tender spot, a diagonal knee is enough to knock an opponent to the ground.

If two fighters are in a tight clinch and don't have the room to execute a straight or diagonal knee, they resort to short, sharp knee strikes to target the inner or front portion of an opponent's thigh. These are called small knee strikes or rabbit knee strikes.

Conversely, from the long range, a fighter can use the deadly flying knee, which is done by "springing off" of one foot, leaping into the air, and landing the knee on an opponent, often in the head. This technique leverages the body's explosive power and the force of gravity, concentrating all of it on the crown of the knee. Landed on the chest, it will feel like being hit by a sledgehammer. Landed on the head, it will most certainly knock someone out, as mixed martial arts (MMA) fighters well know.

On July 6, 2019, at the T-Mobile Arena in Paradise, Nevada, thirty-five-year-old Ben Askren from Iowa fought thirty-four-year-old Jorge Masvidal from Florida. As soon as the referee signaled the start of the match, Masvidal ran toward Askren, leapt into the air, and planted his right knee on Askren's head, knocking Askren unconscious. Masvidal won the match, setting the record for the fastest knockout in the Ultimate Fighting Championship history: a mere five seconds.

A more infamous MMA moment happened on July 16, 2016, at the O2 Arena in London. That day, thirty-eight-year-old Evangelista Santos from Brazil fought twenty-nine-year-old Michael Page from the United Kingdom. During the second round, Page landed a flying knee on Santos's head, literally fracturing his skull. Santos recovered after seven hours of surgery but never again stepped inside an MMA ring.[26]

The knee can hurt, very, very badly.

In many ways, though, Muay Thai is more gut-wrenching. Worldwide, there are only about one thousand professional MMA athletes, but in Thailand alone, some sixty thousand people are registered Muay Thai fighters.[27] On top of *that*, an estimated two to three hundred thousand children under fifteen years of age regularly fight, off the book.[28] While MMA fighters usually start training around the age of fifteen or sixteen, Muay Thai child fighters, boys and girls, start training around six and start competing by the age of seven or eight. Western boxers usually fight no more than a few times a year, while Muay Thai boxers fight as often as twice a month.[29] Training and competition conditions in Thailand are also grim by Western standards: rotting gloves and donated bags, makeshift facilities, leaky roofs, and no air-conditioning in ninety-degree weather.[30]

And, there is no padded headgear or chest guard in Muay Thai, so injury risks are high: 20 percent of fights end in a knockout, and concussions and fractures are common.[31] More devastating may be long-term neurological damages. When 333 Muay Thai child boxers and 200 nonboxers with similar age and socioeconomic backgrounds underwent brain imaging, the Muay Thai kids showed white matter damage, loosening of brain tissues, and decreased brain activity associated with memory.[32] Fighters with two to five years of experience scored an average IQ of 86, compared with nonboxers' scores of 94.6.[33] Fighters with more than five years of experience had a still lower IQ of 83.8, which would make high school a challenge.[34]

By this very brutality, Muay Thai is one way, some say the only way, Thai children and their families can escape a life of poverty. By entering a fight, children earn a fee; by winning a fight, they earn bigger money from gambling. Although gambling is largely illegal in Thailand, Muay Thai fights get a special permit. Gamblers especially like betting on children because they are unpredictable and are unlikely to throw a fight.[35]

In the poverty-stricken area of Isan, eleven-year-old Nhat Tanorak is one of the best child boxers in the region. His entire village would pool their money, money they can ill afford to lose, and bet on him when he fights. If Nhat wins, he gets a cut, which can be more than what his mother brings home in a month.[36] In large cities, stakes are higher. Wagers can jump from $50 to $500 in an instant at the Rajadamnern Stadium in Bangkok, which is packed four times a week with middle-age Thai men and foreign tourists.[37]

Often, children like Nhat, by their own admission, *want* to fight. They train several hours a day before and after school to strengthen their arms and knees. They carry on family traditions, doing what their parents and grandparents once did or dreamt of doing. They relish the wins, the cheering, the fame that they bring to their families and villages. They dream of the bigger fames and glories that will come once they become national and world champions.

Then, there is money—the useful, valuable, desperately needed money. When Nhat was asked, after a particularly grueling training session, "If you didn't get money for Muay Thai, would you do it?" he shook his head wearily and said, "I don't know."[38] When adults make it abundantly clear how much money rides on a fight, when cash unabashedly changes hands around child fighters, how *can* they know?

Some argue that engaging children in Muay Thai fighting is a form of child abuse, something that isn't exactly rare in Thailand. From the sex trade to long hours and heavy labor in the seafood and agriculture sectors, children in Thailand experience some of the worst forms of child labor and exploitation.[39] Muay Thai fighting, however, is technically legal because labor laws only apply to children working in employment relationships. By earning "award" money rather than salaries, child fights exist within the boundaries of the law.

Then, Anucha Tasako happened. In 2018, thirteen-year-old Anucha died of a brain hemorrhage after being punched and knocked out in a Muay Thai fight. Doctors attributed the death to the more than 170 matches Anucha had fought since the age of eight, declaring that his head was hit too many times to survive.[40] Anucha's death was widely publicized and led the Thai government to begin drafting a Thai Boxing Act amendment. The amendment would ban children under twelve years of age from fighting in Muay Thai competitions and would require the use of protective equipment for children between twelve and fifteen years old.[41]

The Thai people fiercely opposed the amendment. "Those who drafted the law do not know anything about the sport of Thai boxing," said Sukrit Parekrithawet, a lawyer who represents several Muay Thai training camps. "If you don't allow younger players to learn their way up, how can they be strong and experienced enough to fight?" asked Parekrithawet. "We call it 'boxing bones.' You need to have boxing bones built from a very young age."[42]

And it's not just well-to-do lawyers who were saying this. "You have to start from seven, eight, nine to develop strong bones," said the proud grandpa of thirteen-year-old Muay Thai fighter Phuripat Poolsuk. "If you start training them at fifteen, you can't survive," grandpa said, shaking his head. "There will be no boxers in Thailand. We will have some, but not good ones."[43]

And protective gears? Fans don't like them. They see it as going against the sport and its tradition.[44] The traditional attire for Muay Thai includes a headpiece called Mongkhon and armbands called Pra Jiad, which are considered sacred objects that bless fighters with power, strength, and good luck. The custom dates back hundreds of years to the warring Siam (present-day Thailand). As warriors headed into battle, they would tie a piece of cloth around their heads and wear armbands made from their mothers' clothing to symbolize courage and protection.

Equally sacred, and ancient, is the prefight ritual Wai Kru Ram Muay, a ceremonial dance performed by fighters in the ring to pay respect to their trainers, parents, and ancestors. The dance is said to be invented by the father of Muay Thai, Nai Khanom Tom, who was captured during the eighteenth-century Burmese invasion of Siam. To celebrate the success of the invasion, the Burmese king called for a festival and ordered the Siamese prisoners to fight Burmese warriors for entertainment. According to the legend, Nai Khanom Tom defeated ten Burmese warriors, greatly impressed the king, and was set free. He returned to Siam and spent his life teaching Muay Thai. In Thailand, this legend is celebrated every year on March 17, the National Muay Thai Day.

Muay Thai, then, is more than brute-force martial skills or personal victory. It is tradition, history, and national pride. If anyone could understand it, Americans ought to, given their national obsession with football, a game of ruthless strength and gladiator-style combat, a game of "beauty and violence."[45]

Like Muay Thai, the violence in American football is not *just* about personal fame, glory, and reward. It is steeped in history and nationalistic sentiment. The National Football League (NFL) may be downplaying the violence of the game in recent years, but it very much celebrated violence in the 1950s and 1960s when Americans were concerned that the postwar boom was making the country soft, that they were losing out to the Soviets.[46]

Football assured the country that American men were anything *but* soft. "We try to hurt everybody. We hit each other as hard as we can. This is a man's game," said Sam Huff, a celebrated linebacker and the first NFL player to appear on the cover of *Time* magazine in 1959.[47] "I just forget about my life when I go in there. I'm not going to worry about what happens to me. It's just going to be a destroy type of deal," mused a rookie player in the 1965 documentary *Mayhem on a Sunday Afternoon*.[48] Today's players may not say these things out loud, and rules may be modified to increase safety, but the violent nature of the game hasn't fundamentally changed. This is why knee injuries are common in American football. This is why players develop chronic traumatic encephalopathy (CTE), a neurodegenerative disease linked to repeated head blows.

In American football, it is not just adults who are hitting, tackling, and hurting. Football is one of the most popular youth sports in the United States, played by more than three million youth players (six to thirteen years old), one million high school players, one hundred thousand college players, and only about two thousand NFL players.[49] In other words, like Muay Thai, there are far more children than professionals in this sport.

Compared with other sports, football claims the most knee casualties in high school: as many as 58 percent of high school male athlete knee injuries are a result of playing football.[50] As players' body mass and speed increase in collegiate football, the risk of knee injury also increases.[51] In addition, an estimated ninety-nine thousand youth players suffer at least one concussion annually, which is one in thirty players.[52] Among former NFL players, those who started playing football before the age of twelve had significantly worse cognitive impairment.[53] For players diagnosed with CTE, the younger they started playing, the sooner they started showing symptoms, with every one year they started earlier equating to two-and-a-half years of advance in symptom onset.[54]

Given these data, a disinterested onlooker might say that Americans having their children play tackle football is no less astonishing than Thais having their children fight in the ring. Yes, in Muay Thai, there is the unsavory gambling that Western documentary makers and journalists have abundantly exposed, but to what extent, I wonder, is that privileged musing? To what extent can wealthy countries justifiably question the morality of a sport when their tourists, money and cameras in hand, pack the stadium to watch people fighting to escape poverty?

Although injuries happen in sports arenas and in martial art rings, there are at least rules, time-outs, and referees. George Floyd got none of that when he was murdered on the street, restrained under the knees of former

Minneapolis police officer Derek Chauvin (see chapter 5). In the aftermath of Floyd's murder, Americans can never talk about knee restraints without feeling a twinge in their conscience.

During his trial, Chauvin's knees were a particular point of interest. The prosecution argued that his left knee was on Floyd's neck more than 90 percent of the time that Floyd was on the ground and his right knee was on Floyd's back at least 57 percent of the time.[55] According to the prosecution's medical examiners,[56] the left knee pressed on Floyd's carotid artery on the side of the neck, impeding blood flow and oxygen supply to his brain. The same knee also restricted air flow to his lung. Meanwhile, the right knee on the back impaired breathing by restricting the movement of the diaphragm, the major respiratory muscle that moves air into and out of the lung.

At the time of Floyd's death, the Minneapolis Police Department permitted neck restraints and chokeholds.[57] When an arrestee is "exhibiting active aggression," can't be controlled with lesser attempts, or for "life saving purposes,"[58] an officer can press the carotid artery on either side of the person's neck to temporarily block blood flow to the brain and render the person unconscious. This is the so-called blood chock. Because a blood choke preserves the airway in the front of the neck, it is considered by law enforcement, and martial artists, as a safer technique than "air choke," which restricts the airway and impedes breathing.[59] Some, however, have long argued that during violent struggles, selectively targeting the carotid artery is tricky and misapplied force on the airway can be lethal.[60] Underlying health conditions, such as vascular diseases, also make the blood choke inherently dangerous and liable to trigger strokes and brain damage.[61]

According to officials from the Minneapolis Police Department, their officers were trained only in using arms, not knees, to execute a blood choke, so they "don't' know what kind of improvised position that [Chauvin's] is."[62] Peculiarly, the Department's training materials did include a demonstration photo of a trainer kneeling on the neck of an arrestee. The person who played the arrestee lay face down on the ground, his neck under the trainer's knee, while two other officers assisted, eerily similar to what happened to Floyd.[63] So, despite sworn testimonies, the Department's training is not as above reproach as its officials wanted to believe.

The Department's saving grace was probable cause and recovery guidelines. Officers can apply a blood choke only when someone shows active aggression, not when they are handcuffed and pleading for their lives. In the world of martial arts, it is well recognized that a blood choke can be safely applied only for very short durations, about ten seconds. Prolonged choking will starve the brain of blood and oxygen and cause irreversible injury and death. Similarly, officers can only restrain someone face down when "handcuffed subjects are combative and still pose a threat,"[64] not when they are unconscious and pulseless. Finally, officers are required to place distressed arrestees on their sides and render first aid, not to continue the use of force to the point of death.[65]

Because of Chauvin's violation of these regulations, a life was lost and a murder committed. That much is clear. What is frustratingly unclear is the exact cause of death.

Dr. Andrew Baker, the Hennepin County Chief Medical Examiner, concluded that Floyd's death was caused by "cardiopulmonary arrest (which is the stop of adequate heart function and breathing) complicating law enforcement subdual restraint and neck compression."[66]

This is a complicated—and semantically awkward—conclusion. By "complicating," Baker could mean that police restraint and neck compression caused medical *complications* that in turn led to cardiopulmonary arrest. Or, I can't help wondering if the word needed here is *implicating*, as in "the cardiopulmonary arrest implicated (i.e., incriminated) police restraint and neck compression."

Whatever the case, Baker's conclusion comes across as far less straightforward and satisfying than that of the independent medical examiners employed by Floyd's family. Their topline conclusion for the cause of death was asphyxia—a lack of oxygen.[67]

A curious, and probably critical, difference in how the two sides reached their conclusions was the video of Chauvin kneeling on Floyd. Baker specifically avoided watching any video to, in his words, avoid bias during the autopsy.[68] The prosecution's medical examiners watched and specifically referenced the video evidence in their reports.

To anyone who did watch, asphyxia seems a sensible conclusion—because Floyd, as the common expression goes, was being strangled under Chauvin's knee. Yet, during the autopsy, Baker found no bruising on Floy's neck or visible damage to his brain; the latter, though, he added, wasn't unusual.[69] What he did find were an enlarged heart that needed more oxygen to function, two clogged heart arteries that decreased blood flow to the heart, and 11 ng/mL of fentanyl, an opioid, in Floyd's bloodstream.[70] "If he were found dead at home alone and no other apparent causes, this could be acceptable to call an OD. Deaths have been certified with levels of three (nanograms of fentanyl)," Baker said.[71] Moreover, "the placement of Chauvin's knee would not have cut off Floyd's airway," Baker believed.[72]

Was Baker trying to help the police get away with murder?

Accustomed to crime TV shows like *Forensic Files*, I imagined that for a murder caught on tape with fresh scenes and known offenders, all sorts of evidence must exist about weight transfer, oxygen intake, and brain imaging that directly link Chauvin's knee to Floyd's death. The knee, as I said repeatedly, can hurt.

Yet, in that expectation, I was sorely disappointed.

Dr. Martin Tobin, a pulmonologist from Chicago, testified that, at one point, Chauvin lifted his left toes off the ground, which means that half of his body weight (91.5 pounds) was directly compressing Floyd's neck.[73] When his toes were on the ground, the weight transferred would be 86.9 pounds, because half of his "shank's weight is supported by the toes."[74]

If this seems scientifically precise, I'm afraid it really is not. In this case, it is impossible to estimate force merely by looking.[75] How much weight one puts on the supporting, stabilizing knee makes a difference. The whole business about "half of the shank's weight" also makes no sense to me, without, at least, supporting details of Tobin's measurement methods.

Not surprisingly, the defense's expert witness Dr. David Flower, retired chief medical examiner for the state of Maryland, estimated that the weight transfer was lower, much lower: 30–35 pounds, about 23 percent of Chauvin's body weight (140 pounds).[76] The accuracy and method of this estimate are likewise unclear. Curiously enough, the prosecution and defense were also using vastly different estimates for Chauvin's baseline weight, something that you'd think would be pretty easy to determine.

Not enlightened by court testimonies, I was lucky to find one single study published in the *American Journal of Forensic Medicine and Pathology* that measured how officers' body weight affects force applied during restraint.[77] A group of law enforcement officers (thirty-six men and five women) knelt on the back of a training mannequin, which was placed face down on an electronic scale to measure weight force transfer. Depending on where the officers put their supporting knees or how they consciously distributed their body weights, putting a single knee on the back of the mannequin exerted between fifty-two and seventy-three pounds of force. Applying both knees exerted an average of 103 pounds.

So, the reality may fall somewhere between the prosecution's and defense's estimates. Either way, that information does little to settle the argument in court. The prosecution believed that the weight applied through Chauvin's knees made it impossible for Floyd to breathe. With the knees on him, "the work that Mr. Floyd has to perform becomes huge. He has to try to lift up the officer's knee with each breath," the prosecution's expert witness Dr. Tobin said.[78]

The defense, in turn, argued that the weight, no matter how much, was irrelevant. "None of the vital structures [of the neck] were in the area where the knee appeared to be from the videos," defense's expert witness Fowler testified.[79] And, "all of his injuries were in areas where the knee was not": the front of the body, the face, and places he was restrained, not the back or the neck.[80]

Who is right? Strictly speaking, neither. We do not know how Chauvin's knees affected Floyd's breathing or heart functions, because we can't pinpoint where on the neck the knee was pressing at all times. A neck is not uniform. Pressure in the lower neck, for example, will restrict blood flow differently than pressure in the middle or higher neck because of the different exposures of carotid and vertebral arteries.[81] More fundamentally, we have no pertinent knee-on-neck medical literature or data to draw upon— at least, I couldn't find any. No sane researcher would design a study in which they would kneel on the neck of voluntary participants for nine minutes at a time. No sane participant would consent to this. No sane institutional review board would approve it.

What I did find are studies on prone restraint—that is, being restrained in the face-down position, as Floyd was. Prone restraint is widely used by law enforcement, medical personnel too, to control someone who may pose a danger to themselves or others. One or more officers may place one or both knees on the person's back or kneel next to the person to gain control for handcuffing.[82] Depending on the nature of resistance, the person's ankles may be tied with a strap; the strap may be further connected to the handcuff, placing the person in the so-called prone maximal restraint position, commonly known as the hogtie position.[83]

The neck doesn't or shouldn't enter this picture. The point of prone restraint is not to limit someone's breathing or render someone unconscious. It is about keeping people on the ground and gaining control of their limbs.

Yet, surprisingly, prone restraint alone is enough to cause death. This was first reported in 1992 by Dr. Donald Reay, chief medical examiner for King County, Washington.[84] Reay described three separate cases in which an individual had died while being restrained in a prone position and transported in a police vehicle. In all cases, autopsies showed no obvious cause

of death, and Reay determined that the men died of positional asphyxia—
as in asphyxia caused by the position of their bodies.

In breathing, we inhale, and we exhale. During inhalation, the diaphragm
drops, the abdomen moves down, and the rib cage expands, drawing air into
the lung. During exhalation, the diaphragm rises, and the rib cage contracts,
pushing air out of the lung. According to the theory of positional asphyxia,
being held face down prevents the movement of the diaphragm and having
shoulders pushed back for handcuffing or hogtying prevents the movement
of the chest wall.[85] As a result, breathing is compromised. If whoever is
holding another person down also applies pressure on the back, it further
restricts the chest and abdominal movement and diminishes breathing.

After Reay, additional cases of positional asphyxia were reported in the
United States, the United Kingdom, and Canada.[86] Deaths occurred under
police custody, in psychiatric hospitals, and in community settings. The victims were predominantly men, often middle-age. Information about race,
when available, showed an equal mix of whites, Blacks, and Hispanics.[87]

In Chauvin's case, the prosecution, in addition to emphasizing the knee
on the neck, was clearly trying to make a case for positional asphyxia: "the
combination of Mr. Chauvin's pressure, the handcuffs pulling Mr. Floyd's
hands behind his back and Mr. Floyd's body being pressed against the
street had caused him to die 'from a low level of oxygen,'" said prosecution's
expert witness Dr. Tobin.[88] It is "positional asphyxia," concluded Williams
Smock, a police surgeon.[89]

These are powerful testimonies, but once again, they are scientifically
ambiguous. There is a serious issue with the theory of positional asphyxia: a
lack of supporting data. In a twelve-month study of seventeen police agencies across six states, 1,085 arrests occurred in which subjects were placed
in the prone position.[90] Handcuffs were used in 96 percent of the cases,
ankle straps were used in 23 percent of the cases, and officers placed their
weight on the arrestees' backs in 70 percent of the cases. In 80 percent of
the cases, arrestees sustained no injury, 16 percent saw mild injuries such
as bruises, and 4 percent had significant injuries, such as fractures. Death
occurred in none of the cases. In short, prone restraint, in and of itself,
does not seem particularly dangerous.

Controlled laboratory studies, too, failed to prove the positional asphyxia
theory. In multiple studies, restraining people face down, putting weight
on their bodies, or hogtying them did impede breathing to some extent but
failed to cause serious damage.

For example, in one study,[91] thirteen healthy men were hogtied for fifteen minutes. Compared with sitting, hogtying reduced the maximal amount of air the men could inhale and exhale by about 23 percent. This value, however, is not clinically significant, and the men showed no signs of ventilatory failure or asphyxiation. To simulate real-world restraint in which physical struggles increase oxygen requirement, researchers had the men exercise on a stationary bicycle and then tied them up. This, too, had no adverse effect on blood oxygen levels.

In another study,[92] researchers applied up to 200–225 pounds of weight on participants' backs while they were lying face down. The weights reduced participants' maximal amount of inhaling and exhaling by about 30 percent, which is still within the normal range for adult men and women. Certainly, if we kept adding weight, things would turn ugly. An adult ribcage could fatally fracture at a weight force of about 570 pounds.[93] This is the equivalent of two or more people, each weighing over 280 pounds, standing and balancing on the back of a facedown subject, a scenario unlikely to happen in real-life restraint. Instead, we are more likely to see this kind of fatality when a soda venting machine (about one thousand pounds) or a vehicle (more than two thousand pounds) tips over a person.[94]

Why, then, do some people die from violent prone restraint, with no apparent cause of death? Researchers speculate that individual vulnerabilities are why. Controlled laboratory studies, for good reasons, enroll only *healthy* adults who are unlikely to succumb to research protocols.[95] But, in real life, not everyone can be a perfect specimen of health.

Obesity, for example, is thought to interfere with breathing in the prone position, because excessive body weight and abdomen fat can limit chest wall and diaphragm movements.[96] Heart disease is another risk factor, given the physical stress and struggle that occur during violent restraint.[97] Psychiatric illness and illicit drug use are also at the top of the list because they can induce extreme agitation, the so-called excited delirium.[98] This condition may impose greater oxygen requirement or disrupt normal brain signaling and shut down cardiorespiratory function.[99] A psychotic or delirious state may also push one to struggle beyond physical limit, creating excessive acid in the body, a condition known as metabolic acidosis.[100] In this condition, an otherwise-small reduction in breathing capacity caused by prone restraint may trigger cardiac arrest.

These medical complications—and unknowns—are why, at the end of the day, Dr. Andrew Baker's autopsy and testimonies, although they *feel* unsatisfying, strike me as the most balanced, conscientious, and ultimately convincing. There is no denying the fact that Floyd had preexisting heart disease and drugs in his system, which could very well have made him more vulnerable to restraint, let alone neck compression. But before the events of the fateful day, Floyd was functioning with his condition and was, as Baker put it, "generally healthy."[101] Chauvin's action, blatantly dangerous and unjustifiably cruel, was "more than Mr. Floyd could take by virtue of that, those heart conditions," the medical examiner concluded.[102]

Would the same action be fatal to a healthy or healthier individual, to you, or me? Maybe, maybe not, we can never know, and we don't really need to know. What was on trial was not one person's well-being, but another person's utmost disregard for it.

Despite my intention, all along in this chapter, to show that the knee can be hurtful, it was distressing to end with death, tinted by police violence and racial profiling. I am, therefore, thankful to learn that a knee restraint, used properly, may also save lives.

Brazilian jiu jitsu (BJJ) is Brazil's transformation of the Japanese martial arts jiu jitsu and judo. In the early twentieth century, Japanese judo expert Mitsuyo Maeda (1878–1941) traveled to Brazil. There, he met and taught a teenager named Carlos Gracie, who then passed his training onto his brothers. Together, the Gracie brothers practiced and innovated, founding the modern martial art known as BJJ.

BJJ focuses on grappling and ground fighting. The goal is to take opponents to the ground, pin them, control them, and apply joint locks or chokeholds to force them into submission. In one of its techniques known as *uki-gatame* or knee-on-belly, you place one knee across the abdomen or chest of a grounded opponent, extend the other knee to the side for balance and support, and control the opponent's upper body with your hands to prevent them from twisting and turning. With this technique, you apply body weight through the pinning knee. To force submission, you point your knee toward the opponent's sternum, lift your butt and heel, and drive the knee into the person's diaphragm to inflict pain and restrict breathing.[103]

As hurtful as this sounds, researchers at University of Washington suggested that this same technique may also save lives.

In traumatic injuries, a major cause of death is blood loss. With civilian incidents, an average of fourteen minutes would pass between times of injury and tourniquet application by trained medical personnel to control bleeding.[104] Even if an injured person or bystander applies a make-shift tourniquet, that action would take seconds, if not minutes. Meanwhile, with an injury to something like the femoral artery, about 17 mL of blood is lost *every* second,[105] so even one minute of delay could mean more than 1,000 mL of blood loss, more than enough to cause serious damage. Although one can use hand compression to control bleeding, that is easily fatiguing and may also deter bystander intervention because of their reluctance to directly touch a bleeding wound.[106]

The *uki-gatame*, according to the researchers, offers a more efficient and effective technique to control blood loss. By kneeling on vascular pressure points, bystanders can apply significant body weight to control bleeding. Ultrasound has shown that mounting a knee on the groin, or the area right between the stomach and the thigh, reduced femoral artery blood flow by 78 percent; pressing in on the shoulder reduced brachial artery blood flow by 97.5 percent.[107] That is potentially lifesaving.

Granted, I can find no corroborating study or further report on using *uki-gatame* for blood loss control, but it is a happy thought to hang on to that a knee that hurts can also save.

8

Race and Money

At the nurses' station in the hospital ward where I work hangs a pinup of British soccer star David Beckham, perhaps the world's most famous sports figure. His annual income is in the region of $25 million. The picture has him wearing a white T-shirt and a pair of designer jeans, carefully torn at the left knee, with a loose thread hanging down. This hint at rags is a fashion, or affectation, that I find offensive. . . . When I think back to the heroic efforts I have witnessed of poor Africans to make themselves clean, smart, and tidy for special occasions—efforts that filled me with admiration— I feel a visceral anger at this frivolous assumption of false poverty by people who have never had to wear rags in their lives.

—THEODORE DALRYMPLE, ENGLISH CULTURAL CRITIC,
AUTHOR, AND PHYSICIAN

More deprived groups accept a greater degree of ill health as normal, and are therefore less likely to consult a general practitioner. . . . [And] poorer quality primary care was to be found in more deprived areas, where the need was greatest.

—NISHI CHATURVEDI AND YOAV BEN-SHLOMO, UNIVERSITY COLLEGE
LONDON MEDICAL SCHOOL

Given the amount of cartilage loss I have in my bad knee (see chapter 2), I wouldn't be surprised—in fact, I fully expect—to develop osteoarthritis (OA) in the coming decades. Although OA affects multiple joints, including the hips, spine, hands, and feet, the knee is the most commonly afflicted. Knee OA accounts for more than 80 percent of total OA and affects an estimated twenty-six million people in the United States alone.[1]

When knee OA sets in, the protective cartilage that cushions the bones inside the knee will be gone. The knee will become, as they say, bone on bone. There will be pain, loss of mobility, and plenty of bitterness.

If you think I'm being overly dramatic, I very much hope you are right, even though I checked, or will check, almost all of the boxes for risk factors. For one, OA is an age-related condition. Being sixty and older increases our risk by two- to threefold; indeed, passing mile marker forty-five (as I soon will) may already push us over into the "high-risk" category.[2] Either way, estimates are that by the time we are fifty-five, one in ten of us will experience pain and mild-to-moderate disability caused by knee OA.[3]

Aside from aging, previous knee injuries, by damaging the joint or changing its mechanics, also speed up cartilage damage. Check for me. Meniscus removal surgery, or meniscectomy, is another risk factor.[4] Check. Women are more likely to develop knee OA than men.[5] Check again. Another commonly mentioned risk factor is obesity, which increases stress at the knee, induces inflammation, and facilitates cartilage breakdown.[6]

Barring these factors, are some of us inherently more prone to develop knee OA than others? It would seem so. In a U.K. study, researchers identified 490 patients who had severe knee OA and then rounded up 737 of their siblings as well as some 1,700 unrelated community members.[7] Radiographs were taken to assess the knee health of everyone. As it turned out, the siblings of patients with knee OA had twice the risk compared with unrelated people in the community. The condition, in other words, seems to run in the family. Among identical twins, the likelihood of sharing knee OA is even higher.[8] Based on these studies, researchers estimate that knee OA is about 40–60 percent inherited.

Check again.

The first time (that I can remember) someone complained to me about their knees, I was in middle school, about fifteen years old. The complainer was my mother, about forty years old.

"My right knee feels tight," my mother would say. "It hurts climbing stairs." She would say.

My mother is, all in all, a lovely person, but she has an above-average tendency to find fault in things and voice her opinion in a moderate yet persistent manner—she nags, that is. The knee was one of the things she nagged about, her teenage daughter being another.

So, as you can understand, I didn't think much of her supposed ailment. The knee hurt; big deal. There was no wound I could see, not even bruises. If it hurt, it didn't stop her from walking about, doing chores, going to work, or going ballroom dancing. If someone could go dance every other weekend, her knees couldn't hurt that bad.

Well, pardon my language, but damn it, karma's a bitch! Some twenty-five years later, I'm my mother's age, and my knee hurts. For the most part, I manage to—I insist on—walking about, doing chores, going to work, and playing the sport I love. But, make no mistake, it can hurt, and sometimes, I just play through the pain. I wonder if that's what my mother did.

My mother is not rough on her knees. She doesn't run, jump, or play ball. The most she ever does is dance—ballroom dancing, strictly recreational, when she was younger; and gentle senior-citizen Zumba, now that she's older. She has the perfect body mass index, never suffered traumatic injuries in her knees, never had knee surgeries, and is overall quite healthy. Yet, she has mild-to-moderate knee OA.

Maybe she simply has inescapable, bad knee genes, genes that I inherited?

To date, researchers have reported finding more than four hundred genes that are associated with OA.[9] The most prominent one is *GDF5*. Animal as well as human studies show that this gene promotes cartilage development and affects bone and joint health. Mice that are genetically engineered to overexpress the gene develop thick and enlarged cartilage.[10] By contrast, mutated *GDF5* in humans causes extremely short limbs, abnormal hand and foot skeletons, and dislocated joints.[11]

These mutations are rare, and chances are that most of us don't have them, but that doesn't mean all of us have the same exact *GDF5*. Like many genes, *GDF5* exists in multiple versions (known as alleles) with slight differences. The different versions are all technically normal, but some of them may confer a higher, or a lower, disease risk. When researchers compared over five thousand patients with knee OA and more than eight thousand healthy individuals, they found that having a particular version of *GDF5* (known as rs143383) increased the odds of having knee OA.[12]

Disappointingly (if that's the right word), the odds didn't increase all that much: a mere 1.15 times. The effects of other genetic variations that allegedly increase knee OA risk are similarly small.[13] This, some researchers believe, reflects the reality that OA is a complex disease that is triggered not by singular genetic variants but rather by the interaction of multiple variants, each making a small contribution.[14]

That may well be true, but there is something else to consider. As it's common with human genetic studies, the research that has been done to date on OA, and the data that have been carefully collected, all focus on white populations. Or, to be more biologically accurate, they all focus on people of European descent.

When scientists try to determine which genetic variations are associated with particular diseases, they scan the entire sets of DNA, or genomes, of many people. Some of the people will have the disease in question;

some don't. Comparing their genomes, scientists can then identify which genetic variants are associated with the disease—or, in other words, which genetic variants increase disease risk. This is the so-called genome-wide association study.

If we look at the number of people included in genome-wide association studies worldwide, 78 percent were of European descent, 10 percent were Asian, 2 percent were African, and 1 percent were Hispanic or Latin American.[15] Situations in the United States were even more one-sided: some 90 percent of the study participants were of European descent, followed by African Americans at about 3 percent.[16] The Asians fell off the map because, worldwide, they were accounted for only by studies conducted in Asia, in countries like Japan and China, and not in countries like the United States where they are a minority group. As for people of African and Latin American ancestry, Hispanic people, native people, and indigenous people, they are not accounted for anywhere.[17]

The reasons for this disparity are cold and simple. Genome-wide association studies are expensive efforts. They require equipment and operations upward of millions of dollars. They require expertise that comes from hefty investment in education and training. Emerging economies and low-income countries can't afford them and don't have the infrastructure to run them.

In high-income and racially diverse countries like the United States, the reasons for the lack of diversity in genetic studies are more complex. First, many of these studies use established cohorts as their initial populations, and many of these cohorts were created back in the days when we didn't know better.[18] In addition, large cohort data are often generated by well-established medical centers, which are more likely to be located in places that primarily serve high-income, white populations.[19] Then, there is the mistrust, especially among African American communities, toward biomedical research. Some ninety years later, the Tuskegee syphilis study remains in the minds of many African Americans, fueled by recent research exploitations and health-care disparities.[20] Asking people who have been repeatedly betrayed by science to give up their DNA for scientific research does not sound very convincing, does it?

Certainly, as it is well known, humans share 99.9 percent of their DNA. Black, white, brown, or yellow, underneath the skin, we are essentially the same. Races are not meaningful biological categories. They are social constructs built upon superficial differences—and deep-seated prejudices.

This, however, doesn't mean that genetic findings from one human population apply equally well to all populations. For example, genetic variants that were found to be associated with type 2 diabetes, body mass index, and lipid levels in people of European descent often do not bear out in non-European populations.[21] What this means is that if you are of, say, African descent, you are more likely to receive ambiguous results from genetic testing, to be told that you have genetic variants of "unknown significance."

Genetic variations between populations also affect drug metabolism, so drugs may be safer and more effective for populations that have been intensively studied—and riskier and less effective for others. Examples of such drugs include those developed to treat cancer, reduce blood pressure, and prevent blood clots.[22]

Knee OA may be yet another condition for which the impact of genetics varies across populations. Among people of European descent, the *GDF5* gene variant mentioned above, rs143383, increased the odds of knee OA by 1.13 times.[23] In a Japanese population, the odds increased to 1.3 times.[24] In a Han Chinese population (the ethnic majority of China), the risk conferred was higher yet, at 1.54 times.[25]

Still other genetic risk factors of knee OA seem to affect only Asian populations, such as the *ASPN* gene. This gene codes for the production of a protein called asporin. The asporin protein binds to and inhibits another protein (called TGF-β) that is in turn essential for cartilage formation. Different variations of the *ASPN* gene, by being a stronger or weaker inhibitor of TGF-β, affect cartilage health.[26] In a Japanese population, a particular variant of the *ASPN* gene (called D14) increased the odds of knee OA by up to 2.6 times.[27] In a Han Chinese population, the same variant increased the odds twofold.[28] Both the U.S. and U.K. Caucasian populations, however, are unfazed by this variant.[29]

If all of our genetic architectures are 99.9 percent similar, where do all these population differences come from? Some of them are caused by random changes: a certain genetic variant just *happened* to become more commonly inherited in a population.[30] Or, they may be caused by local adaptations: a certain variant may be selected because it provides protection again some local diseases.[31] Or, they may interact with environmental factors: for example, some genetic risks may be triggered by certain diets or local environmental toxins.[32] Whichever the case, we need to stop assuming that the genomes of people of European descent are the standard that applies to everyone. Diversifying our genetic studies matters.

For changes to happen, researchers need to target diverse and underrepresented populations worldwide, and funding agencies need to prioritize grant requests that propose to do so. Even so, gaining access and building infrastructures in low-income countries are challenging. Changing funding priority may also mean reducing the amount of money spent on people of European descent,[33] which may not sit well with researchers whose studies focus on those populations.

In the United States, an even tougher job is to build trust among minority populations. Frankly, no one quite knows precisely how to do that. Possible ideas include establishing clinics in Black or Hispanic neighborhoods to provide genetic testing, creating partnerships with historically Black colleges and universities, and recruiting biomedical researchers who are, themselves, minorities.[34] None of these measures, as you can imagine, will be easy or quick fixes and will require large-scale systemic changes that are the hardest kind to make.

Even if we manage to do all that and do it well, it is, unfortunately, not going to be enough. A pure biomedical approach, a narrow genetic focus ignores socioeconomic factors that find their way into our knee pain.

African Americans, especially African American women, are more likely than their Caucasian counterparts to have knee OA.[35] The difference is especially notable for severe knee OA. A study conducted in Johnston County, North Carolina, found that African Americans forty-five years and older were twice as likely to have severe knee OA as their Caucasian counterparts.[36]

Lower educational levels, a commonly used marker for socioeconomic status, are also associated with increased risk for knee OA.[37] Among white adults with knee OA, those living below the poverty line have significantly more pain and worse physical function than those living above the poverty line, despite having similar levels of joint deterioration as shown in radiographs.[38] Moreover, Black patients living *both above and below* the poverty line have more pain and worse function than white patients living above the poverty line, despite similar joint deterioration.

In other words, when it comes to knee OA, money matters, race matters, and socioeconomic status matters.

Why would this be? Why would a seemingly innocent biological phenomenon like knee pain be tangled up with messy social factors that we would just as soon pretend didn't exist?

Researchers speculate that there are many possible reasons.[39] Lifestyle choices related to diet and physical activity may contribute to a higher body mass index among Black adults, which in turn exacerbates knee OA symptoms. Remember, though, that diet and physical activity are not simply individual choices. They have to do with access to affordable fresh produce and exercise facilities, or the lack thereof, in deprived neighborhoods. Psychosocial factors may also exacerbate knee OA symptoms. These include a lack of social support, the stress associated with living in poverty, and real or perceived discrimination.

On top of having worse symptoms, minority groups and people with lower socioeconomic status also receive less medical relief for their knee OA.

An effective surgical option for advanced knee OA is total knee arthroplasty, commonly known as total knee replacement.[40] During the procedure, a surgeon shaves off damaged cartilage and bone in the knee joint, including at the lower end of the thigh bone, the top of the shinbone, and possibly under the kneecap. Exposed bone ends are then covered with artificial parts to resemble the anatomy of a natural knee. The part on the thigh bone has two round metal knobs to resemble the condyles. The part on the shinbone has a metal tray to resemble the plateau. In between the metals is a bearing made of high-grade plastic with indentations for the condyles to grind into (figure 8.1). That there gives us a new hinge, a new knee.

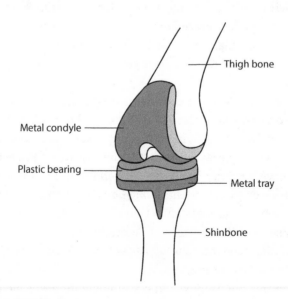

FIGURE 8.1 Artificial knee

With this artificial new knee comes less pain, increased mobility, and improved quality of life. Total knee replacement has a relatively low risk of complication and a greater than 90 percent patient satisfaction.[41] With the increased shelf-life of today's knee implants, the procedure is no longer restricted to older populations and is performed in younger and more active age-groups. Estimates are that by 2030, some 3.5 million total knee replacements will be performed each year in the United States.[42]

The underside of this medical triumph is that, in the United States, knee replacement surgery is utilized far less by minority groups and people of lower socioeconomic status, despite them having similar or higher medical needs.[43] Black men, for example, are 2.8 times less likely than white men to have total knee replacement, and Asian women are 3 times less likely than white women to do so.[44] Compared with people who have a college education, those who don't are 1.5 times less likely to receive a joint replacement.[45]

Why? My immediate guess was financial burden. When I was a lone foreign student in this country trying to make ends meet and had next to nothing in the department of medical insurance, I, like many others in my situation, skipped medical care. And I guessed right: lack of insurance and low income contributed to the low rates of knee surgery among deprived groups, especially Hispanics.[46] Even for older adults who have Medicare, concerns about deductibles, copays, and lost wages can be barriers to surgery.[47]

But that's not all. Inequalities in the U.S. health-care system, researchers believe, are more profound reasons. These include lack of access to orthopedic care, lack of informative educational resources, and poor patient–doctor interactions caused by language and cultural barriers.[48] This last deterrent is especially felt by Hispanics and Asians.

Patients' preferences also come into play. African Americans' prevailing mistrust of the medical system doesn't exactly inspire confidence in surgery. Instead, the Black communities have a stronger belief in self-care and alternative treatments like reducing activities, herbal medicine, and, yes, faith and prayers.[49]

And, you know what, they may just have a point: African Americans *do* have inferior surgery outcomes. Compared with white patients, Blacks, especially those living in impoverished communities, have worse pain and function two years after total knee replacement surgery.[50] They also suffer higher rates of surgery failure. A definitive marker of replacement failure is the need for revision surgery. Generally speaking, 2 percent to 5.7 percent of patients require revision surgery within five years of total knee replacement.[51] Common reasons for revision include loosening of the implant, knee instability, and infection. Across multiple studies that together included about 452,000 patients who underwent total knee replacement, African Americans were 38 percent more likely than Caucasians to require revision surgeries.[52]

Why *that* should happen is also a mystery. Part of it may be that African Americans are more likely to undergo surgeries in low-quality hospitals with inexperienced surgeons.[53] Part of it may be that they receive less post-surgical rehabilitation care.[54] And part of it may be that they put off the surgery for too long, which has a negative impact on surgery outcomes.[55]

The saddest part is that all these issues create a vicious circle. Seeing poor surgical outcomes in their Black friends, families, and neighbors reduces African Americans' willingness to seek surgery.[56] Having few people in their social circles who had the surgery also means that African Americans are less likely to even know about this treatment option.[57] Overtime, surgery rates will continue to be low, and the idea that surgery doesn't work becomes a self-fulfilling prophecy.

Elsewhere in the world, knee OA poses different kinds of puzzles. Despite their relatively lower body mass index, Asian populations have more knee OA than Western populations.[58] In Malaysia, for example, an estimated 25 percent of people over the age of fifty-five suffer from knee OA.[59] In Korea, 38 percent of people over the age of sixty-five do.[60]

Aside from the genetic factors noted earlier, cultural and lifestyle factors—namely, the Asian "floor culture"—are thought to play a role. Chinese and Japanese, as we've learned, have a long tradition of kneeling on the floor. Other Asian populations likewise habitually kneel, sit, or squat on the floor in postures that require deep knee flexion—in other words, postures that require them to bend their knees a great deal.

In Thailand, for example, people, especially rural residents, spend considerable time on the floor for work and leisure activities, sitting cross-legged (figure 8.2A) or with knees bent to the side (figure 8.2B).[61] In the cross-legged position, the knees are bent to about 130 degrees; in the side-knee position, the knees are bent more than 120 degrees.[62] In China and Korea, deep squatting, at least in certain regions, is a common posture in daily living (figure 8.2C).[63] After all, squatting-style toilets are still prevalent in Asia. In deep squatting, the knees are bent to about 120–140 degrees, more if the heels of the feet rise from the ground and touch the butt.[64]

Cross-legged sitting   Side-knee bending   Deep squatting   Maximum-flexion kneeling

(A)   (B)   (C)   (D)

FIGURE 8.2 Common postures with deep knee flexion

In Muslim communities, prayers involve maximum-flexion kneeling (figure 8.2D). In this position, the top of one foot is on the ground, the other foot is upright, and the knees are bent up to 160 degrees.[65] From this position, a person bows forward into prostration, with the forehead touching the ground. The maximum-flexion kneeling is similar to the *seiza* sitting that continues to be part of the Japanese tradition and important for ceremonial rituals like martial arts and tea ceremonies. In seiza, the tops of both feet are flat on the ground, and the knees are bent 150–180 degrees.[66]

These kinds of knee bending are not common in daily activities in the West. Going downstairs, for example, requires only about 90 degrees of knee flexion, rising from a chair requires about 93 degrees, and lifting an object requires about 117 degrees.[67] Still, these daily activities are *already* stressing our knees. When we go up and down stairs, for example, the contact stress between the thigh bone and shinbone goes up 50 percent, and the stress between the kneecap and the thigh bone goes up 100 percent.[68]

As you can imagine, greater knee flexion stresses the joint even more. Among Western populations, frequent kneeling and squatting, often as a result of occupational activities, were found to increase cartilage damage and the risk of knee OA.[69] In a comparison of 84 Saudi Arabian knees and 106 North American knees that were scheduled for total knee replacement surgeries, the former showed cartilage wear patterns consistent with kneeling and squatting.[70] In a study of 576 Thais, habitual cross-legged sitting, side-knee sitting, and squatting increased the risk of knee OA.[71] Likewise, in a study of 1,850 Chinese, frequent squatting increased the risk of knee OA.[72]

Despite all of these studies, age-old customs and traditions don't simply go away. Plus, risk is such an abstract thing. It's not like you can see, overnight, the direct damage of squatting on your knees and resolve to do better tomorrow. Plus, floor-living is not without possible health benefits.

Some scholars have suggested that kneeling and squatting, compared with chair sitting, involve higher levels of lower-limb muscle activation. As a kind of "active rest," these postures may be beneficial to our metabolism, blood flow, and heart and vascular health.[73] Either way, in the foreseeable future, Asia's floor culture will continue to exist, not to mention the ardent Muslim prayers.

Given this reality, some surgeons suggest that knee replacement ought to consider the special needs of Asian populations. In the West, successful knee replacement allows individuals to perform the usual array of daily activities: climb stairs, eat at a table, use a sitting toilet. By that logic, in the East, an ideal knee replacement should allow Japanese women to kneel on the floor for a tea ceremony, devoted Muslims to offer daily prayers, and rural Chinese residents to use squatting toilets.

To meet this demand, implant manufacturers have now developed the so-called high-flexion knee implants. In addition to targeting patients whose cultural customs require deep flexion, they are marketed to people whose recreational or work activities require greater ranges of motion in the knee.

These implants often feature an "extended posterior femoral condyle," which is to say the metal knobs attached to the end of the thigh bone (see figure 8.1) are enlarged in the rear.[74] This design allows more room for the metal condyles to roll on the plastic bearing to achieve deeper knee flexion. With this enlarged contact area, there is less stress and less risk of the metal condyles digging into the plastic bearing. A deeper cutout is then made on the front of the plastic bearing, creating more space for the patellar tendon (which connects the kneecap and the shinbone) to settle in during deep knee flexion.[75]

With these designs, the implants promise 155 degrees of safe knee flexion, compared with 125 degrees enabled by traditional implants.[76] In laboratory testing, high-flexion knee implants withstood stresses that equated to doing thirty squats a day for twenty years.[77]

But alas, engineering specifications aren't the same thing as real-life surgical outcomes. 155 degrees of knee flexion is rare, if at all possible, in total knee replacement patients. In Taiwan, among twenty-five patients who received high-flexion implants, the average angle of knee flexion two years after the surgery was 138 degrees.[78] 80 percent of the patients were able to squat on command, although none was actually able to squat for daily activities.

In Thailand, among 156 patients who received high-flexion implants, maximum knee flexion at the six-year follow-up was 135 degrees.[79] Only about 15 percent of the patients were able to kneel, 37 percent were able to sit with legs bent to the side, and 43 percent were able to sit cross-legged.

In Korea, among forty-seven patients who received high-flexion implants, average knee flexion eight years after surgery was 132 degrees.[80] Only half of the new knees allowed for activities such as squatting, kneeling, or sitting cross-legged.

So, it seems fair to say that high-flexion knee implants underperform in reality. This, to an extent, is inevitable. The knee is not a simple mechanical hinge. When cemented and screwed into a living body, it is not going to perform as efficiently as it does in a pristine engineering lab. Various elements can influence surgery outcomes, including patients' presurgical range of motion, surgical techniques received, and postsurgical rehabilitation.[81]

That said, some surgeons fear that a high-flexion implant is inherently prone to failure. In the same Korean study cited earlier, 46 percent of the knees had loosened and required revision an average of three-and-a-half years after the initial surgery.[82] It would be terribly ironic if, in trying to accommodate Asian patients, we are actually dooming their knees. Fortunately, other reports coming in from Korea and elsewhere in Asia contested this gloomy view. In another Korean study that followed 172 patients with high-flexion implants for an average of four years, no patients experienced implant loosening that required revision.[83] Longer follow-up studies in Korean and Thai patients yielded similarly favorable results—or, at least, they showed that high-flexion implants were no more likely to fail than conventional ones.[84]

Perhaps, some Korean researchers hinted, the high failure rate managed by their fellow surgeons was simply due to their poor surgical techniques.[85] Ouch.

Race and money matter not only in knee health, but also in knee fashion. In chapter 3, we saw women fight for the right to expose their knees in short skirts and the right to liberate those knees in bifurcated "male attire." Similar (but also very different) struggles are faced by people with darker skin and without material wealth. And, compared with 1920s, these struggles are more poignant today with fashion being the humongous transnational enterprise that it is.

Certainly, not all people get to embrace that which we call fashion. In the poverty-stricken countries of Africa—Congo, for example, or Uganda—many people don't get to choose the "in" looks. Clothes are primarily used to shelter the body. So they need to be clean, practical, and sturdy. Tattered clothes are a sign of inadequacy, of painful necessity.

In wealthy countries—the United States, for example, or the United Kingdom—where consumers can afford endless choices of garments, a, or *the*, primary function of clothes is to declare who people are or who they want to be. With this change in function, tattered clothes can somehow become items of joy and pride. Men wear them; women wear them. Everyday people wear them; celebrities wear them. You see them in neighborhood pharmacies, and you see them on the Paris catwalk. They are, yes, the ripped jeans—or, as fashion likes to call them, distressed jeans.

Distressed jeans are, literally, jeans that have been distressed, damaged, and battered. They use tears, rips, scrapes, and frays to create a worn-in look. The knee is—and has always been—a favorite spot to create this distress. "[T]he more ragged the rip, the better. It is especially chic if the knees are exposed," exclaimed the *Bradenton Herald* in 1987.[86] While one can "go for multiple horizontal slashes a few inches apart, giving a kind of fish-gill look," the newspaper continued, "[t]he more conservative types seem to prefer a single slash at the knee."

Thirty years later, fashion has not changed—or rather, it has circled back (figure 8.3 A and B).

Then, as now, the tears made at the knees must go horizontally. A vertical tear is considered silly because it is not *natural*. A little ironic, isn't it? Anyways, the idea is that the fabric at the knee "naturally" rips as a result of a wearer repeatedly bending the knees, so the pattern will always be horizontal. Fashion also dictates that the tear should never be wider than the width of the leg,[87] possibly because we'll otherwise risk the leg falling off. It is also a good idea to keep the slit no more than an inch tall when you stand up.[88] This way, the knee will peek through tastefully when you stand up and come into full view when you sit down.

The symbolic meaning of distressed jeans is an interesting one to ponder. Unlike the 1920s flapper dress, these jeans, clearly, do not intend to leverage one's feminine knees as tantalizing attraction. On the contrary, they intimate physical labor and rugged masculinity: as if the pants were stretched, thinned, and torn by someone who's constantly on his knees cutting wood, mining coal, or tending cattle.

At least, that's how the earliest jeans fell apart. While the word "jeans" may have come from "Genoa," which referred to pants made of fustian and worn by sailors from the Italian port of Genoa around the seventeenth century, the garments that we recognize today as "blue jeans" were first made for laborers in the late-nineteenth-century American West.[89] These workers—woodcutters,

(A)   (B)

FIGURE 8.3 Distressed jeans today: (A) Wesner Rodrigues, *Woman Looking Up*, July 17, 2018, photograph, Pexels, https://www.pexels.com/photo/woman-looking-up-1892779/; (B) Milad Heran, *Man in Brown Jacket and Blue Denim Jeans*, March 27, 2020, photograph, Pexels, https://www.pexels.com/photo/man-in-brown-jacket-and-blue-denim-jeans-4069117/.

miners, farmers, mechanics—wanted something that could stand up to a fair amount of wear and tear on the frontier.

No, these workers weren't responsible for making ripped jeans a fashion. We have the punk rockers of the 1970s to thank for that. One of the most popular bands at the time was Television, from New York City. Its bassist, Richard Hell, used to rip his T-shirts—and his jeans.[90] The scrappy look was promptly picked up by others, including the English punk-rock band from London, the Sex Pistols, and another band from New York city, the Ramones.[91]

On these musicians, ripped jeans and exposed knees were reborn as a symbol of rebellion and free spirit. Here we find a kindred connection with the flappers, although the rockers, in ripping their jeans, also supposedly announced their contempt for consumerism, wealth, and social status,[92] which the flappers rather relished.

Importantly, ripping apart *any* pants won't do. Ripped jeans are fashion and chic; ripped trousers would just be rags and poor. Jeans are special because they are a self-contradictory fashion and cultural icon. They stand for physical labor, but in a way that is romanticized by white collars. They are manufactured to last, but when torn, they underscore the garment's original durability.[93] They say, as it were, "we may be torn, but we hold it together and we don't care."

Of course, people who *actually* do physical labor, like miners, would think twice about going to work wearing jeans that expose their knees. That's practically asking for injury. People who *actually* lack material wealth and social status would also think twice about wearing jeans tattered at the knee. It borders on insulting. It is only the privileged who can afford to publicly perform their disregard for wealth and status.

And, accordingly (or ironically), ripped jeans cost more than intact jeans. At Nordstrom, a pair of women's Paige Verdugo skinny jeans retails $199, while a near-identical pair with frayed ankles and a slashed knee lists for $225. A pair of men's Blake slim fit straight jeans costs $185, and a same design in a faded look with a prominent hole in one knee costs $235. Then, there are luxury brands that easily sell upward of $1,000—today, as well as twenty years ago. In a 2004 episode of *60 Minutes*, the famous Russian soprano Anna Netrebko was seen wearing a pair of Dolce & Gabbana jeans with "considerable gashes in the knees." She explained that they cost $1,200, to which the *60 Minutes* correspondent Bob Simon grumbled, "For twelve hundred dollars, you'd think they wouldn't have holes in them."[94]

But, Bob, distressed jeans *need* to be more expensive because they take more trouble to make. The tears may look accidental, but they are anything but. They are carefully created either by laser machines or by hand using shears and drills—the latter method being more intricate, time-confusing, and preferred by premium designers.[95] The threads are then pulled apart and frayed using a fabric picker. That's not all. For the tears and threads to look authentic, the jeans need to have a faded, aged appearance. To achieve that, manufacturers invented various finishing methods, from chemical treatment and stone washing, to brushing and manual sanding. Denim is tough stuff, and destroying it is tough work.

Most of this work is outsourced to countries where labor is cheap and legislation weak. Whether it's chemical treatment or hand sanding, the distressing processes are potentially hazardous. Exposure to the chemicals (often potassium permanganate) can damage multiple organs from eyes, skin, to liver, while using brushes and sandpaper can cause asthma due to dust exposure.[96]

The worst process is sandblasting, where workers use high-pressure guns to blast sand onto jeans. The abrasive sand softens and lightens the denim to create a faded, worn-in look. This is a popular method because it is cheap. There is no requirement for sophisticated equipment, and sand is inexpensive. Sandblasting is also efficient. While a worker can produce twenty to thirty pieces of garments an hour doing hand sanding, that same worker can produce thirty-five to sixty pieces with manual sandblasting.[97] The only thing that gives is workers' health. Manual sandblasting generates a large amount of silica dust, silica being the major compound in sand. The tiny particles lodge in workers' lungs, causing difficulty breathing, chronic cough, chest pain, and, in the worst cases, death. The condition is known as silicosis, and it can't be cured.

Before 2009, a lot of denim sandblasting happened in Turkey. An estimated three hundred thousand people in Turkey worked in the denim industry, and ten to fifteen thousand of them worked with sandblasting.[98] Most of these were young men in their early twenties or younger.[99] Many started sandblasting around the age of twelve to seventeen years old; some started as early as ten. They worked ten hours a day, six days a week in unlicensed workshops, and some slept in the dusty workshop when their shift was over. Most of these workshops had no ventilation. In fact, to avoid wasting sand, some kept their doors and windows tightly closed. The only, if any, protective equipment used was a simple mask, usually rationed to one a day.

From 2005 to 2008, a series of silicosis cases among Turkey's sandblasting workers were reported.[100] An estimated five thousand people were diagnosed, and more than fifty died.[101] The speed at which the disease developed was astonishing. In other industries like mining and quarrying where silicosis happens, the disease is chronic and develops after decades of exposure, but the denim sandblasters became ill within about three years.[102] Researchers believed that this acute onset was caused by heavy exposure due to the massive amounts of dust, poor ventilation, long work hours, and a lack of protection.[103] Alarmed, in 2009, Turkey prohibited denim sandblasting, some sixty years after it was prohibited in the United Kingdom.[104]

But the practice didn't simply go away. After the ban, the sandblasting industry simply moved farther out from the fashion centers of Europe and North America to countries like China, Bangladesh, Pakistan, and parts of North Africa.[105] In Bangladesh, the work conditions faced by sandblasters were equally deplorable. Workers described coughing out "balls of sand," feeling a "constant biting sensation" in the chest, and having their eyes "filled with sand" when they woke up.[106]

It is true that many fashion brands in the West have publicly issued bans on sandblasting their jeans, but such bans have little effect on what goes on in remote, illegal workshops several contractors down the supply chain and several continents over. When brand auditors came, sandblast equipment was simply hidden and the work moved to night shifts to avoid detection.[107] When denim sandblasters in Bangladesh were shown brand logos, they recognized many that had supposedly banned sandblasting as being manufactured in their workshops: H&M, Levi Strauss, Lee, and Diesel.[108] In China, Levi Strauss, Gap, and Wrangler, all of which had banned sandblasting, were found manufactured in factories that used sandblasting.[109]

As long as we adore ripped jeans as fashion, as long as companies design them, and as long as consumers buy them, we may never be able to stop cheap labor being endangered in the process. Unless, perhaps, we all become distressing do-it-yourselfers with scissors, utility knifes, sandpaper, and wire brushes.

At the sprawling Xipamanine market in Maputo, Mozambique, where thirty-some-year-old Mario makes a living selling secondhand jeans, no one likes to buy them ripped.[110] Mario purchases his jeans in one-hundred-pound bales coming in as container shipments from North America and Europe. The bales are sold "blind": sealed and unopened. If Mario is lucky, he gets gently used ones that will fetch good prices. If not, he gets them ripped, scuffed, and stained. The dirtiest and most torn ones won't sell, and Mario will end up losing money if his bales contain too many pairs like that.

Might some of these unwanted pairs be distressed jeans that were deliberately ripped, torn, and stained, I can't help wondering. And if so, can Mario tell, and will his Mozambican customers appreciate them as fashion? Probably not.

For Western consumers to deliberately distress clothes, using cheap labor from developing countries, and then dispense with the garments at a charity store, the irony would go full circle. Chances are that your unwanted jeans won't make their way to the local shelter. Instead, they will be sorted, packaged, and sold back to developing countries, for a profit. Globally, about four million metric tons of old clothes are exported annually, which are equivalent to seven-and-a-half billion pairs of jeans and valued at more than $4 billion.[111]

Many of these clothes find their way to Africa, to places like the Xipamanine market in Mozambique. This country alone imports $55 million of used clothes a year, which is fifty-two thousand metric tons of garments, or the equivalent of ninety-nine million pairs of jeans.[112]

These secondhand clothes provide impoverished Africans with a cheap source of clothing, at the expense of style and preference. And so it is that in Maputo, it is common for men to wear ladies' trousers, and in Uganda, a Karamoja elder would attend a tribal ceremony wearing nothing but a woman's tweed, oatmeal-colored winter overcoat.[113]

These secondhand clothes provide Mario and thousands of used-clothes traders in Africa a precarious small business to keep their heads above water—but never to escape poverty. The traders face the constant dilemma of risking money on another blind bale or using it to provide for their families, or, indeed, to save money to pay for education toward a better-paying job.[114]

These secondhand clothes, along with cheap new garments coming in from East Asia, outcompete Africa's domestically manufactured garments, chocking to death the local industry and the prospect of economic growth, the hope of escaping distressed clothes.[115]

In fashion as in health, money, race, power, and status all matter. Our knees, curiously enough, provide a glimpse of that reality.

9

Last Words

hen I conceived the idea for this book and proposed it to my editor at Columbia University Press, I had to explain what I wanted it to achieve. That's what presses ask, you know. They always want you to define your "purpose." If an author can't articulate a purpose, chances are that there isn't a book in it, at least not a very good one.

I explained, or thought I did, in the form of a book proposal. Reasonably pleased, my editor sent the proposal out for anonymous reviews by knee experts. Who are "experts" on the topic of human knees, you ask? Well, I imagine anatomists, orthopedic doctors, or anthropologists. Like I said, the reviews were anonymous (to promote candid feedback), so I really don't have the details.

Alas, not all these experts loved my idea or my purpose. Quite a few did, but one in particular was, to put it mildly, dissatisfied. The topics proposed are too disparate, he (or she) opined, what with those distracting details about women's dresses and people kneeling. This is not the proper way to educate public readers about the scientific facts of the knee, he (or she) declared, what with the sensational talks of race and money. I should add that there was also the unsubtle overtone that *I* was not the proper person to educate the public, what with me not being an anatomist or medical doctor.

The word "educate" made me cringe. Scrolling through social media posts and comments, we see the phrase often: "Educate yourself!" The very act of asking someone to "educate" themselves automatically gives the asker authority over whatever topic is in question, regardless of their actual authority. Conversely, the person who is ordered to be educated is left feeling, and presumed to be, pathetically ignorant. Wary of the word, nowhere in my proposal had I used the word "educate." My dissatisfied reviewer, in his (or her) haste to educate me, failed to realize that.

Yes, yes, I am, by profession, an educator: I teach in a university for a living. But that is a very different context, isn't it? In classrooms, I teach a narrow set of topics that, because of my years of training and work in those topics, I have developed a decent amount of knowledge and skill. My students, diverse and accomplished though they are in various aspects of life and knowledge domains, are generally not as experienced in that narrow set of topics. So, I feel comfortable educating my students on those topics—and every semester tends to boost my confidence, some smart (and stubborn) students notwithstanding.

But, the public community you and I live in is not a classroom. This may come as a shock to some "professional academics" (as my dissatisfied reviewer self-identified), but it is rare that the average reader leisurely browsing the neighborhood bookstore or library would pick up a title and exclaim, "Wow, I must read this, because I want to be educated!"

Well, at least they won't with a book like this, a book that dares to cover a diverse range of topics. No single author, I venture, can simultaneously have advanced training (a triple PhD?!) in, say, fashion, anatomy, *and* history, so there can be no "expert" in the absolute sense of the word. More to the point, public readers are more numerous and diverse than any student bodies that can fit into a classroom. I bet many of you, in some way, shape, or form, have more experiences with knee pain, knee surgery, knee fashion, or knee anatomy than I do. So, who am I to educate?

This is why "to educate" was never my purpose in writing this book. In fact, I do not believe it should be a primary purpose for *any* popular science book—unless we don't mind making something fairly boring, like a textbook students consult to pass quizzes. I'm inclined to believe that, outside of a classroom, most people want to read a book because they can relate to the topic, they anticipate it being fun, they are curious about what an author has to say, and they want to compare what they already know with what the author has to say. If they come out feeling that they have learned something useful, all the better!

And *that*, as I tried to explain to my anonymous reviewers, is my purpose in writing about the human knee. I want to intrigue my readers, to surprise them, to take them on a journey of discovery about this supposedly lowly joint that really is rather fascinating. I want readers to see that science and medicine, which we are often told are objective, are in fact deeply intertwined with diverse human lives, experiences, and values. I want to invite readers (especially those who self-identify as *not* being "sciency") to explore the sciences about the knee, including their complexities and unknowns. I'm of the opinion that public readers are more intelligent than academic experts often give them credit for and that readers don't need to be fed a simplified story of scientific consensuses.

Equally importantly, I want my readers to value history and culture, including and especially their deep-seated injustice. There can be no science or medicine (or the need for them) without people, communities, and societies, all of which are steeped in historical and cultural contexts. Whenever we presume to talk about science and medicine in the public sphere, we *have to* talk about contexts. Some of these contexts are not pretty, but we can't pretend that they don't exist. Being an immigrant woman of color, I also want to share my slice of historical baggage, cultural heritage, and multiplied identities, so that the world can know more about "my kind," so that there can be hope for genuine connections.

Ultimately, with all the above, I want my readers—I want you—to appreciate the curiosity that is the human knee and, by extension, the curiosity that is the human life, both of which are messy yet wonderful moving targets.

This is why the book covers what it does: not only evolution, anatomy, and medicine, but also pain, fashion, and personal struggles; not only *Australopithecus*, the anterior cruciate ligament (ACL), and knee braces, but also American flappers, Chinese kowtow, Japanese knee walking, and Thai boxers; not only legendary paleoanthropologists and renowned scientists, but also unnamed knee patients, Colin Kaepernick, and George Floyd.

As I draw this journey to an end, now seems the time to offer some sort of grand conclusion about the knee. And believe me, I tried. But, as I look back to the stories I told, I'm rather afraid that any conclusion I try to come up with will be woefully inadequate. I mean, how can I offer pithy conclusions about unknown human ancestries, incomplete scientific understandings, pain and lost abilities, women's struggle in society, medical inequality, and social injustice in more places than one?

My only hope, I suppose, is that the next time you see a pair of exposed knees, the next time you feel a tinge of pain or relief in your own knees, the next time you witness someone bend their knees for whatever reasons, you catch yourself dwelling on that moment and observing the perspectives crossing your mind. You don't take it for granted. You appreciate. You don't judge. You think. Better yet, you remember that you too have curious stories to tell about the knee. I'd love to hear those stories—drop me a note or leave a review online.

Notes

1. KNEES BEFORE THE BRAIN

The quote in the chapter epigraph comes from S. L. Washburn, "Speculations on the Interrelations of the History of Tools and Biological Evolution," *Human Biology* 31, no. 1 (1959): 23.

1. "Apes" mean members of the Hominoidea family other than humans.
2. Some mammals, such as kangaroos, hop bipedally.
3. Philadelphia Zoo, "Snacking on the Run," March 5, 2018, https://www.facebook.com/watch/?v=10155414572887934.
4. Sarah Gibbens, "Why This Gorilla Prefers to Walk Upright," *National Geographic*, March 19, 2018, https://www.nationalgeographic.com/news/2018/03/gorilla-walking-upright-bipedal-philadelphia-zoo-spd/.
5. Ivor Janković, "Certain Medical Problems Resulting from Evolutionary Processes: Bipedalism as an Example," *Periodicum Biologorum* 117, no. 1 (2015): 17–26.
6. Bruce Latimer, "The Perils of Being Bipedal," *Annals of Biomedical Engineering* 33, no. 1 (2005): 3–6, https://doi.org/10.1007/s10439-005-8957-8.
7. Latimer, "The Perils."
8. Donald C. Johanson and Maitland Edey, *Lucy: The Beginnings of Humankind* (New York: Simon & Schuster, 1981).
9. Johanson and Edey, *Lucy*.
10. "Gibbons," *National Geographic*, accessed November 11, 2021, https://www.nationalgeographic.com/animals/mammals/group/gibbons/.
11. Thure E. Cerling, Yang Wang, and Jay Quade, "Expansion of C4 Ecosystems as an Indicator of Global Ecological Change in the Late Miocene," *Nature* 361, no. 6410 (1993): 344–45, https://doi.org/10.1038/361344a0.
12. Gordon W. Hewes, "Food Transport and the Origin of Hominid Bipedalism," *American Anthropologist* 63, no. 4 (1961): 687–710, https://doi.org/10.1525/aa.1961.63.4.02a00020.
13. Christine Tardieu and Erik Trinkaus, "Early Ontogeny of the Human Femoral Bicondylar Angle," *American Journal of Physical Anthropology* 95, no. 2 (1994): 183–95, https://doi.org/10.1002/ajpa.1330950206.

14. C. Owen Lovejoy, "The Natural History of Human Gait and Posture," *Gait and Posture* 25, no. 3 (2007): 325–41, https://doi.org/10.1016/j.gaitpost.2006 .05.001.

15. Tardieu and Trinkaus, "Early Ontogeny." People with congenital conditions who never stood or walked upright don't have a bicondylar angle either.

16. S. Hayama, M. Nakatsukasa, and Y. Kunimatsu, "Monkey Performance: The Development of Bipedalism in Trained Japanese Monkeys," *Acta Anat Nippon* 67, no. 3 (1992): 169–85; C. Tardieu, Y. Glard, E. Garron, C. Boulay, J. L. Jouve, O. Dutour, G. Boetsch, and G. Bollini, "Relationship Between Formation of the Femoral Bicondylar Angle and Trochlear Shape: Independence of Diaphyseal and Epiphyseal Growth," *American Journal of Physical Anthropology* 130, no. 4 (2006): 491–500, https://doi.org/10.1002/ajpa.20373.

17. Lovejoy, "The Natural History."

18. Lovejoy, "The Natural History."

19. The following account of the discovery of the knee joint draws from Johanson and Edey, *Lucy*.

20. Johanson and Edey, *Lucy*, 156.

21. Johanson and Edey, *Lucy*, 159.

22. The following account of the discovery of Lucy draws from Johanson and Edey, *Lucy*.

23. Johanson and Edey, *Lucy*, 15.

24. Donald C. Johanson and Kate Wong, *Lucy's Legacy: The Quest for Human Origins* (New York: Three Rivers Press, 2009).

25. Donald. C. Johanson, Lovejoy C. Owen, A. H. Burstein, and K. G. Heiple, "Functional Implications of the Afar Knee Joint," *American Journal of Physical Anthropology* 44, no. 1 (1976): 188; Jack T. Stern, "Climbing to the Top: A Personal Memoir of *Australopithecus afarensis*," *Evolutionary Anthropology: Issues, News, and Reviews* 9, no. 3 (2000): 113–33, https://doi.org/10.1002/1520 -6505(2000)9:3<113::AID-EVAN2>3.0.CO;2-W; Carol V. Ward, "Interpreting the Posture and Locomotion of *Australopithecus afarensis*: Where Do We Stand?," *Yearbook of Physical Anthropology* 45 (2002): 185–215.

26. Ward, "Interpreting"; Yoel Rak, "Lucy's Pelvic Anatomy: Its Role in Bipedal Gait," *Journal of Human Evolution* 20, no. 4 (1991): 283–90, https://doi.org/10 .1016/0047-2484(91)90011-J; C. Owen Lovejoy, "Evolution of Human Walking," *Scientific American* 259, no. 5 (1988): 118–25, https://doi.org/10.1038 /scientificamerican1188-118.

27. William H. Kimbel and Lucas K. Delezene, "'Lucy' Redux: A Review of Research on *Australopithecus afarensis*," *American Journal of Physical Anthropology* 140, no. S49 (2009): 2–48, https://doi.org/10.1002/ajpa.21183.

28. See Stern, "Climbing."

29. Ward, "Interpreting."

30. Some studies suggest that Sadiman may not be the volcano that provided the ash for the Laetoli footprints. See Anatoly N. Zaitsev, Thomas Wenzel, John Spratt, Terry C. Williams, Stanislav Strekopytov, Victor V. Sharygin, Sergey V.

Petrov, Tamara A. Golovina, Elena O. Zaitseva, and Gregor Markl, "Was Sadiman Volcano a Source for the Laetoli Footprint Tuff?," *Journal of Human Evolution* 61, no. 1 (2011): 121–24, https://doi.org/10.1016/j.jhevol.2011.02.004.

31. Johanson and Edey, *Lucy.*

32. Smithsonian National Museum of Natural History, "Laetoli Footprint Trails," updated December 17, 2020, http://humanorigins.si.edu/evidence/behavior/footprints/laetoli-footprint-trails.

33. M. R. Bennett, S. C. Reynolds, S. A. Morse, and M. Budka, "Laetoli's Lost Tracks: 3D Generated Mean Shape and Missing Footprints," *Scientific Reports* 6, no. 1 (2016): 21916, https://doi.org/10.1038/srep21916.

34. Robin H. Crompton, Todd C. Pataky, Russell Savage, Kristiaan D'AoÛt, Matthew R. Bennett, Michael H. Day, Karl Bates, Sarita Morse, and William I. Sellers, "Human-Like External Function of the Foot, and Fully Upright Gait, Confirmed in the 3.66 Million Year Old Laetoli Hominin Footprints by Topographic Statistics, Experimental Footprint-Formation and Computer Simulation," *Journal of the Royal Society, Interface* 9, no. 69 (2012): 707–19, https://doi.org/10.1098/rsif.2011.0258.

35. Johanson and Edey, *Lucy*, 250.

36. David A. Raichlen, Adam D. Gordon, William E. H. Harcourt-Smith, Adam D. Foster, and Wm. Randall Haas Jr., "Laetoli Footprints Preserve Earliest Direct Evidence of Human-Like Bipedal Biomechanics," *PLoS One* 5, no. 3 (2010): e9769, https://doi.org/10.1371/journal.pone.0009769; Crompton et al., "Human-Like"; Bennett et al., "Laetoli's Lost Tracks."

37. Johanson and Edey, *Lucy.*

38. Robert C. Walter, "Age of Lucy and the First Family: Single-Crystal ^{40}Ar/^{39}Ar Dating of the Denen Dora and Lower Kada Hadar Members of the Hadar Formation, Ethiopia," *Geology* 22, no. 1 (1994): 6–10, https://doi.org/10.1130/0091-7613(1994)022<0006:AOLATF>2.3.CO;2; Alan L. Deino, "40ar/39ar Dating of Laetoli, Tanzania," in *Paleontology and Geology of Laetoli: Human Evolution in Context*, vol. 1, *Geology, Geochronology, Paleoecology and Paleoenvironment*, ed. Terry Harrison (Dordrecht: Springer Netherlands, 2011), 77–97.

39. S. Semaw, P. Renne, J. W. K. Harris, C. S. Feibel, R. L. Bernor, N. Fesseha, and K. Mowbray, "2.5-Million-Year-Old Stone Tools from Gona, Ethiopia," *Nature* 385, no. 6614 (1997): 333–36, https://doi.org/10.1038/385333a0.

40. The earliest stone tools currently known are dated 3.3 million years ago and were found in West Turkana, Kenya. We don't know which hominin species made them, though *Kenyanthropus platyops*, whose fossils were found in the same region and time period, is a possible guess. See Sonia Harmand, Jason E. Lewis, Craig S. Feibel, Christopher J. Lepre, Sandrine Prat, Arnaud Lenoble, Xavier Boës, et al., "3.3-Million-Year-Old Stone Tools from Lomekwi 3, West Turkana, Kenya," *Nature* 521, no. 7552 (2015): 310–15, https://doi.org/10.1038/nature14464.

41. Tim D. White, C. Owen Lovejoy, Berhane Asfaw, Joshua P. Carlson, and Gen Suwa, "Neither Chimpanzee nor Human, Ardipithecus Reveals the Surprising

Ancestry of Both," *Proceedings of the National Academy of Sciences of the United States of America* 112, no. 16 (2015): 4877–84, https://doi.org/10.1073/pnas.1403659111.

42. Ann Gibbons, "A New Kind of Ancestor: Ardipithecus Unveiled," *Science* 326, no. 5949 (2009): 36–40, https://doi.org/10.1126/science.326.5949.36; White et al., "Neither Chimpanzee nor Human."

43. C. Owen Lovejoy, "Reexamining Human Origins in Light of *Ardipithecus ramidus*," *Science* 326, no. 5949 (2009): 74, 74e1-e8, https://doi.org/10.1126/science.1175834; White et al., "Neither Chimpanzee nor Human."

44. Gerard D. Gierliński, Grzegorz Niedźwiedzki, Martin G. Lockley, Athanassios Athanassiou, Charalampos Fassoulas, Zofia Dubicka, Andrzej Boczarowski, Matthew R. Bennett, and Per Erik Ahlberg, "Possible Hominin Footprints from the Late Miocene (c. 5.7 Ma) of Crete?," *Proceedings of the Geologists' Association* 128, nos. 5–6 (2017): 697–710, https://doi.org/10.1016/j.pgeola.2017.07.006.

45. Gierliński et al., "Possible Hominin Footprints"; Robin H. Crompton, "Making the Case for Possible Hominin Footprints from the Late Miocene (c. 5.7 Ma) of Crete?," *Proceedings of the Geologists' Association* 128, nos. 5–6 (2017): 692–93, https://doi.org/10.1016/j.pgeola.2017.09.001.

46. Brian G. Richmond and William L. Jungers. "*Orrorin tugenensis* Femoral Morphology and the Evolution of Hominin Bipedalism," *Science* 319, no. 5870 (2008): 1662–65, https://doi.org/10.1126/science.1154197.

47. Richmond and Jungers, "*Orrorin tugenensis*"; David R. Begun, "The Earliest Hominins: Is Less More?," *Science* 303, no. 5663 (2004): 1478–80.

48. Lee R. Berger, Darryl J. De Ruiter, Steven E. Churchill, Peter Schmid, Kristian J. Carlson, Paul H. G. M. Dirks, and Job M. Kibii, "*Australopithecus sediba*: A New Species of *Homo*-Like Australopith from South Africa," *Science* 328, no. 5975 (2010): 195–204, https://doi.org/10.1126/science.1184944; Jeremy M. DeSilva, Kenneth G. Holt, Steven E. Churchill, Kristian J. Carlson, Christopher S. Walker, Bernhard Zipfel, and Lee R. Berger, "The Lower Limb and Mechanics of Walking in *Australopithecus sediba*," *Science* 340, no. 6129 (2013): 1232999, https://doi.org/10.1126/science.1232999.

49. Tim Bertelsman and Brandon Steele, "Foot Hyperpronation," Illinois Chiropractic Society, accessed November 11, 2021, https://ilchiro.org/foot-hyperpronation/.

50. DeSilva et al., "The Lower Limb."

51. Lee R. Berger, John Hawks, Darryl J. de Ruiter, Steven E. Churchill, Peter Schmid, Lucas K. Delezene, Tracy L. Kivell, et al., "*Homo naledi*, A New Species of the Genus *Homo* from the Dinaledi Chamber, South Africa," *eLife* 4 (2015), https://doi.org/10.7554/eLife.09560; Florent Détroit, Armand Salvador Mijares, Julien Corny, Guillaume Daver, Clément Zanolli, Eusebio Dizon, Emil Robles, Rainer Grün, and Philip J. Piper, "A New Species of *Homo* from the Late Pleistocene of the Philippines," *Nature* 568, no. 7751 (2019): 181–86, https://doi.org/10.1038/s41586-019-1067-9.

52. Détroit et al., "A New Species."

53. Jeremy DeSilva, *First Steps: How Walking Upright Made Us Human* (New York: HarperCollins, 2021).

54. Hewes, "Food Transport"; Gordon W. Hewes, "Hominid Bipedalism: Independent Evidence for the Food-Carrying Theory," *Science* 146, no. 3642 (1964): 416–18, https://doi.org/10.1126/science.146.3642.416.

55. Hewes, "Food Transport"; Hewes," Hominid Bipedalism."

56. Johanson and Edey, *Lucy*; Lovejoy, "Reexamining Human Origins."

57. Goodall, quoted in C. Owen Lovejoy, "The Origin of Man," *Science* 211, no. 4480 (1981): 341–50, 343.

58. Goodall, quoted in Lovejoy, "The Origin," 344.

59. Lovejoy, "Reexamining Human Origins."

60. Donna Hart and Robert Sussman, *Man the Hunted: Primates, Predators, and Human Evolution* (Boulder, CO: Westview Press, 2009).

61. Hart and Sussman, *Man the Hunted*.

62. Johanson and Edey, *Lucy*, 334; Lovejoy, "Reexamining Human Origins."

63. Johanson and Edey, *Lucy*, 339–40.

64. Glynn L. Isaac, "Models of Human Evolution," *Science* 217, no. 4557 (1982): 295; Johanson and Wong, *Lucy's Legacy*.

65. Steven R. Leigh and Brian T. Shea, "Ontogeny and the Evolution of Adult Body Size Dimorphism in Apes," *American Journal of Primatology* 36, no. 1 (1995): 37–60, https://doi.org/10.1002/ajp.1350360104.

66. Ryan Schacht and Karen L. Kramer, "Are We Monogamous? A Review of the Evolution of Pair-Bonding in Humans and Its Contemporary Variation Cross-Culturally," *Frontiers in Ecology and Evolution* 7 (2019), https://doi.org/10.3389/fevo.2019.00230.

67. Philip L. Reno, Richard S. Meindl, Melanie A. McCollum, and Owen Lovejoy, "Sexual Dimorphism in *Australopithecus afarensis* Was Similar to That of Modern Humans," *Proceedings of the National Academy of Sciences of the United States* 100, no. 16 (2003): 9404–409, https://doi.org/10.1073/pnas.1133180100.

68. Reno et al., "Sexual Dimorphism"; Charles A. Lockwood, Brian G. Richmond, William L. Jungers, and William H. Kimbel, "Randomization Procedures and Sexual Dimorphism in *Australopithecus afarensis*," *Journal of Human Evolution* 31, no. 6 (1996): 537–48, https://doi.org/10.1006/jhev.1996.0078; Adam D. Gordon, David J. Green, and Brian G. Richmond, "Strong Postcranial Size Dimorphism in *Australopithecus afarensis*: Results from Two New Resampling Methods for Multivariate Data Sets with Missing Data," *American Journal of Physical Anthropology* 135, no. 3 (2008): 311–28, https://doi.org/10.1002/ajpa.20745.

69. White, "Neither Chimpanzee nor Human"; Lovejoy, "Reexamining Human Origins."

70. T. V. Venkateswaran, "Did Ardi Walk for Sex? Gender, Science and World Views," *Economic and Political Weekly* 46, no. 3 (2011): 19–23, 21.

71. Venkateswaran, "Did Ardi Walk," 21.

72. Carsten Niemitz, "The Evolution of the Upright Posture and Gait—a Review and a New Synthesis," *Die Naturwissenschaften* 97, no. 3 (2010): 241–63, https://doi.org/10.1007/s00114-009-0637-3.

73. Nina G. Jablonski and George Chaplin, "Origin of Habitual Terrestrial Bipedalism in the Ancestor of the Hominidae," *Journal of Human Evolution* 24, no. 4 (1993): 259–80, https://doi.org/10.1006/jhev.1993.1021.

74. Hewes, "Food Transport"; Niemitz, "The Evolution."

75. "Gorilla Strolls on Hind Legs," NBC News, January 27, 2011, http://www.nbcnews.com/id/41292533/ns/technology_and_science-science/t/gorilla-strolls-hind-legs/#.XrBe1mXQguX.

76. Niemitz, "The Evolution."

77. Hewes, "Food Transport."

78. Kevin D. Hunt, "The Postural Feeding Hypothesis: An Ecological Model for the Evolution of Bipedalism," *South African Journal of Science* 92, no. 2 (1996): 77–90.

79. Hunt, "The Postural Feeding Hypothesis."

80. Jablonski and Chaplin, "Origin."

81. Jablonski and Chaplin, "Origin," 274.

82. Niemitz, "The Evolution," 248.

83. Niemitz, "The Evolution"; Michael D. Sockol, David A. Raichlen, and Herman Pontzer, "Chimpanzee Locomotor Energetics and the Origin of Human Bipedalism," *Proceedings of the National Academy of Sciences of the United States of America* 104, no. 30 (2007): 12265–69, https://doi.org/10.1073/pnas.0703267104.

84. Johanson and Edey, *Lucy*, 338.

2. CONFUSED ANATOMY

The quote in the chapter epigraph comes from C. Owen Lovejoy, "The Natural History of Human Gait and Posture," *Gait and Posture* 25, no. 3 (2007): 326, https://doi.org/10.1016/j.gaitpost.2006.05.001.

1. C. F. Cox, M. A. Sinkler, and J. B. Hubbard, "Anatomy, Bony Pelvis and Lower Limb, Knee Patella," in *StatPearls* (Treasure Island, FL: StatPearls, 2020), 30137819.

2. Jawad Abulhasan and Michael Grey, "Anatomy and Physiology of Knee Stability," *Journal of Functional Morphology and Kinesiology* 2, no. 4 (2017): 34, https://doi.org/10.3390/jfmk2040034.

3. Jennifer Evans and Jeffery I. Nielson, "Anterior Cruciate Ligament Knee Injuries," in *StatPearls* (Treasure Island, FL: StatPearls, 2020), 29763023.

4. Hannah Yoo and Raghavendra Marappa-Ganeshan, "Anatomy, Bony Pelvis and Lower Limb, Knee Anterior Cruciate Ligament," in *StatPearls* (Treasure Island, FL: StatPearls, 2020), 32644659; Evans and Nielson, "Anterior Cruciate Ligament."

5. Marc A. Raj, Ahmed Mabrouk, and Matthew Varacallo, "Posterior Cruciate Ligament Knee Injuries," in *StatPearls* (Treasure Island, FL: StatPearls, 2020), 28613477.

6. Raj, Mabrouk, and Varacallo, "Posterior Cruciate Ligament."

7. The MCL connects the thigh bone to the shinbone; the LCL connects the thigh bone to the calf bone.

8. Reed J. Yaras, Nicholas O'Neill, and Amjad M. Yaish, "Lateral Collateral Ligament Knee Injuries," in *StatPearls* (Treasure Island, FL: StatPearls, 2020), 32809682.

9. Usker Naqvi and Andrew I. Sherman, "Medial Collateral Ligament Knee Injuries," in *StatPearls* (Treasure Island, FL: StatPearls, 2020), 28613747.

10. Alice J. S. Fox, Asheesh Bedi, and Scott A. Rodeo, "The Basic Science of Human Knee Menisci: Structure, Composition, and Function," *Sports Health* 4, no. 4 (2012): 340–51, https://doi.org/10.1177/1941738111429419.

11. Jennifer C. Jones, Robert Burks, Brett D. Owens, Rodney X. Sturdivant, Steven J. Svoboda, and Kenneth L. Cameron, "Incidence and Risk Factors Associated with Meniscal Injuries Among Active-Duty US Military Service Members," *Journal of Athletic Training* 47, no. 1 (2012): 67–73.

12. Chandler F. Cox and John B. Hubbard, "Anatomy, Bony Pelvis and Lower Limb, Knee Lateral Meniscus," in *StatPearls* (Treasure Island, FL: StatPearls, 2020), 30137778; Connor Farrell, Alan G. Shamrock, and John Kiel, "Anatomy, Bony Pelvis and Lower Limb, Medial Meniscus," in *StatPearls* (Treasure Island, FL: StatPearls, 2020), 30725961.

13. Simon C. Mordecai, Nawfal Al-Hadithy, Howard E. Ware, and Chinmay M. Gupte, "Treatment of Meniscal Tears: An Evidence Based Approach," *World Journal of Orthopedics* 5, no. 3 (2014): 233–41, https://doi.org/10.5312/wjo.v5.i3.233.

14. Abulhasan and Grey, "Anatomy."

15. Shantanu Sudhakar Patil, Anshu Shekhar, and Sachin Ramchandra Tapasvi, "Meniscal Preservation Is Important for the Knee Joint," *Indian Journal of Orthopaedics* 51, no. 5 (2017): 576–87, https://doi.org/10.4103/ortho.ijortho_247_17.

16. T. P. McMurray, "The Semilunar Cartilages," *British Journal of Surgery* 29, no. 116 (1942): 407–14. https://doi.org/10.1002/bjs.18002911612.

17. T. J. Fairbank, "Knee Joint Changes After Meniscectomy," *Journal of Bone and Joint Surgery* 30-B, no. 4 (1948): 670, https://doi.org/10.1302/0301-620x.30b4.664.

18. Jack C. Hughston, "A Simple Meniscectomy," *Journal of Sports Medicine* 3, no. 4 (1975): 179–87, https://doi.org/10.1177/036354657500300406.

19. Fox, Bedi, and Rodeo, "The Basic Science."

20. Frank R. Noyes, Ryan C. Chen, Sue D. Barber-Westin, and Hollis G. Potter, "Greater Than 10-Year Results of Red-White Longitudinal Meniscal Repairs in Patients 20 Years of Age or Younger," *American Journal of Sports Medicine* 39, no. 5 (2011): 1008–17, https://doi.org/10.1177/0363546510392014; Patil, Shekhar, and Tapasvi, "Meniscal Preservation."

21. P. D. Gallacher, R. E. Gilbert, G. Kanes, S. N. J. Roberts, and D. Rees, "White on White Meniscal Tears to Fix or Not to Fix?," *The Knee* 17, no. 4 (2010): 270–73, https://doi.org/10.1016/j.knee.2010.02.016.

22. Steven P. Arnoczky and Russell F. Warren, "Microvasculature of the Human Meniscus," *American Journal of Sports Medicine* 10, no. 2 (1982): 90–95, https://doi.org/10.1177/036354658201000205.

23. Noyes et al., "Greater Than 10-Year Results."

24. Gallacher et al., "White on White."

25. Gallacher et al., "White on White."

26. Michiel F. van Trommel, Peter T. Simonian, Hollis G. Potter, and Thoma L. Wickiewicz, "Arthroscopic Meniscal Repair with Fibrin Clot of Complete Radial Tears of the Lateral Meniscus in the Avascular Zone," *Arthroscopy* 14, no. 4 (1998): 360–65, https://doi.org/10.1016/S0749-8063(98)70002-7; Mordecai et al., "Treatment of Meniscal Tears."

27. Mordecai et al., "Treatment of Meniscal Tears."

28. Matt Daggett, Steven Claes, Camilo P. Helito, Pierre Imbert, Edoardo Monaco, Christian Lutz, and Bertrand Sonnery-Cottet, "The Role of the Anterolateral Structures and the ACL in Controlling Laxity of the Intact and ACL-Deficient Knee: Letter to the Editor," *American Journal of Sports Medicine* 44, no. 4 (2016): NP14–NP15, https://doi.org/10.1177/0363546516638069.

29. Christoph Kittl, Hadi El-Daou, Kiron K. Athwal, Chinmay M. Gupte, Andreas Weiler, Andy Williams, and Andrew A. Amis, "The Role of the Anterolateral Structures and the ACL in Controlling Laxity of the Intact and ACL-Deficient Knee: Response," *American Journal of Sports Medicine* 44, no. 4 (2016): NP15–NP18, https://doi.org/10.1177/0363546516638070; Jacob L. Cartner, Zane M. Hartsell, William M. Ricci, and Paul Tornetta III, "Can We Trust Ex Vivo Mechanical Testing of Fresh-Frozen Cadaveric Specimens? The Effect of Postfreezing Delays," *Journal of Orthopaedic Trauma* 25, no. 8 (2011): 459–61, https://doi.org/10.1097/BOT.0b013e318225b875.

30. Michael T. Hirschmann and Werner Müller, "Complex Function of the Knee Joint: The Current Understanding of the Knee," *Knee Surgery, Sports Traumatology, Arthroscopy* 23, no. 10 (2015): 2780–88, https://doi.org/10.1007/s00167-015-3619-3.

31. Hirschmann and Müller, "Complex Function."

32. Armin Runer, Stephan Birkmaier, Mathias Pamminger, Simon Reider, Elmar Herbst, Karl-Heinz Künzel, Erich Brenner, and Christian Fink, "The Anterolateral Ligament of the Knee: A Dissection Study," *The Knee* 23, no. 1 (2016): 8–12, https://doi.org/10.1016/j.knee.2015.09.014.

33. Anthony M. J. Bull and Andrew A. Amis, "The Pivot-Shift Phenomenon: A Clinical and Biomechanical Perspective," *The Knee* 5, no. 3 (1998): 141–58, https://doi.org/10.1016/s0968-0160(97)10027-8.

34. Rob J. P. M. Scholten, Wim Opstelten, Cees G. van der Plas, Dick Bijl, Walter L. J. M. Deville, and Lex M. Bouter, "Accuracy of Physical Diagnostic Tests

for Assessing Ruptures of the Anterior Cruciate Ligament: A Meta-Analysis," *Journal of Family Practice* 52, no. 9 (2003): 689–94; Richard Nickinson, Clare Darrah, and Simon Donell, "Accuracy of Clinical Diagnosis in Patients Undergoing Knee Arthroscopy," *International Orthopaedics* 34, no. 1 (2010): 39–44, https://doi.org/10.1007/s00264-009-0760-y.

35. Bull and Amis, "The Pivot-Shift"; E. Monaco, A. Ferretti, L. Labianca, B. Maestri, A. Speranza, M. J. Kelly, and C. D'Arrigo, "Navigated Knee Kinematics After Cutting of the ACL and Its Secondary Restraint," *Knee Surgery, Sports Traumatology, Arthroscopy* 20, no. 5 (2011): 870–77, https://doi.org/10.1007/s00167 -011-1640-8.

36. Bull and Amis, "The Pivot-Shift."

37. Nicholas J. Vaudreuil, Benjamin B. Rothrauff, Darren de Sa, and Volker Musahl, "The Pivot Shift: Current Experimental Methodology and Clinical Utility for Anterior Cruciate Ligament Rupture and Associated Injury," *Current Reviews in Musculoskeletal Medicine* 12, no. 1 (2019): 41–49, https://doi .org/10.1007/s12178-019-09529-7. Granted, some of the inconsistencies are due to varying examiner skills and patients' guarding against pain. See Vaudreuil et al., "The Pivot Shift"; Takehiko Matsushita, Shinya Oka, Kouki Nagamune, Tomoyuki Matsumoto, Yuichiro Nishizawa, Yuichi Hoshino, Seiji Kubo, Masahiro Kurosaka, and Ryosuke Kuroda, "Differences in Knee Kinematics Between Awake and Anesthetized Patients During the Lachman and Pivot-Shift Tests for Anterior Cruciate Ligament Deficiency," *Orthopaedic Journal of Sports Medicine* 1, no. 1 (2013): 2325967113487855–55, https://doi.org/10.1177 /2325967113487855.

38. Steven Claes, Evie Vereecke, Michael Maes, Jan Victor, Peter Verdonk, and Johan Bellemans, "Anatomy of the Anterolateral Ligament of the Knee," *Journal of Anatomy* 223, no. 4 (2013): 321, https://doi.org/10.1111/joa.12087.

39. Claes et al., "Anatomy of the Anterolateral Ligament," 321.

40. Daniel Guenther, Amir A. Rahnemai-Azar, Kevin M. Bell, Sebastián Irarrázaval, Freddie H. Fu, Volker Musahl, and Richard E. Debski, "The Anterolateral Capsule of the Knee Behaves Like a Sheet of Fibrous Tissue," *American Journal of Sports Medicine* 45, no. 4 (2016): 849–55, https://doi.org/10.1177/0363546516674477; Malcolm E. Dombrowski, Joanna M. Costello, Bruno Ohashi, Christopher D. Murawski, Benjamin B. Rothrauff, Fabio V. Arilla, Nicole A. Friel, et al., "Macroscopic Anatomical, Histological and Magnetic Resonance Imaging Correlation of the Lateral Capsule of the Knee," *Knee Surgery, Sports Traumatology, Arthroscopy* 24, no. 9 (2016): 2854–60, https://doi.org/10.1007/s00167 -015-3517-8; Elmar Herbst, Marcio Albers, Andreas Imhoff, Freddie Fu, and Volker Musahl, "The Anterolateral Complex of the Knee," *Orthopaedic Journal of Sports Medicine* 5, no. 10 (2017): 2325967117730805, https://doi.org/10.1177 /2325967118S00031.

41. Claes et al., "Anatomy of the Anterolateral Ligament"; Jean-Philippe Vincent, Robert A. Magnussen, Ferittu Gezmez, Arnaud Uguen, Matthias Jacobi, Florent

Weppe, Ma'ad F. Al-Saati, et al., "The Anterolateral Ligament of the Human Knee: An Anatomic and Histologic Study," *Knee Surgery, Sports Traumatology, Arthroscopy* 20, no. 1 (2011): 147–52, https://doi.org/10.1007/s00167-011 -1580-3; Runer et al., "The Anterolateral Ligament." The average width at the femoral origin is seven to eight millimeters; the average width at the tibia insertion is eleven to twelve millimeters. See Claes et al., "Anatomy of the Anterolateral Ligament"; Runer et al., "The Anterolateral Ligament."

42. See, for example, Vincent et al., "The Anterolateral Ligament"; Guenther et al., "The Anterolateral Capsule."

43. Herbst et al., "The Anterolateral Complex."

44. Vincent et al., "The Anterolateral Ligament."

45. Runer et al., "The Anterolateral Ligament."

46. Claes et al., "Anatomy of the Anterolateral Ligament."

47. Kittl et al., "The Role of the Anterolateral Structures"; Ran Thein, James Boorman-Padgett, Kyle Stone, Thomas L. Wickiewicz, Carl W. Imhauser, and Andrew D. Pearle, "Biomechanical Assessment of the Anterolateral Ligament of the Knee: A Secondary Restraint in Simulated Tests of the Pivot Shift and of Anterior Stability," *Journal of Bone and Joint Surgery* 98, no. 11 (2016): 937–43, https://doi.org/10.2106/JBJS.15.00344.

48. Matthew T. Rasmussen, Marco Nitri, Brady T. Williams, Samuel G. Moulton, Raphael Serra Cruz, Grant J. Dornan, Mary T. Goldsmith, and Robert F. LaPrade, "An in Vitro Robotic Assessment of the Anterolateral Ligament, Part 1: Secondary Role of the Anterolateral Ligament in the Setting of an Anterior Cruciate Ligament Injury," *American Journal of Sports Medicine* 44, no. 3 (2015): 585–92, https://doi.org/10.1177/0363546515618387.

49. Marco Nitri, Matthew T. Rasmussen, Brady T. Williams, Samuel G. Moulton, Raphael Serra Cruz, Grant J. Dornan, Mary T. Goldsmith, and Robert F. LaPrade, "An in Vitro Robotic Assessment of the Anterolateral Ligament, Part 2: Anterolateral Ligament Reconstruction Combined with Anterior Cruciate Ligament Reconstruction," *American Journal of Sports Medicine* 44, no. 3 (2016): 593–601, https://doi.org/10.1177/0363546515620183.

50. Bertrand Sonnery-Cottet, Mathieu Thaunat, Benjamin Freychet, Barbara H. B. Pupim, Colin G. Murphy, and Steven Claes, "Outcome of a Combined Anterior Cruciate Ligament and Anterolateral Ligament Reconstruction Technique with a Minimum 2-Year Follow-Up," *American Journal of Sports Medicine* 43, no. 7 (2015): 1598–605, https://doi.org/10.1177/0363546515571571.

51. Claes et al., "Anatomy of the Anterolateral Ligament"; Vincent et al., "The Anterolateral Ligament"; Kevin G. Shea, Matthew D. Milewski, Peter C. Cannamela, Theodore J. Ganley, Peter D. Fabricant, Elizabeth B. Terhune, Alexandra C. Styhl, Allen F. Anderson, and John D. Polousky, "Anterolateral Ligament of the Knee Shows Variable Anatomy in Pediatric Specimens," *Clinical Orthopaedics and Related Research* 475, no. 6 (2017): 1583–91, https://doi.org/10.1007 /s11999-016-5123-6.

52. Humberto G. Rosas, "Unraveling the Posterolateral Corner of the Knee," *Radiographics* 36, no. 6 (2016): 1776–91, https://doi.org/10.1148/rg.2016160027;

Jorge Chahla, Gilbert Moatshe, Chase S. Dean, and Robert F. LaPrade, "Posterolateral Corner of the Knee: Current Concepts," *Archives of Bone and Joint Surgery* 4, no. 2 (2016): 97–103.

53. Andreas Diamantopoulos, Anastasios Tokis, Matheus Tzurbakis, Iraklis Patsopoulos, and Anastasios Georgoulis, "The Posterolateral Corner of the Knee: Evaluation Under Microsurgical Dissection." *Arthroscopy* 21, no. 7 (2005): 826–33, https://doi.org/10.1016/j.arthro.2005.03.021.

54. John Nyland, Narusha Lachman, Yavuz Kocabey, Joseph Brosky, Remziye Altun, and David Caborn, "Anatomy, Function, and Rehabilitation of the Popliteus Musculotendinous Complex," *Journal of Orthopaedic and Sports Physical Therapy* 35, no. 3 (2005): 165–79, https://doi.org/10.2519/jospt.2005.35.3.165.

55. Nyland et al., "Anatomy, Function, and Rehabilitation"; Alfred J. Tria, Christopher D. Johnson, and Joseph P. Zawadsky, "The Popliteus Tendon," *Journal of Bone and Joint Surgery* 71, no. 5 (1989): 714–16, https://doi.org/10.2106/00004623-198971050-00011.

56. Daniel Geiger, Eric Y. Chang, Mini N. Pathria, and Christine B. Chung, "Posterolateral and Posteromedial Corner Injuries of the Knee," *Magnetic Resonance Imaging Clinics of North America* 22, no. 4 (2014): 581–99, https://doi.org/10.1016/j.mric.2014.08.001.

57. Geiger et al., "Posterolateral and Posteromedial."

58. Russell Flato, Giovanni J. Passanante, Matthew R. Skalski, Dakshesh B. Patel, Eric A. White, and George R. Matcuk, "The Iliotibial Tract: Imaging, Anatomy, Injuries, and Other Pathology," *Skeletal Radiology* 46, no. 5 (2017): 605–22, https://doi.org/10.1007/s00256-017-2604-y.

59. Irene Sher, Hilary Umans, Sherry A. Downie, Keith Tobin, Ritika Arora, and Todd R. Olson, "Proximal Iliotibial Band Syndrome: What Is It and Where Is It?," *Skeletal Radiology* 40, no. 12 (2011): 1553–56, https://doi.org/10.1007/s00256-011-1168-5.

60. Carolyn M. Eng, Allison S. Arnold, Andrew A. Biewener, and Daniel E. Lieberman, "The Human Iliotibial Band Is Specialized for Elastic Energy Storage Compared with the Chimp Fascia Lata," *Journal of Experimental Biology* 218, no. 15 (2015): 2382–93, https://doi.org/10.1242/jeb.117952; Carolyn M. Eng, Allison S. Arnold, Daniel E. Lieberman, and Andrew A. Biewener, "The Capacity of the Human Iliotibial Band to Store Elastic Energy During Running," *Journal of Biomechanics* 48, no. 12 (2015): 3341–48, https://doi.org/10.1016/j.jbiomech.2015.06.017.

61. J. W. Renne, "The Iliotibial Band Friction Syndrome," *Journal of Bone and Joint Surgery* 57, no. 8 (1975): 1110–11, https://doi.org/10.2106/00004623-197557080-00014.

62. Renne, "The Iliotibial Band," 1110.

63. Some believe that a fluid-filled sac known as a bursa, which normally helps tendons and ligaments glide over bones, is present at the friction site. In IT band syndrome, the bursa gets irritated and inflamed, causing pain. However, multiple imaging and cadaver studies have failed to locate such a bursa.

See E. C. Falvey, R. A. Clark, A. Franklyn-Miller, A. L. Bryant, C. Briggs, and P. R. McCrory, "Iliotibial Band Syndrome: An Examination of the Evidence Behind a Number of Treatment Options," *Scandinavian Journal of Medicine and Science in Sports* 20, no. 4 (2010): 580–87, https://doi.org/10.1111/j.1600-0838 .2009.00968.x.; Flato et al., "The Iliotibial Tract."

64. Richard Ellis, Wayne Hing, and Duncan Reid, "Iliotibial Band Friction Syndrome— A Systematic Review," *Manual Therapy* 12, no. 3 (2007): 200–208, https://doi .org/10.1016/j.math.2006.08.004.

65. Renne, "The Iliotibial Band," 1111.

66. Ellis, Hing, and Reid, "Iliotibial Band."

67. Paul E. Niemuth, Robert J. Johnson, Marcella J. Myers, and Thomas J. Thieman, "Hip Muscle Weakness and Overuse Injuries in Recreational Runners," *Clinical Journal of Sport Medicine* 15, no. 1 (2005): 14–21, https://doi.org/10.1097 /00042752-200501000-00004.

68. J. E. Taunton, M. B. Ryan, D. B. Clement, D. C. McKenzie, D. R. Lloyd-Smith, and B. D. Zumbo, "A Retrospective Case-Control Analysis of 2002 Running Injuries," *British Journal of Sports Medicine* 36, no. 2 (2002): 95–101, https:// doi.org/10.1136/bjsm.36.2.95.

69. P. Gunter and M. P. Schwellnus, "Local Corticosteroid Injection in Iliotibial Band Friction Syndrome in Runners: A Randomised Controlled Trial," *British Journal of Sports Medicine* 38, no. 3 (2004): 269–72, https://doi.org/10.1136 /bjsm.2003.000283.

70. Gunter and Schwellnus, "Local Corticosteroid"; M. P. Schwellnus, L. Theunissen, T. D. Noakes, and S. G. Reinach, "Anti-Inflammatory and Combined Anti-Inflammatory/Analgesic Medication in the Early Management of Iliotibial Band Friction Syndrome. A Clinical Trial," *South African Medical Journal* 79, no. 10 (1991): 602–6.

71. John Fairclough, Koji Hayashi, Hechmi Toumi, Kathleen Lyons, Graeme Bydder, Nicola Phillips, Thomas M. Best, and Mike Benjamin, "Is Iliotibial Band Syndrome Really a Friction Syndrome?," *Journal of Science and Medicine in Sport* 10, no. 2 (2007): 74–76, https://doi.org/10.1016/j.jsams.2006.05.017; John Fairclough, Koji Hayashi, Hechmi Toumi, Kathleen Lyons, Graeme Bydder, Nicola Phillips, Thomas M. Best, and Mike Benjamin, "The Functional Anatomy of the Iliotibial Band During Flexion and Extension of the Knee: Implications for Understanding Iliotibial Band Syndrome," *Journal of Anatomy* 208, no. 3 (2006): 309–16, https://doi.org/10.1111/j.1469-7580 .2006.00531.x.

72. Elena J. Jelsing, Jonathan T. Finnoff, Andrea L. Cheville, Bruce A. Levy, and Jay Smith, "Sonographic Evaluation of the Iliotibial Band at the Lateral Femoral Epicondyle: Does the Iliotibial Band Move?," *Journal of Ultrasound in Medicine* 32, no. 7 (2013): 1199–206, https://doi.org/10.7863/ultra.32.7.1199.

73. Jelsing et al., "Sonographic Evaluation."

74. Stephen P. Messier, David G. Edwards, David F. Martin, Robert B. Lowery, D. Wayne Cannon, Margaret K. James, Walton W. Curl, Hank M. Read Jr., and

D. Monte Hunter, "Etiology of Iliotibial Band Friction Syndrome in Distance Runners," *Medicine and Science in Sports and Exercise* 27, no. 7 (1995): 951–60, https://doi.org/10.1249/00005768-199507000-00002; Maarten P. van der Worp, Nick van der Horst, Anton de Wijer, Frank J. G. Backx, and Maria W. G. Nijhuis-van der Sanden, "Iliotibial Band Syndrome in Runners: A Systematic Review," *Sports Medicine* 42, no. 11 (2012): 969–92, https://doi.org/10.1007/BF03262306; Maryke Louw and Clare Deary, "The Biomechanical Variables Involved in the Aetiology of Iliotibial Band Syndrome in Distance Runners—A Systematic Review of the Literature," *Physical Therapy in Sport* 15, no. 1 (2014): 64–75, https://doi.org/10.1016/j.ptsp.2013.07.002.

75. Messier et al., "Etiology of Iliotibial Band."

76. John W. Orchard, Peter A. Fricker, Anna T. Abud, and Bruce R. Mason, "Biomechanics of Iliotibial Band Friction Syndrome in Runners," *American Journal of Sports Medicine* 24, no. 3 (1996): 375–79, https://doi.org/10.1177/036354659602400321.

77. Niemuth et al., "Hip Muscle Weakness."

78. Michael Fredericson, Curtis L. Cookingham, Ajit M. Chaudhari, Brian C. Dowdell, Nina Oestreicher, and Shirley A. Sahrmann, "Hip Abductor Weakness in Distance Runners with Iliotibial Band Syndrome," *Clinical Journal of Sport Medicine* 10, no. 3 (2000): 169–75, https://doi.org/10.1097/00042752-200007000-00004.

79. Fredericson et al., "Hip Abductor Weakness."

80. Amanda Beers, Michael Ryan, Zenya Kasubuchi, Scott Fraser, and Jack E. Taunton, "Effects of Multi-Modal Physiotherapy, Including Hip Abductor Strengthening, in Patients with Iliotibial Band Friction Syndrome," *Physiotherapy Canada* 60, no. 2 (2008): 180–88, https://doi.org/10.3138/physio.60.2.180.

81. Fredericson et al., "Hip Abductor Weakness."

82. S. Grau, I. Krauss, C. Maiwald, R. Best, and T. Horstmann, "Hip Abductor Weakness Is Not the Cause for Iliotibial Band Syndrome," *International Journal of Sports Medicine* 29, no. 7 (2008): 579–83, https://doi.org/10.1055/s-2007-989323; Brian Noehren, Anne Schmitz, Ross Hempel, Carolyn Westlake, and William Black, "Assessment of Strength, Flexibility, and Running Mechanics in Men with Iliotibial Band Syndrome," *Journal of Orthopaedic and Sports Physical Therapy* 44, no. 3 (2014): 217–22, https://doi.org/10.2519/jospt.2014.4991.

83. Falvey et al., "Iliotibial Band Syndrome."

84. Hsing-Kuo Wang, Tiffany Ting-Fang Shih, Kwan-Hwa Lin, and Tyng-Guey Wang, "Real-Time Morphologic Changes of the Iliotibial Band During Therapeutic Stretching; An Ultrasonographic Study," *Manual Therapy* 13, no. 4 (2007): 334–40, https://doi.org/10.1016/j.math.2007.03.002; Tyng-Guey Wang, Mei-Hwa Jan, Kwan-Hwa Lin, and Hsing-Kuo Wang, "Assessment of Stretching of the Iliotibial Tract with Ober and Modified Ober Tests: An Ultrasonographic Study," *Archives of Physical Medicine and Rehabilitation* 87, no. 10 (2006): 1407–11, https://doi.org/10.1016/j.apmr.2006.06.007.

85. Michael Fredericson, Jeremy J. White, John M. MacMahon, and Thomas P. Andriacchi, "Quantitative Analysis of the Relative Effectiveness of 3 Iliotibial Band Stretches," *Archives of Physical Medicine and Rehabilitation* 83, no. 5 (2002): 589–92, https://doi.org/10.1053/apmr.2002.31606.

86. Mark Wilhelm, Omer Matthijs, Kevin Browne, Gesine Seeber, Anja Matthijs, Phillip S. Sizer, Jean-Michel Brismée, C. Roger James, and Kerry K. Gilbert, "Deformation Response of the Iliotibial Band-Tensor Fascia Lata Complex to Clinical-Grade Longitudinal Tension Loading in-Vitro," *International Journal of Sports Physical Therapy* 12, no. 1 (2017): 16–24.

87. Jan Wilke, Anna-Lena Müller, Florian Giesche, Gerard Power, Hamid Ahmedi, and David G. Behm, "Acute Effects of Foam Rolling on Range of Motion in Healthy Adults: A Systematic Review with Multilevel Meta-Analysis," *Sports Medicine* 50, no. 2 (2020): 387–402, https://doi.org/10.1007/s40279-019 -01205-7; David J. Bradbury-Squires, Jennifer C. Noftall, Kathleen M. Sullivan, David G. Behm, Kevin E. Power, and Duane C. Button, "Roller-Massager Application to the Quadriceps and Knee-Joint Range of Motion and Neuromuscular Efficiency During a Lunge," *Journal of Athletic Training* 50, no. 2 (2015): 133–40, https://doi.org/10.4085/1062-6050-49.5.03; Graham Z. MacDonald, Michael D. H. Penney, Michelle E. Mullaley, Amanda L. Cuconato, Corey D. J. Drake, David G. Behm, and Duane C. Button, "An Acute Bout of Self-Myofascial Release Increases Range of Motion Without a Subsequent Decrease in Muscle Activation or Force," *Journal of Strength and Conditioning Research* 27, no. 3 (2013): 812–21, https://doi.org/10.1519/JSC.0b013e31825c2bc1; Jake Phillips, David Diggin, Deborah L. King, and Gary A. Sforzo, "Effect of Varying Self-Myofascial Release Duration on Subsequent Athletic Performance," *Journal of Strength and Conditioning Research* 35, no. 3 (2018): 746–53, https://doi.org /10.1519/JSC.0000000000002751.

88. Talin M. Pepper, Jean-Michel Brismée, Phillip S. Sizer Jr., Jeegisha Kapila, Gesine H. Seeber, Christopher A. Huggins, and Troy L. Hooper, "The Immediate Effects of Foam Rolling and Stretching on Iliotibial Band Stiffness: A Randomized Controlled Trial," *International Journal of Sports Physical Therapy* 16, no. 3 (2021): 651–61, https://doi.org/10.26603/001c.23606.

89. David G. Behm and Jan Wilke, "Do Self-Myofascial Release Devices Release Myofascia? Rolling Mechanisms: A Narrative Review," *Sports Medicine (Auckland)* 49, no. 8 (2019): 1173–81, https://doi.org/10.1007/s40279-019 -01149-y.

90. Melissa Nicol Conte and Peter Kessler, "Method and device for therapeutic treatment of iliotibial band syndrome, myofascial and musculoskeletal dysfunctions," US Patent 20,150,313,788, filed November 5, 2015, 1, http://appft .uspto.gov/netacgi/nph-Parser?Sect1=PTO1&Sect2=HITOFF&p=1&u=/netahtml /PTO/srchnum.html&r=1&f=G&l=50&d=PG01&s1=20150313788.

91. Conte and Kessler, Method and device, 1.

92. Paul Walbron, Adrien Jacquot, Jean-Marc Geoffroy, François Sirveaux, and Daniel Molé, "Iliotibial Band Friction Syndrome: An Original Technique of

Digastric Release of the Iliotibial Band from Gerdy's Tubercle," *Orthopaedics and Traumatology, Surgery and Research* 104, no. 8 (2018): 1209–13, https://doi.org/10.1016/j.otsr.2018.08.013.

93. Walbron et al., "Iliotibial Band Friction."
94. Walbron et al., "Iliotibial Band Friction."
95. Jonathan A. Godin, Jorge Chahla, Gilbert Moatshe, Bradley M. Kruckeberg, Kyle J. Muckenhirn, Alexander R. Vap, Andrew G. Geeslin, and Robert F. LaPrade, "A Comprehensive Reanalysis of the Distal Iliotibial Band: Quantitative Anatomy, Radiographic Markers, and Biomechanical Properties," *American Journal of Sports Medicine* 45, no. 11 (2017): 2595–603, https://doi.org/10.1177/0363546517707961.

3. BARE KNEES, DICEY POWER

1. Associated Press, "Flappers-Puritans Declare Open War in Pennsylvania," *Daily Tribune* (Wisconsin Rapids, WI), August 24, 1923, 1.
2. Associated Press, "Flappers-Puritans."
3. Associated Press, "Flappers in Verse Defend Near Nudity," *Morning News* (Wilmington, DE), August 25, 1923, 1.
4. Blanche Payne, *History of Costume: From the Ancient Egyptians to the Twentieth Century* (New York: Harper & Row, 1965).
5. Mary Harlow, *A Cultural History of Dress and Fashion*, vol. 1 (New York: Bloomsbury, 2017).
6. Annette Becker, "The Body," in *A Cultural History of Dress and Fashion in the Age of Empire*, ed. Denise Amy Baxter (New York: Bloomsbury, 2017), 67.
7. Payne, *History of Costume*; Harlow, *A Cultural History*
8. "Not for Old Fogies" is the tagline of the *Flapper* magazine.
9. "Big Business Banishes the Flapper," *Morning Tulsa Daily World* (Tulsa, OK), July 16, 1922, 26.
10. "Wants School Girls to Hide Their Knees," *New York Times*, January 27, 1922, 10.
11. "Chicago Pastor Does Not Like Flappers," *Springfield News-Leader* (Springfield, MO), May 12, 1922, 5.
12. Theresa L. Lennon, Sharron J. Lennon, and Kim K. P. Johnson, "Is Clothing Probative of Attitude or Intent? Implications for Rape and Sexual Harassment Cases," *Law and Inequality: A Journal of Theory and Practice* 11, no. 2 (1993): 391.
13. Bhuvanesh Awasthi, "From Attire to Assault: Clothing, Objectification, and De-Humanization—a Possible Prelude to Sexual Violence?," *Frontiers in Psychology* 8 (2017): 338, https://doi.org/10.3389/fpsyg.2017.00338.
14. Lennon, Lennon, and Johnson, "Is Clothing Probative of Attitude or Intent?"
15. "Against Immodest Fashions," *Pittsburgh Press* (Pittsburgh, PA), March 20, 1921, 93.

16. "Against Immodest Fashions."
17. "Wants School Girls to Hide Their Knees."
18. "Against Immodest Fashions."
19. "Against Immodest Fashions."
20. "Big Business Banishes the Flapper."
21. "Big Business Banishes the Flapper."
22. "Big Business Banishes the Flapper."
23. Kenneth A. Yellis, "Prosperity's Child: Some Thoughts on the Flapper," *American Quarterly* 21, no. 1 (1969): 44–64, https://doi.org/10.2307/2710772.
24. Karen J. Kriebl, "From Bloomers to Flappers: The American Women's Dress Reform Movement, 1840–1920" (PhD diss., Ohio State University, 1998), ProQuest.
25. Kriebl, "From Bloomers."
26. Kriebl, "From Bloomers."
27. Payne, *History of Costume*, 524.
28. Becker, "The Body."
29. Joshua Zeitz, *Flapper: A Madcap Story of Sex, Style, Celebrity, and the Women Who Made America Modern* (New York: Broadway Books, 2006).
30. Kriebl, "From Bloomers."
31. Becker, "The Body," 67.
32. Becker, "The Body."
33. Becker, "The Body."
34. Kriebl, "From Bloomers."
35. Gilbert, Theodosia, "An Eye Sore," *Water-Cure Journal* 11, no. 5 (1851): 117.
36. Gilbert, "An Eye Sore," 117.
37. M. S. Gove Nichols, "Woman the Physician," *Water-Cure Journal* 12, no. 4 (1851): 75.
38. Elizabeth Cady Stanton, Susan B. Anthony, Emmeline Pankhurst, Anna Howard Shaw, Millicent Garrett Fawcett, Jane Addams, Lucy Stone, Carrie Chapman Catt, and Alice Paul, *The Women of the Suffrage Movement: Autobiographies and Biographies of the Most Influential Suffragettes* (n.p.: Musaicum Books, 2018).
39. Stanton et al., *The Women*.
40. Kriebl, "From Bloomers."
41. Ella H. Cooper, "Bicycle skirt," US Patent 555,211, filed July 18, 1895, and issued February 25, 1896.
42. "Much Smartness in School Rainment," *Journal and Tribune* (Knoxville, TN), January 17, 1915, 36.
43. "What Is a Flapper? The Critics Disagree," *Kansas City Times* (Kansas City, MO), March 18, 1922, 9.
44. "The Psychology of Knees," *The Flapper*, June 1922.
45. Zoë Beery, "Flappers Didn't Really Wear Fringed Dresses," Racked, May 19, 2017, https://www.racked.com/2017/5/19/15612000/flappers-fringe-myth.
46. Bruce Bliven, "Flapper Jane," *New Republic*, September 9, 1925, 65.
47. "Rubber Roll Garters," *El Paso Herald* (El Paso, TX), August 3, 1928, 12.

48. Bill Murray, "Roll 'em Girls," *Dance Music of 1925*, Jack Shilkret Orchestra (1925).
49. "Women Now Rouging Their Knees, Says a N.Y. Beauty Parlor Manager," *Birmingham News* (Birmingham, AL), May 26, 1921, 9.
50. "Rouged Knee Mode Interests Paris," *Boston Post* (Boston, MA), July 14, 1921, 24.
51. "Rouged Knee Mode Interests Paris."
52. Einav Rabinovitch-Fox, "Fabricating Black Modernity: Fashion and African American Womanhood During the First Great Migration," *International Journal of Fashion Studies* 6, no. 2 (2019): 239–60.
53. Rabinovitch-Fox, "Fabricating Black Modernity."
54. Einav Rabinovitch-Fox, "This Is What a Feminist Looks Like: The New Woman Image, American Feminism, and the Politics of Women's Fashion 1890–1930" (PhD diss., New York University, 2014), ProQuest.
55. "Design Painted on Knee of Stocking Newest in London," *St. Louis Star and Times* (St. Louis, MO), April 20, 1928, 39.
56. "Hand-Painted Knees Latest Beauty Stunt," *Chattanooga News* (Chattanooga, OK), August 27, 1925, 5.
57. "Hand-Painted Knees Latest Beauty Stunt," 5.
58. "Girls' Painted Knees Bothering Professor," *San Francisco Examiner* (San Francisco, CA), May 14, 1925, 4.
59. "Painted Knee Fad Hits," *Nebraska State Journal* (Lincoln, NE), July 23, 1925, 3.
60. Serviss, Myrna, "Form a Flapper Flock of Your Own," *The Flapper*, June 1922, 20.
61. Serviss, Myrna, "News of the Flapper Flocks," *The Flapper*, November 1922, 32–33.
62. "Announcing Flapper Beauty Contest," *The Flapper*, June 1922, 2–3.
63. Zeitz, *Flapper: A Madcap Story*.
64. "Underpinning the 1920s: Brassieres, Bandeaux, and Bust Flatteners," witness-2fashion, April 27, 2014, https://witness2fashion.wordpress.com/tag/breast-binding-1920s/.
65. Zeitz, *Flapper: A Madcap Story*.
66. Ariel Levy, *Female Chauvinist Pigs: Women and the Rise of Raunch Culture* (New York: Free Press, 2006).
67. Lynne Richards, "The Rise and Fall of It All: The Hemlines and Hiplines of the 1920s," *Clothing and Textiles Research Journal* 2, no. 1 (1983): 42–48.
68. Hall, Linda, "Fashion and Style in the Twenties: The Change," *Historian* 34, no. 3 (1972): 485–97, https://doi.org/10.1111/j.1540-6563.1972.tb00424.x.
69. Hall, "Fashion and Style," 488.
70. Hall, "Fashion and Style."
71. "Are Bloomers Ugly?," *Hanford Semi-Weekly Journal* (Hanford, CA), November 5, 1895, 2; "Discuss the Bloomer," *Chicago Tribune* (Chicago, IL), September 8, 1895, 39.
72. "For the Knickerbocker," *San Francisco Examiner* (San Francisco, CA), June 23, 1895, 13.
73. "Discuss the Bloomer," 39.

74. "Does It Pay to Visit Yo Semite?," *Leavenworth Times* (Leavenworth, KS), October 4, 1870, 2.
75. "Much Smartness in School Rainment," 36.
76. "Much Smartness in School Rainment," 36.
77. "Flapper Suits by Lady Duff-Gordon," *Ogden Standard-Examiner* (Ogden, Utah), March 26, 1922, 31.
78. Clare Rose, *Making, Selling and Wearing Boys' Clothes in Late-Victorian England* (Surrey, England: Ashgate, 2010), 158.
79. "What Shall the New Woman Wear, Skirts or Bloomers," *Los Angeles Herald* (Los Angeles, CA), September 15, 1895, 16.
80. Kriebl, "From Bloomers."
81. Bonnie Tsui, *She Went to the Field: Women Soldiers of the Civil War* (Guilford, CT: TwoDot, 2006).
82. "Arrest for Wearing 'Male Attire,'" *Times-Picayune* (New Orleans, LA), June 23, 1866, 1.
83. Zeitz, *Flapper: A Madcap Story*, 30. Cities are defined as those with a population of at least eight thousand.
84. Mary V. Dempsey, *The Occupational Progress of Women, 1910 to 1930* (Washington, DC: U.S. Government Printing Office, 1933), 7.
85. Zeitz, *Flapper: A Madcap Story*, 66.
86. Rabinovitch-Fox, "This Is What a Feminist Looks Like."
87. Rabinovitch-Fox, "This Is What a Feminist Looks Like."
88. Julie Marks, "What Caused the Stock Market Crash of 1929?," History, updated April 27, 2021, https://www.history.com/news/what-caused-the-stock-market-crash-of-1929.
89. Marks, "What Caused."
90. Zeitz, *Flapper: A Madcap Story*.
91. "Stock Market Crash of 1929," History, updated April 27, 2021, https://www.history.com/topics/great-depression/1929-stock-market-crash.
92. "Stock Market Crash of 1929."
93. Nigel Barber, "Women's Dress Fashions as a Function of Reproductive Strategy," *Sex Roles* 40, no. 5 (1999): 459–71, https://doi.org/10.1023/A:1018823727012.
94. Barber, "Women's Dress."
95. Mary Ann Mabry, "The Relationship Between Fluctuations in Hemlines and Stock Market Averages from 1921 to 1971" (master's thesis, University of Tennessee, 1971), https://trace.tennessee.edu/utk_gradthes/1121; Marjolein van Baardwijk and Philip Hans Franses, "The Hemline and the Economy: Is There Any Match?" (report no. EI 2010-40, Econometric Institute, Erasmus University Rotterdam, 2010), https://repub.eur.nl/pub/20147; Soohyun Kim and Insook Ahn, "Impact of Macro-Economic Factors on the Hemline Cycles," in *ITAA Proceedings*, 72 (Santa Fe, NM: International Textile and Apparel Association, 2015), 1–2, https://iastatedigitalpress.com/itaa/article/2476/galley/2349/view/.
96. Isadore Barmash, "Furor Over the Mini-Skirt," *Des Moines Register* (Des Moines, IA), December 7, 1966, 9.

97. David A. Jewell, "Exit the Mini-Skirts Enter the Micros," *News-Journal* (Mansfield, OH), May 14, 1967, 24.

98. "Weep Not, Girls, Breeze'll Make Them Red, Soon," *Capital Times* (Madison, WI), August 1, 1925, 13.

99. Rubye Graham, "Beauty Industry Goes out on a Limb," *Philadelphia Inquirer* (Philadelphia, PA), July 3, 1966, 37.

100. "Mini-Skirt Comes to Mt. Carmel In 'Modified' Form, Study Reveals," *Daily Republican-Register* (Mount Carmel, IL), September 30, 1966, 1.

101. "Mini-Skirt Comes to Mt. Carmel"; David Smothers, "Leg-islators Find Short Skirts No Mini-Controversy," *Wisconsin State Journal* (Madison, WI), April 13, 1969, 73.

102. Bob Krauss, "Mini? Not Many Downtown," *Honolulu Advertiser* (Honolulu, HI), December 1, 1967, 35.

103. Krauss, "Mini?," 35.

104. Smothers, "Leg-islators," 73.

105. Smothers, "Leg-islators," 73.

106. Brigitte Studer, "'1968' and the Formation of the Feminist Subject," *Twentieth Century Communism*, no. 3 (2011): 38–69.

107. Studer, "'1968,'" 48–49.

4. THE WEAKER SEX?

The quote in the chapter epigraph comes from Jules Boykoff, "Tokyo Olympics Head Yoshiro Mori Called Out by Naomi Osaka and Others for Sexism," NBC News, February 10, 2021, https://www.nbcnews.com/think/opinion/tokyo-olympics-head-yoshiro-mori-called-out-naomi-osaka-others-ncna1257163.

1. Mary Anne Case, "Heterosexuality as a Factor in the Long History of Women's Sports," *Law and Contemporary Problems* 80, no. 4 (2018): 25–46.

2. Case, "Heterosexuality as a Factor," 34.

3. Boykoff, "Tokyo Olympics."

4. Roberta Park, "Sport, Gender and Society in a Transatlantic Victorian Perspective," in *From "Fair Sex" to Feminism: Sport and the Socialization of Women in the Industrial and Post-Industrial Eras*, ed. J. A. Mangan and Roberta Park (Totowa, NJ: Frank Cass, 1987), 61.

5. Park, "Sport, Gender and Society," 85.

6. John Temesi, Pierrick J. Arnal, Thomas Rupp, Léonard Féasson, Régine Cartier, Laurent Gergelé, Samuel Verges, Vincent Martin, and Guillaume Y. Millet, "Are Females More Resistant to Extreme Neuromuscular Fatigue?," *Medicine and Science in Sports and Exercise* 47, no. 7 (2015): 1372–82, https://doi.org/10.1249/MSS.0000000000000540; Sophie Williams, "Are Women Better Ultra-Endurance Athletes Than Men?," BBC News, August 11, 2019, https://www.bbc.com/news/world-49284389.

7. Beth A. Brooke-Marciniak and Donna de Varona, "Amazing Things Happen When You Give Female Athletes the Same Funding as Men," World Economic

Forum, August 25, 2016, https://www.weforum.org/agenda/2016/08/sustaining -the-olympic-legacy-women-sports-and-public-policy/; Timothy E. Hewett and Gregory D. Myer, "Reducing Knee and Anterior Cruciate Ligament Injuries Among Female Athletes: A Systematic Review of Neuromuscular Training Interventions," *Journal of Knee Surgery* 18, no. 1 (2005): 82–88, https://doi.org /10.1055/s-0030-1248163.

8. John W. Powell and Kim D. Barber-Foss, "Sex-Related Injury Patterns Among Selected High School Sports," *American Journal of Sports Medicine* 28, no. 3 (2000): 385–91, https://doi.org/10.1177/03635465000280031801.

9. Powell and Barber-Foss, "Sex-Related Injury."

10. Elizabeth Arendt and Randall Dick, "Knee Injury Patterns Among Men and Women in Collegiate Basketball and Soccer: NCAA Data and Review of Literature," *American Journal of Sports Medicine* 23, no. 6 (1995): 694–701, https://doi.org/10.1177/036354659502300611.

11. Arendt and Dick, "Knee Injury"; Scott L. Zuckerman, Adam M. Wegner, Karen G. Roos, Aristarque Djoko, Thomas P. Dompier, and Zachary Y. Kerr, "Injuries Sustained in National Collegiate Athletic Association Men's and Women's Basketball, 2009/2010–2014/2015," *British Journal of Sports Medicine* 52, no. 4 (2018): 261–68, https://doi.org/10.1136/bjsports-2016-096005.

12. Andrea Ferretti, Paola Papandrea, Fabio Conteduca, and Pier Paolo Mariani, "Knee Ligament Injuries in Volleyball Players," *American Journal of Sports Medicine* 20, no. 2 (1992): 203–7, https://doi.org/10.1177/036354659202000219.

13. More precisely, the Q angle is the angle between two lines: the line from the anterior superior iliac spine to the center of the patella and the line from the center of the patella to the center of the tibial tubercle.

14. All human knees are angulated (see chapter 1).

15. Melissa G. Horton and Terry L. Hall, "Quadriceps Femoris Muscle Angle: Normal Values and Relationships with Gender and Selected Skeletal Measures," *Physical Therapy* 69, no. 11 (1989): 897–901, https://doi.org/10.1093/ptj /69.11.897; Ramada R. Khasawneh, Mohammed Z. Allouh, and Ejlal Abu-El-Rub, "Measurement of the Quadriceps (Q) Angle with Respect to Various Body Parameters in Young Arab Population," *PLoS One* 14, no. 6 (2019): e0218387 -e87, https://doi.org/10.1371/journal.pone.0218387.

16. Cara M. Wall-Scheffler and Marcella J. Myers, "The Biomechanical and Energetic Advantages of a Mediolaterally Wide Pelvis in Women," *Anatomical Record* 300, no. 4 (2017): 764–75, https://doi.org/10.1002/ar.23553.

17. Wall-Scheffler and Myers, "The Biomechanical."

18. Lia Betti and Andrea Manica, "Human Variation in the Shape of the Birth Canal Is Significant and Geographically Structured," *Proceedings of the Royal Society B: Biological Sciences* 285, no. 1889 (2018): 20181807, https://doi.org /10.1098/rspb.2018.1807.

19. Horton and Hall, "Quadriceps."

20. Veeramani Raveendranath, Shankar Nachiket, Narayanan Sujatha, Ranganath Priya, and Devi Rema, "The Quadriceps Angle (Q Angle) in Indian Men and Women," *European Journal of Anatomy* 13, no. 3 (2009): 105–9.

21. R. P. Grelsamer, A. Dubey, and C. H. Weinstein, "Men and Women Have Similar Q Angles: A Clinical and Trigonometric Evaluation," *Journal of Bone and Joint Surgery* 87, no. 11 (2005): 1498–501, https://doi.org/10.1302/0301-620X .87B11.16485.

22. J. E. Taunton, M. B. Ryan, D. B. Clement, D. C. McKenzie, D. R. Lloyd-Smith, and B. D. Zumbo, "A Retrospective Case-Control Analysis of 2002 Running Injuries," *British Journal of Sports Medicine* 36, no. 2 (2002): 95–101, https:// doi.org/10.1136/bjsm.36.2.95.

23. Erik Witvrouw, Roeland Lysens, Johan Bellemans, Dirk Cambier, and Guy Vanderstraeten, "Intrinsic Risk Factors for the Development of Anterior Knee Pain in an Athletic Population: A Two-Year Prospective Study," *American Journal of Sports Medicine* 28, no. 4 (2000): 480–89, https://doi.org/10.1177 /03635465000280040701.

24. Mohammad-Jafar Emami, Mohammad-Hossein Ghahramani, Farzad Abdine-jad, and Hamid Namazi, "Q-Angle: An Invaluable Parameter for Evaluation of Anterior Knee Pain," *Archives of Iranian Medicine* 10, no. 1 (2007): 24–26; Amir Haim, Moshe Yaniv, Samuel Dekel, and Hagay Amir, "Patellofemoral Pain Syndrome: Validity of Clinical and Radiological Features," *Clinical Orthopaedics and Related Research* 451 (2006): 223–28, https://doi.org/10.1097/01 .blo.0000229284.45485.6c.

25. Witvrouw et al., "Intrinsic Risk Factors"; V. Lun, W. H. Meeuwisse, P. Stergiou, and D. Stefanyshyn, "Relation Between Running Injury and Static Lower Limb Alignment in Recreational Runners," *British Journal of Sports Medicine* 38, no. 5 (2004): 576–80, https://doi.org/10.1136/bjsm.2003.005488.

26. Mitchell J. Rauh, Thomas D. Koepsell, Frederick P. Rivara, Stephen G. Rice, and Anthony J. Margherita, "Quadriceps Angle and Risk of Injury Among High School Cross-Country Runners." *Journal of Orthopaedic and Sports Physical Therapy* 37, no. 12 (2007): 725–33, https://doi.org/10.2519/jospt.2007.2453; Philip J. Shambaugh, Andrew Klein, and John H. Herbert, "Structural Measures as Predictors of Injury in Basketball Players," *Medicine and Science in Sports and Exercise* 23, no. 5 (1991): 522–27, https://doi.org/10.1249/00005768 -199105000-00003.

27. Janice K. Loudon, Walter Jenkins, and Karen L. Loudon, "The Relationship Between Static Posture and ACL Injury in Female Athletes," *Journal of Orthopaedic and Sports Physical Therapy* 24, no. 2 (1996): 91–97, https://doi.org /10.2519/jospt.1996.24.2.91; E. E. Mohamed, U. Useh, and B. F. Mtshali, "Q-Angle, Pelvic Width, and Intercondylar Notch Width as Predictors of Knee Injuries in Women Soccer Players in South Africa," *African Health Sciences* 12, no. 2 (2012): 174–80, https://doi.org/10.4314/ahs.v12i2.15.

28. Stephen Lombardo, Paul M. Sethi, and Chad Starkey, "Intercondylar Notch Stenosis Is Not a Risk Factor for Anterior Cruciate Ligament Tears in Professional Male Basketball Players: An 11-Year Prospective Study," *American Journal of Sports Medicine* 33, no. 1 (2005): 29–34, https://doi.org/10.1177 /0363546504266482; Michael Dienst, Guenther Schneider, Katrin Altmeyer, Kristina Voelkering, Thomas Georg, Bernhard Kramann, and Dieter Kohn,

"Correlation of Intercondylar Notch Cross Sections to the ACL Size: A High Resolution MR Tomographic in Vivo Analysis," *Archives of Orthopaedic and Trauma Surgery* 127, no. 4 (2007): 253–60, https://doi.org/10.1007/s00402-006-0177-7.

29. Chao Zeng, Shu-guang Gao, Jie Wei, Tu-bao Yang, Ling Cheng, Wei Luo, Min Tu, et al., "The Influence of the Intercondylar Notch Dimensions on Injury of the Anterior Cruciate Ligament: A Meta-Analysis," *Knee Surgery, Sports Traumatology, Arthroscopy* 21, no. 4 (2013): 804–15, https://doi.org/10.1007/s00167-012-2166-4; Zheng Li, Changshu Li, Li Li, and Ping Wang, "Correlation Between Notch Width Index Assessed Via Magnetic Resonance Imaging and Risk of Anterior Cruciate Ligament Injury: An Updated Meta-Analysis," *Surgical and Radiologic Anatomy* 42, no. 10 (2020): 1209–17, https://doi.org/10.1007/s00276-020-02496-6. Contradictory findings do exist. See, for example, Lombardo, Sethi, and Starkey, "Intercondylar Notch Stenosis"; Floor M. van Diek, Megan R. Wolf, Christopher D. Murawski, Carola F. van Eck, and Freddie H. Fu, "Knee Morphology and Risk Factors for Developing an Anterior Cruciate Ligament Rupture: An MRI Comparison Between ACL-Ruptured and Non-Injured Knees," *Knee Surgery, Sports Traumatology, Arthroscopy* 22, no. 5 (2014): 987–94, https://doi.org/10.1007/s00167-013-2588-7.

30. Hakon Lund-Hanssen, James Gannon, Lars Engebretsen, Ketil J. Holen, Svein Anda, and Lars Vatten, "Intercondylar Notch Width and the Risk for Anterior Cruciate Ligament Rupture: A Case-Control Study in 46 Female Handball Players," *Acta Orthopaedica* 65, no. 5 (1994): 529–32, https://doi.org/10.3109/17453679409000907.

31. Tomás Fernández-Jaén, Juan Manuel López-Alcorocho, Elena Rodriguez-Iñigo, Fabián Castellán, Juan Carlos Hernández, and Pedro Guillén-García, "The Importance of the Intercondylar Notch in Anterior Cruciate Ligament Tears," *Orthopaedic Journal of Sports Medicine* 3, no. 8 (2015): 2325967115597882–82, https://doi.org/10.1177/2325967115597882; Lars Good, Magnus Odensten, and Jan Gillquist, "Intercondylar Notch Measurements with Special Reference to Anterior Cruciate Ligament Surgery," *Clinical Orthopaedics and Related Research*, no. 263 (1991): 185–89, https://doi.org/10.1097/00003086-199102000-00022.

32. Lazar Stijak, Vidosava Radonjić, Valentina Nikolić, Zoran Blagojević, Milan Aksić, and Branislav Filipović, "Correlation Between the Morphometric Parameters of the Anterior Cruciate Ligament and the Intercondylar Width: Gender and Age Differences," *Knee Surgery, Sports Traumatology, Arthroscopy* 17, no. 7 (2009): 812–17, https://doi.org/10.1007/s00167-009-0807-z.

33. Allen F. Anderson, David C. Dome, Shiva Gautam, Mark H. Awh, and Gregory W. Rennirt, "Correlation of Anthropometric Measurements, Strength, Anterior Cruciate Ligament Size, and Intercondylar Notch Characteristics to Sex Differences in Anterior Cruciate Ligament Tear Rates," *American Journal of Sports Medicine* 29, no. 1 (2001): 58–66, https://doi.org/10.1177/03635465010290011501.

34. Robert F. LaPrade, Glenn C. Terry, Ronald D. Montgomery, David Curd, and David J. Simmons, "The Effects of Aggressive Notchplasty on the Normal Knee in Dogs," *American Journal of Sports Medicine* 26, no. 2 (1998): 193–200, https://doi.org/10.1177/03635465980260020801.

35. LaPrade et al., "The Effects."

36. LaPrade et al., "The Effects"; Shintaro Asahina, Takeshi Muneta, and Yoichi Ezura, "Notchplasty in Anterior Cruciate Ligament Reconstruction: An Experimental Animal Study," *Arthroscopy* 16, no. 2 (2000): 165–72, https://doi.org/10.1016/s0749-8063(00)90031-8.

37. Francesco Ranuccio, Filippo Familiari, Giuseppe Tedesco, Francesco La Camera, and Giorgio Gasparini, "Effects of Notchplasty on Anterior Cruciate Ligament Reconstruction: A Systematic Review," *Joints* 5, no. 3 (2017): 173–79, https://doi.org/10.1055/s-0037-1605551.

38. Ross Wilson and Alan A. Barhorst, "Intercondylar Notch Impingement of the Anterior Cruciate Ligament: A Cadaveric in Vitro Study Using Robots," *Journal of Healthcare Engineering* 2018 (2018): 8698167–27, https://doi.org/10.1155/2018/8698167.

39. Dienst et al., "Correlation of Intercondylar Notch."

40. Stijak et al., "Correlation Between the Morphometric Parameters."

41. Takeshi Muneta, Kazuo Takakuda, and Haruyasu Yamamoto, "Intercondylar Notch Width and Its Relation to the Configuration and Cross-Sectional Area of the Anterior Cruciate Ligament: A Cadaveric Knee Study," *American Journal of Sports Medicine* 25, no. 1 (1997): 69–72, https://doi.org/10.1177/036354659702500113; Anderson et al., "Correlation of Anthropometric Measurements."

42. Donald K. Shelbourne, Thorp J. Davis, and Thomas E. Klootwyk, "The Relationship Between Intercondylar Notch Width of the Femur and the Incidence of Anterior Cruciate Ligament Tears: A Prospective Study," *American Journal of Sports Medicine* 26, no. 3 (1998): 402–8.

43. M. Steiner, "Editorial Commentary: Size Does Matter—Anterior Cruciate Ligament Graft Diameter Affects Biomechanical and Clinical Outcomes," *Arthroscopy* 33, no. 5 (May 2017): 1014–15, https://doi.org/10.1016/j.arthro.2017.01.020.

44. Jennifer M. Medina McKeon and Jay Hertel, "Sex Differences and Representative Values for 6 Lower Extremity Alignment Measures," *Journal of Athletic Training* 44, no. 3 (2009): 249–55, https://doi.org/10.4085/1062-6050-44.3.249; Anh-Dung Nguyen and Sandra J. Shultz, "Sex Differences in Clinical Measures of Lower Extremity Alignment," *Journal of Orthopaedic and Sports Physical Therapy* 37, no. 7 (2007): 389–98, https://doi.org/10.2519/jospt.2007.2487; Samuel C. Wordeman, Carmen E. Quatman, Christopher C. Kaeding, and Timothy E. Hewett, "In Vivo Evidence for Tibial Plateau Slope as a Risk Factor for Anterior Cruciate Ligament Injury: A Systematic Review and Meta-Analysis," *American Journal of Sports Medicine* 40, no. 7 (2012): 1673–81, https://doi.org/10.1177/0363546512442307.

45. Catherine Y. Wild, Julie R. Steele, and Bridget J. Munro, "Why Do Girls Sustain More Anterior Cruciate Ligament Injuries Than Boys? A Review of the

Changes in Estrogen and Musculoskeletal Structure and Function During Puberty," *Sports Medicine* 42, no. 9 (2012): 733–49, https://doi.org/10.1007/BF03262292.

46. Stephen H. Liu, Raad Al-Shaikh, Vahé Panossian, Rong-Sen Yang, Scott D. Nelson, Neptune Soleiman, Gerald A. M. Finerman, and Joseph M. Lane, "Primary Immunolocalization of Estrogen and Progesterone Target Cells in the Human Anterior Cruciate Ligament," *Journal of Orthopaedic Research* 14, no. 4 (1996): 526–33, https://doi.org/10.1002/jor.1100140405.

47. Wild, Steele, and Munro, "Why Do Girls."

48. Wild, Steele, and Munro, "Why Do Girls."

49. Warren D. Yu, Stephen H. Liu, Joshua D. Hatch, Vahé Panossian, and Gerald A. M. Finerman, "Effect of Estrogen on Cellular Metabolism of the Human Anterior Cruciate Ligament," *Clinical Orthopaedics and Related Research* 366 (1999): 229–38, https://doi.org/10.1097/00003086-199909000-00030.

50. Warren D. Yu, Vahé Panossian, Joshua D. Hatch, Stephen H. Liu, and Gerald A. M. Finerman, "Combined Effects of Estrogen and Progesterone on the Anterior Cruciate Ligament," *Clinical Orthopaedics and Related Research* 383 (2001): 268–81, https://doi.org/10.1097/00003086-200102000-00031.

51. Aruna Seneviratne, Erik Attia, Riley J. Williams, Scott A. Rodeo, and Jo A. Hannafin, "The Effect of Estrogen on Ovine Anterior Cruciate Ligament Fibroblasts: Cell Proliferation and Collagen Synthesis," *American Journal of Sports Medicine* 32, no. 7 (2004): 1613–18, https://doi.org/10.1177/0363546503262179; Sabrina M. Strickland, Thomas W. Belknap, Simon A. Turner, Timothy M. Wright, and Jo A. Hannafin, "Lack of Hormonal Influences on Mechanical Properties of Sheep Knee Ligaments," *American Journal of Sports Medicine* 31, no. 2 (2003): 210–15, https://doi.org/10.1177/03635465030310020901.

52. Stuart J. Warden, Leanne K. Saxon, Alesha B. Castillo, and Charles H. Turner, "Knee Ligament Mechanical Properties Are Not Influenced by Estrogen or Its Receptors," *American Journal of Physiology Endocrinology and Metabolism* 290, no. 5 (2006): 1034–40, https://doi.org/10.1152/ajpendo.00367.2005.

53. Timothy E. Hewett, Bohdanna T. Zazulak, and Gregory D. Myer, "Effects of the Menstrual Cycle on Anterior Cruciate Ligament Injury Risk: A Systematic Review," *American Journal of Sports Medicine* 35, no. 4 (2007): 659–68, https://doi.org/10.1177/0363546506295699; N. Lefevre, Y. Bohu, S. Klouche, J. Lecocq, and S. Herman, "Anterior Cruciate Ligament Tear During the Menstrual Cycle in Female Recreational Skiers," *Orthopaedics and Traumatology: Surgery and Research* 99, no. 5 (September 2013): 571–75, https://doi.org/10.1016/j.otsr.2013.02.005; Gerhard Ruedl, Patrick Ploner, Ingrid Linortner, Alois Schranz, Christian Fink, Renate Sommersacher, Elena Pocecco, Werner Nachbauer, and Martin Burtscher, "Are Oral Contraceptive Use and Menstrual Cycle Phase Related to Anterior Cruciate Ligament Injury Risk in Female Recreational Skiers?," *Knee Surgery, Sports Traumatology, Arthroscopy* 17, no. 9 (2009): 1065–69, https://doi.org/10.1007/s00167-009-0786-0; Simone D. Herzberg, Makalapua L. Motu'apuaka, William Lambert, Rongwei

Fu, Jacqueline Brady, and Jeanne-Marie Guise, "The Effect of Menstrual Cycle and Contraceptives on ACL Injuries and Laxity: A Systematic Review and Meta-Analysis," *Orthopaedic Journal of Sports Medicine* 5, no. 7 (2017), https://doi.org/10.1177/2325967117718781; Noriko Adachi, Koji Nawata, Michio Maeta, and Youichi Kurozawa, "Relationship of the Menstrual Cycle Phase to Anterior Cruciate Ligament Injuries in Teenaged Female Athletes," *Archives of Orthopaedic and Trauma Surgery* 128, no. 5 (2008): 473–78, https://doi.org/10.1007/s00402-007-0461-1.

54. Edward M. Wojtys, Laura J. Huston, Melbourne D. Boynton, Kurt P. Spindler, and Thomas N. Lindenfeld, "The Effect of the Menstrual Cycle on Anterior Cruciate Ligament Injuries in Women as Determined by Hormone Levels," *American Journal of Sports Medicine* 30, no. 2 (2002): 182–88, https://doi.org/10.1177/03635465020300020601.

55. Ovulation days are also linked with increased knee laxity. See Bohdanna T. Zazulak, Mark Paterno, Gregory D. Myer, William A. Romani, and Timothy E. Hewett, "The Effects of the Menstrual Cycle on Anterior Knee Laxity: A Systematic Review," *Sports Medicine* 36, no. 10 (2006): 847–62, https://doi.org/10.2165/00007256-200636100-00004; Hewett, Zazulak, and Myer, "Effects of the Menstrual Cycle"; Herzberg et al., "The Effect of Menstrual Cycle." Conflicting findings exist, see Zazulak et al., "The Effects of the Menstrual Cycle."

56. Wild, Steele, and Munro, "Why Do Girls."

57. H. L. Fevold, Frederick L. Hisaw, and R. K. Meyer, "The Relaxative Hormone of the Corpus Luteum. Its Purification and Concentration," *Journal of the American Chemical Society* 52, no. 8 (1930): 3340–48, https://doi.org/10.1021/ja01371a051.

58. Fevold, Hisaw, and Meyer, "The Relaxative Hormone."

59. Laura T. Goldsmith and Gerson Weiss, "Relaxin in Human Pregnancy," *Annals of the New York Academy of Sciences* 1160, no. 1 (2009): 130–35, https://doi.org/10.1111/j.1749-6632.2008.03800.x.

60. Goldsmith and Weiss, "Relaxin."

61. Charles W. Schauberger, Brenda L. Rooney, Laura Goldsmith, David Shenton, Paul D. Silva, and Ana Schaper, "Peripheral Joint Laxity Increases in Pregnancy but Does Not Correlate with Serum Relaxin Levels," *American Journal of Obstetrics and Gynecology* 174, no. 2 (1996): 667–71, https://doi.org/10.1016/S0002-9378(96)70447-7.

   There is not, however, an exact correlation between relaxin levels and degrees of laxity. Authors speculate that the effect of relaxin is not acute. Rather, prolonged exposure during pregnancy causes increased joint laxity late in pregnancy.

62. Zazulak et al., "The Effects of the Menstrual Cycle."

63. Carmen E. Quatman, Kevin R. Ford, Gregory D. Myer, Mark V. Paterno, and Timothy E. Hewett, "The Effects of Gender and Pubertal Status on Generalized Joint Laxity in Young Athletes," *Journal of Science and Medicine in Sport* 11, no. 3 (2007): 257–63, https://doi.org/10.1016/j.jsams.2007.05.005.

64. Jason L. Dragoo, Richard S. Lee, Prosper Benhaim, Gerald A. M. Finerman, and Sharon L. Hame, "Relaxin Receptors in the Human Female Anterior Cruciate Ligament," *American Journal of Sports Medicine* 31, no. 4 (2003): 577–84, https://doi.org/10.1177/03635465030310041701.

65. Elaine N. Unemori, L. Steven Beck, Wyne Pun Lee, Yvette Xu, Mark Siegel, Gilbert Keller, H. Denny Liggitt, Eugene A. Bauer, and Edward P. Amento, "Human Relaxin Decreases Collagen Accumulation in Vivo in Two Rodent Models of Fibrosis," *Journal of Investigative Dermatology* 101, no. 3 (1993): 280–85, https://doi.org/10.1111/1523-1747.ep12365206.

66. Dennis R. Stewart, Abbie C. Celniker, Clinton A. Taylor, Jeffrey R. Cragun, James W. Overstreet, and Bill L. Lasley, "Relaxin in the Peri-Implantation Period," *Journal of Clinical Endocrinology and Metabolism* 70, no. 6 (1990): 1771–73, https://doi.org/10.1210/jcem-70-6-1771.

67. Stewart et al., "Relaxin."

68. Jason L. Dragoo, Tiffany N. Castillo, Hillary J. Braun, Bethany A. Ridley, Ashleigh C. Kennedy, and S. Raymond Golish, "Prospective Correlation Between Serum Relaxin Concentration and Anterior Cruciate Ligament Tears Among Elite Collegiate Female Athletes," *American Journal of Sports Medicine* 39, no. 10 (2011): 2175–80, https://doi.org/10.1177/0363546511413378.

69. Zazulak et al., "The Effects of the Menstrual Cycle."

70. Dragoo et al., "Prospective Correlation"; Christine D. Pollard, Barry Braun, and Joseph Hamill, "Influence of Gender, Estrogen and Exercise on Anterior Knee Laxity," *Clinical Biomechanics* 21, no. 10 (2006): 1060–66, https://doi.org/10.1016/j.clinbiomech.2006.07.002.

71. Danielle Cooper and Heba Mahdy, "Oral Contraceptive Pills," in *StatPearls* (Treasure Island, FL: StatPearls, 2020), 28613632; Dragoo et al., "Prospective Correlation."

72. Lene Rahr-Wagner, Theis Muncholm Thillemann, Frank Mehnert, Alma Becic Pedersen, and Martin Lind, "Is the Use of Oral Contraceptives Associated with Operatively Treated Anterior Cruciate Ligament Injury? A Case-Control Study from the Danish Knee Ligament Reconstruction Registry," *American Journal of Sports Medicine* 42, no. 12 (2014): 2897–905, https://doi.org/10.1177/0363546514557240.

73. Rahr-Wagner et al., "Is the Use."

74. Ruedl et al., "Are Oral Contraceptive Use."

75. Julie Agel, Boris Bershadsky, and Elizabeth A. Arendt, "Hormonal Therapy: ACL and Ankle Injury," *Medicine and Science in Sports and Exercise* 38, no. 1 (2006): 7–12, https://doi.org/10.1249/01.mss.0000194072.13021.78.

76. Emma Woodhouse, Gregory A. Schmale, Peter Simonian, Allan Tencer, Phillipe Huber, and Kristy Seidel, "Reproductive Hormone Effects on Strength of the Rat Anterior Cruciate Ligament," *Knee Surgery, Sports Traumatology, Arthroscopy* 15, no. 4 (2007): 453–60, https://doi.org/10.1007/s00167-006-0237-0.

77. Woodhouse et al., "Reproductive Hormone."

78. Andrew Chamberlain, Daniel Zhao, and Amanda Stansell, *Progress on the Gender Pay Gap: 2019*, Glassdoor Economic Research, March 27, 2019. https://www.glassdoor.com/research/gender-pay-gap-2019/#.

79. See Lise Eliot, "Neurosexism: The Myth That Men and Women Have Different Brains," *Nature* 566 (2019): 453–54.

80. Bruce Goldman, "Two Minds: The Cognitive Differences Between Men and Women," *Stanford Medicine*, Spring 2017, https://stanmed.stanford.edu/2017spring/how-mens-and-womens-brains-are-different.html; Larry Cahill, "Denying the Neuroscience of Sex Differences," Quillette, March 29, 2019, https://quillette.com/2019/03/29/denying-the-neuroscience-of-sex-differences/; Diane F. Halpern, Camilla P. Benbow, David C. Geary, Ruben C. Gur, Janet Shibley Hyde, and Morton Ann Gernsbacher, "The Science of Sex Differences in Science and Mathematics," *Psychological Science in the Public Interest* 8, no. 1 (2007): 1–51, https://doi.org/10.1111/j.1529-1006.2007.00032.x.

81. Susan L. Rozzi, Scott M. Lephart, William S. Gear, and Freddie H. Fu, "Knee Joint Laxity and Neuromuscular Characteristics of Male and Female Soccer and Basketball Players," *American Journal of Sports Medicine* 27, no. 3 (1999): 312–19, https://doi.org/10.1177/03635465990270030801.

82. Takashi Nagai, Timothy C. Sell, John P. Abt, and Scott M. Lephart, "Reliability, Precision, and Gender Differences in Knee Internal/External Rotation Proprioception Measurements," *Physical Therapy in Sport* 13, no. 4 (2011): 233–37, https://doi.org/10.1016/j.ptsp.2011.11.004.

83. Cecilia Fridén, Angelica Lindén Hirschberg, Tönu Saartok, and Per Renström, "Knee Joint Kinaesthesia and Neuromuscular Coordination During Three Phases of the Menstrual Cycle in Moderately Active Women," *Knee Surgery, Sports Traumatology, Arthroscopy* 14, no. 4 (2006): 383–89, https://doi.org/10.1007/s00167-005-0663-4.

84. Jay Hertel, Nancy I. Williams, Lauren C. Olmsted-Kramer, Heather J. Leidy, and Margot Putukian, "Neuromuscular Performance and Knee Laxity Do Not Change Across the Menstrual Cycle in Female Athletes," *Knee Surgery, Sports Traumatology, Arthroscopy* 14, no. 9 (2006): 817–22, https://doi.org/10.1007/s00167-006-0047-4.

85. Laura J. Huston and Edward M. Wojtys, "Neuromuscular Performance Characteristics in Elite Female Athletes," *American Journal of Sports Medicine* 24, no. 4 (1996): 427–36, https://doi.org/10.1177/036354659602400405.

86. P. Renström, S. W. Arms, T. S. Stanwyck, R. J. Johnson, and M. H. Pope, "Strain Within the Anterior Cruciate Ligament During Hamstring and Quadriceps Activity," *American Journal of Sports Medicine* 14, no. 1 (1986): 83–87, https://doi.org/10.1177/036354658601400114.

87. Renström et al., "Strain Within the Anterior." Some conflicting findings exist. See Arne K. Aune, Patrick W. Cawley, and Arne Ekeland, "Quadriceps Muscle Contraction Protects the Anterior Cruciate Ligament During Anterior Tibial Translation," *American Journal of Sports Medicine* 25, no. 2 (1997): 187–90, https://doi.org/10.1177/036354659702500208.

88. Huston and Wojtys, "Neuromuscular Performance."
89. Robert A. Malinzak, Scott M. Colby, Donald T. Kirkendall, Bing Yu, and William E. Garrett, "A Comparison of Knee Joint Motion Patterns Between Men and Women in Selected Athletic Tasks," *Clinical Biomechanics* 16, no. 5 (2001): 438–45, https://doi.org/10.1016/S0268-0033(01)00019-5; Kevin R. Ford, Gregory D. Myer, Laura C. Schmitt, Timothy L. Uhl, and Timothy E. Hewett, "Preferential Quadriceps Activation in Female Athletes with Incremental Increases in Landing Intensity," *Journal of Applied Biomechanics* 27, no. 3 (2011): 215–22, https://doi.org/10.1123/jab.27.3.215; Ashley M. Hanson, Darin A. Padua, J. Troy Blackburn, William E. Prentice, and Christopher J. Hirth, "Muscle Activation During Side-Step Cutting Maneuvers in Male and Female Soccer Athletes," *Journal of Athletic Training* 43, no. 2 (2008): 133–43, https://doi.org/10.4085/1062-6050-43.2.133.
90. Renström et al., "Strain Within the Anterior."
91. Malinzak et al., "A Comparison."
92. T. E. Hewett, J. S. Torg, and B. P. Boden, "Video Analysis of Trunk and Knee Motion During Non-Contact Anterior Cruciate Ligament Injury in Female Athletes: Lateral Trunk and Knee Abduction Motion Are Combined Components of the Injury Mechanism," *British Journal of Sports Medicine* 43, no. 6 (2009): 417–22, https://doi.org/10.1136/bjsm.2009.059162.
93. Hewett, Torg, and Boden, "Video Analysis."
94. Kyla A. Russell, Riann M. Palmieri, Steven M. Zinder, and Christopher D. Ingersoll, "Sex Differences in Valgus Knee Angle During a Single-Leg Drop Jump," *Journal of Athletic Training* 41, no. 2 (2006): 166–71.
95. Kevin R. Ford, Gregory D. Myer, and Timothy E. Hewett, "Valgus Knee Motion During Landing in High School Female and Male Basketball Players," *Medicine and Science in Sports and Exercise* 35, no. 10 (2003): 1745–50, https://doi.org/10.1249/01.MSS.0000089346.85744.D9.
96. M. Z. Bendjaballah, A. Shirazi-Adl, and D. J. Zukor, "Finite Element Analysis of Human Knee Joint in Varus-Valgus," *Clinical Biomechanics* (Bristol) 12, no. 3 (1997): 139–48, https://doi.org/10.1016/S0268-0033(97)00072-7.
97. Rozzi et al., "Knee Joint Laxity."
98. Timothy E. Hewett, Amanda L. Stroupe, Thomas A. Nance, and Frank R. Noyes, "Plyometric Training in Female Athletes: Decreased Impact Forces and Increased Hamstring Torques," *American Journal of Sports Medicine* 24, no. 6(1996): 765–73, https://doi.org/10.1177/036354659602400611.
99. One of the eleven participants did not decrease her landing force.
100. Bert R. Mandelbaum, Holly J. Silvers, Diane S. Watanabe, John F. Knarr, Stephen D. Thomas, Letha Y. Griffin, Donald T. Kirkendall, and William Garrett, Jr., "Effectiveness of a Neuromuscular and Proprioceptive Training Program in Preventing Anterior Cruciate Ligament Injuries in Female Athletes," *American Journal of Sports Medicine* 33, no. 7 (2005): 1003, https://doi.org/10.1177/0363546504272261.

101. Hewett and Myer, "Reducing Knee"; J. Herbert Stevenson, Chad S. Beattie, Jennifer B. Schwartz, and Brian D. Busconi, "Assessing the Effectiveness of Neuromuscular Training Programs in Reducing the Incidence of Anterior Cruciate Ligament Injuries in Female Athletes: A Systematic Review," *American Journal of Sports Medicine* 43, no. 2 (2015): 482–90, https://doi.org/10.1177/0363546514523388.
102. Stevenson et al., "Assessing."
103. Stevenson et al., "Assessing."
104. Dai Sugimoto, Gregory D. Myer, Heather M. Bush, Maddie F. Klugman, Jennifer M. Medina McKeon, and Timothy E. Hewett, "Compliance with Neuromuscular Training and Anterior Cruciate Ligament Injury Risk Reduction in Female Athletes: A Meta-Analysis," *Journal of Athletic Training* 47, no. 6 (2012): 714–23, https://doi.org/10.4085/1062-6050-47.6.10.

5. TO KNEEL, OR NOT TO KNEEL

The quote in the first chapter epigraph comes from George Staunton, *An Authentic Account of an Embassy from the King of Great Britain to the Emperor of China*, vol. 2 (London: W. Bulmer for G. Nicol, 1797), 129–130.

The quote in the second chapter epigraph comes from *West Virginia State Board of Education et al. v. Barnette et al.*, 319 U.S. 624 (1943), Legal Information Institute, https://www.law.cornell.edu/supremecourt/text/319/624.

1. Dongqing Wang, "Representing Kowtow: Civility and Civilization in Early Sino-British Encounters," *The Eighteenth Century* 60, no. 3 (2019): 269–92, https://doi.org/10.1353/ecy.2019.0022.
2. Paul Gillingham, "The Macartney Embassy to China, 1792–94," *History Today*, 43, no. 11 (1993): 28–34.
3. George Staunton, *An Authentic Account of an Embassy from the King of Great Britain to the Emperor of China*, vol. 1 (London: W. Bulmer for G. Nicol, 1797), 54.
4. Frances Wood, "Britain's First View of China: The Macartney Embassy 1792–1794," *RSA Journal* 142, no. 5447 (1994): 59–68.
5. Gillingham, "The Macartney Embassy"; Wood, "Britain's First View."
6. Gillingham, "The Macartney Embassy."
7. Xingwu Dan, "从马葛尔尼使华看国际体系之争 [A Study of Conflicting International Systems Through the Macartney Embassy to China]," 国际政治科学 2 (2006): 1–27.
8. Staunton, *An Authentic Account*, vol. 1, 484–85.
9. Wang, "Representing Kowtow."
10. Staunton, *An Authentic Account*, vol. 2.
11. Earl H. Pritchard, "The Kotow in the Macartney Embassy to China in 1793," *Far Eastern Quarterly* 2, no. 2 (1943): 163–203, https://doi.org/10.2307/2049496.
12. Staunton, *An Authentic Account*, vol. 2, 213.

13. Staunton, *An Authentic Account*, vol. 2, 213.

14. Yinong Huang, "印象与真相：清朝中英两国的观礼之争 [Impression and Reality: The Sino-British Ritual War During the Qing Dynasty]," 中央研究院历史语言研究所集刊 78, no. 1 (2007): 35–106; Pritchard, "The Kotow."

15. Huang, "Impression and Reality."

16. Huang, "Impression and Reality."

17. Hao Gao, "The 'Inner Kowtow Controversy' During the Amherst Embassy to China, 1816–1817," *Diplomacy and Statecraft* 27, no. 4 (2016): 595–614, https://doi.org/10.1080/09592296.2016.1238691.

18. Pritchard, "The Kotow."

19. Wood, "Britain's First View."

20. Wei Wang, 椅子改变中国 [*Chairs Changed China*] (Beijing, China: China International Radio Press, 2009).

21. Jiji Kyodo, "Traditional Japanese Sitting Style to Be Recognized as Punishment Under New Law," *Japan Times*, December 4, 2019, https://www.japantimes.co.jp/news/2019/12/04/national/social-issues/japanese-sitting-style-recognized-punishment-new-law/.

22. Wang, *Chairs Changed China*.

23. Wang, "Representing Kowtow," 273.

24. Wang, *Chairs Changed China*.

25. Wang, *Chairs Changed China*.

26. The following account on the rise of chairs in China draws from Wang, *Chairs Changed China*.

27. Practically speaking, a person who is kneeling would also be less of a physical threat.

28. Wang, *Chairs Changed China*; Wang, "Representing Kowtow."

29. "横扫一切牛鬼蛇神 [Sweep Away All Evils]," editorial, *People's Daily*, June 1, 1966.

30. F. Adnan, "Chinese Customers Furious after Samsung Executives Knelt to Apologize for the Galaxy Note 7," SamMobile, Last updated November 2, 2016, https://www.sammobile.com/2016/11/02/chinese-customers-furious-after-samsung-executives-knelt-to-apologize-for-the-galaxy-note-7/.

31. Lin Tan and Wei Zhang, "网友质疑2000中学生集体跪拜父母; 我们需要何种感恩教育 [2,000 Middle School Students Kowtowing to Parents Invites Critique; What Kind of Gratitude Education Do We Need]," People's Daily Online, November 23, 2018, http://www.people.com.cn/n1/2018/1123/c347407-30418728.html.

32. Lili Xiang and Yan Jiang, "李阳博客贴出三千学生集体跪拜老师照片 [Li Yang's Blog Posted Photos of Three Thousand Students Kneeling to their Teachers]," Sina News, September 11, 2007, http://news.sina.com.cn/c/2007-09-11/014113860064.shtml.

33. Steve Wyche, "Colin Kaepernick Explains Why He Sat During National Anthem," NFL, August 27, 2016, https://www.nfl.com/news/colin-kaepernick-explains-why-he-sat-during-national-anthem-0ap3000000691077.

34. Rhiannon Walker, "One Year Later, Steve Wyche Reflects on Breaking the Colin Kaepernick Story," The Undefeated, August 28, 2017, https://theundefeated .com/features/one-year-later-steve-wyche-colin-kaepernick-story/.
35. Will Brinson, "Here's How Nate Boyer Got Colin Kaepernick to Go from Sitting to Kneeling," CBS/NFL, September 27, 2016, https://www.cbssports.com/nfl/news /heres-how-nate-boyer-got-colin-kaepernick-to-go-from-sitting-to-kneeling/.
36. Eric Reid, "Eric Reid: Why Colin Kaepernick and I Decided to Take a Knee," New York Times, September 25, 2017, https://www.nytimes.com/2017/09/25 /opinion/colin-kaepernick-football-protests.html.https://www.nytimes.com /2017/09/25/opinion/colin-kaepernick-football-protests.html?_r=0&referer =https://www.google.com/
37. "The Image of the Supplicant Slave: Advert or Advocate?," 1807 Commemo-rated (Institute for the Public Understanding of the Past, University of York, 2007), https://archives.history.ac.uk/1807commemorated/discussion /supplicant_slave.html.
38. Adam Howard, "Colin Kaepernick National Anthem Protest Catches on in NFL," NBC News, September 12, 2016, https://www.nbcnews.com/news/us -news/colin-kaepernick-national-anthem-protest-catches-nfl-n646671.
39. Howard, "Colin Kaepernick."
40. Kelly Grovier, "The Surprising Power of Kneeling," BBC Culture, September 29, 2017, https://www.bbc.com/culture/article/20170929-the-surprising-power-of -kneeling.
41. Jeremy Adam Smith and Dacher Keltner, "The Psychology of Taking a Knee," Scientific American, September 29, 2017, https://blogs.scientificamerican.com /voices/the-psychology-of-taking-a-knee/. Keltner was born in Mexico and moved to the United States at a young age.
42. I consider psychology a science because of its use of empirical research.
43. Mark Maske, "Americans Generally Disapprove of Players' Anthem Pro-tests but Opinion Divided Along Racial Lines, Poll Finds," Washington Post, October 11, 2016, https://www.washingtonpost.com/news/sports/wp/2016/10 /11/americans-generally-disapprove-of-players-anthem-protests-but-opinion -divided-along-racial-lines-poll-finds/.
44. Carrie Dann, "NBC/WSJ Poll: Majority Say Kneeling During Anthem 'Not Appropriate,'" NBC News, August 31, 2018, https://www.nbcnews.com/politics /first-read/nbc-wsj-poll-majority-say-kneeling-during-anthem-not-appropriate -n904891.
45. Maske, "Americans Generally Disapprove"; Dann, "NBC/WSJ Poll."
46. Kelsey Dallas, "The Religious Significance of Taking a Knee," Deseret News, August 14, 2020, https://www.deseret.com/indepth/2020/8/14/21362248 /athletes-kneeling-national-anthem-colin-kaepernick-eric-reid-sam-coonrod -patriotism-religion.
47. Dallas, "The Religious Significance."
48. Henry Schulman, "Giants' Sam Coonrod Cites His Faith, Issues with Black Lives Matter for Decision Not to Kneel," San Francisco Chronicle, July 24,

2020, https://www.sfchronicle.com/giants/article/Giants-Coonrod-cites-his
-faith-issues-with-15430927.php.

49. *Men's Health*, "Survey: Men Bend Knee to Tradition," February 13, 2012, https://
www.menshealth.com/sex-women/a19515770/wedding-proposals/.

50. Dallas, "The Religious Significance."

51. Reid, "Eric Reid."

52. Wyche, "Colin Kaepernick."

53. Billy Witz, "This Time, Colin Kaepernick Takes a Stand by Kneeling," *New York Times*, September 1, 2016, https://www.nytimes.com/2016/09/02/sports/football/colin-kaepernick-kneels-national-anthem-protest.html.

54. Aimee Lewis, "Colin Kaepernick: A Cultural Star Fast Turning Into a Global Icon," CNN, September 10, 2018, https://www.cnn.com/2018/09/07/sport/colin-kaepernick-protest-taking-the-knee-nate-boyer-spt-intl/index.html.

55. Wang, "Representing Kowtow."

56. James Baldwin, *Notes of a Native Son* (Boston, Beacon, 1955), 9.

57. Eric Burin, "Race, Dissent, and Patriotism in 21st Century America," in *Protesting on Bended Knee: Race, Dissent, and Patriotism in 21st Century America*, ed. Eric Burin (Grand Forks, ND: Digital Press at the University of North Dakota, 2018), 1–83.

58. Reid, "Eric Reid."

59. David F. Krugler, "African American Patriotism During the World War I Era," in *Protesting on Bended Knee: Race, Dissent, and Patriotism in 21st Century America*, ed. Eric Burin (Grand Forks, ND: Digital Press at The University of North Dakota, 2018), 189.

60. NBC News, "NBC News Poll: June 1995" (Cornell University, Ithaca, NY: Roper Center for Public Opinion Research, 1995), https://ropercenter.cornell.edu/ipoll/study/31106496. Apparently, patriotism is also a male trait. Only 3 percent of white respondents pictured a white woman when hearing someone being "patriotic," and 0 percent of white respondents pictured a Black woman; 5 percent of Black respondents pictured a white woman, and 5 percent of Black respondents pictured a Black woman.

61. Michael Tesler, "To Many Americans, Being Patriotic Means Being White," *Washington Post*, October 13, 2017, https://www.washingtonpost.com/news/monkey-cage/wp/2017/10/13/is-white-resentment-about-the-nfl-protests-about-race-or-patriotism-or-both/.

62. Burin, "Race, Dissent, and Patriotism."

63. M. L. Nestel, "Trump Says Issue of NFL Players Kneeling 'Has Nothing to Do with Race'," ABC News, September 25, 2017, https://abcnews.go.com/US/trump-issue-nfl-players-kneeling-race/story?id=50074211.

64. Burin, "Race, Dissent, and Patriotism."

65. Tesler, "To Many Americans."

66. JEFF (@jeffisrael25), "Thinking NFL players are 'protesting the flag' is like thinking Rosa Parks was protesting public transportation," Twitter, September 24, 2017, 3:24 p.m., https://twitter.com/jeffisrael25/status/9120651340335390

73?lang=en; Bree Newsome (@BreeNewsome), "Don't allow racists to reframe #TakeAKnee as being a debate about anthem & flag. It's a protest of police brutality & racism," Twitter, September 23, 2017, 7:45 a.m., https://twitter.com/breenewsome/status/911587445296254982.

67. Dianna Cahn, "VFW, American Legion: NFL Protests Disrespectful to Vets; Others Disagree," Stripes, 2017, accessed April 29, 2021, https://www.stripes.com/news/us/vfw-american-legion-nfl-protests-disrespectful-to-vets-others-disagree-1.489529.

68. Cahn, "VFW, American Legion."

69. Matt Eidson, "The Veteran View of Colin Kaepernick," in *Protesting on Bended Knee: Race, Dissent, and Patriotism in 21st Century America*, ed. Eric Burin (Grand Forks, ND: Digital Press at The University of North Dakota, 2018), 253–54.

70. VoteVets (@votevets), "As veterans, we swore an oath to support and defend the Constitutional rights of all citizens to speak freely and protest," Twitter, November 16, 2019, 8:17 p.m., https://twitter.com/votevets/status/1195903699706634240; Jo Wright (@JoWright59), "I'm sick of these people who believe the military is disrespected by Kaepernick & others kneeling. I'm a veteran and I support the right to kneel," Twitter, September 7, 2018, 9:42 p.m., https://mobile.twitter.com/JoWright59/status/1038271217596334080?cxt=HHwWgIC8lfyK1-gcAAAA.

71. Brian Stelter, "NFL Aired Unity Ad in Prime Time on Sunday," CNN, September 25, 2017, https://money.cnn.com/2017/09/24/media/nfl-unity-ad-trump/.

72. In 2018, Reid was signed by the Carolina Panthers.

73. A. J. Perez, "Report: Colin Kaepernick, Eric Reid Got Less Than $10 Million in NFL Collusion Settlement," *USA Today*, March 21, 2019, https://www.usatoday.com/story/sports/nfl/2019/03/21/colin-kaepernick-eric-reid-nfl-collusion-settlement/3237678002/.

74. The following account surrounding Floyd's death draws from "George Floyd: What Happened in the Final Moments of His Life," BBC News, July 16, 2020, https://www.bbc.com/news/world-us-canada-52861726; Evan Hill, Ainara Tiefenthäler, Christiaan Triebert, Drew Jordan, Haley Willis, and Robin Stein, "How George Floyd Was Killed in Police Custody," *New York Times*, May 31, 2020, https://www.nytimes.com/2020/05/31/us/george-floyd-investigation.html; *State of Minnesota vs. Thomas Kiernan Lane*, Court File No. 27-CR-20-12951, Minnesota Judicial Branch, July 7, 2020, https://www.mncourts.gov/mncourtsgov/media/High-Profile-Cases/27-CR-20-12951-TKL/Memorandum07072020.pdf.

75. On April 20, 2021, Chauvin was convicted of second-degree unintentional murder, third-degree murder, and second-degree manslaughter. On February 24, 2022, the other three ex-officers were found guilty of violating George Floyd's civil rights, among other charges.

76. "George Floyd: What Happened."

77. Derrick Bryson Taylor, "George Floyd Protests: A Timeline," *New York Times*, March 28, 2021, https://www.nytimes.com/article/george-floyd-protests-timeline.html; Roudabeh Kishi and Sam Jones, "Demonstrations and Political

Violence in America: New Data for Summer 2020," Armed Conflict Location and Event Data Project (ACLED), September 2020, https://acleddata.com/2020/09/03/demonstrations-political-violence-in-america-new-data-for-summer-2020/.

78. "Protests Across the Globe After George Floyd's Death," CNN, June 13, 2020, https://www.cnn.com/2020/06/06/world/gallery/intl-george-floyd-protests/index.html.

79. Kurt Streeter, "Kneeling, Fiercely Debated in the N.F.L., Resonates in Protests," *New York Times*, June 5, 2020, https://www.nytimes.com/2020/06/05/sports/football/george-floyd-kaepernick-kneeling-nfl-protests.html.

80. Caitlin O'Kane, "Police Officers Kneel in Solidarity with Protesters in Several U.S. Cities," CBS News, June 1, 2020, https://www.cbsnews.com/news/protesters-police-kneel-solidarity-george-floyd/.

81. Streeter, "Kneeling, Fiercely Debated."

82. Around the NFL, "Roger Goodell: NFL 'Wrong' for Not Listening to Protesting Players Earlier," NFL, June 5, 2020, https://www.nfl.com/news/roger-goodell-nfl-wrong-for-not-listening-to-protesting-players-earlier.

83. "George Floyd: US Soccer Overturns Ban on Players Kneeling," BBC News, June 11, 2020, https://www.bbc.com/news/world-us-canada-53003816.

84. Rick Maese and Emily Guskin, "Most Americans Support Athletes Speaking Out, Say Anthem Protests Are Appropriate, Post Poll Finds," *Washington Post*, September 10, 2020, https://www.washingtonpost.com/sports/2020/09/10/poll-nfl-anthem-protests/.

85. Dann, "NBC/WSJ Poll."

86. Rachel Hill (@r_hill3), "Unity," Twitter, June 30, 2020, 7:56 p.m., https://twitter.com/r_hill3/status/1278145465406627841.

87. John Branch, "The Anthem Debate Is Back. But Now It's Standing That's Polarizing," *New York Times*, July 4, 2020, https://www.nytimes.com/2020/07/04/sports/football/anthem-kneeling-sports.html.

6. TREATMENT, OR PLACEBO

The quote in the first chapter epigraph comes from Gabe Mirkin, "Why Ice Delays Recovery," September 16, 2015, https://www.drmirkin.com/fitness/why-ice-delays-recovery.html.

The quote in the second chapter epigraph comes from Y. Tegner and R. Lorentzon, "Evaluation of Knee Braces in Swedish Ice Hockey Players," *British Journal of Sports Medicine* 25, no. 3 (1991): 159, https://doi.org/10.1136/bjsm.25.3.159.

1. Giovanni Lombardi, Ewa Ziemann, and Giuseppe Banfi, "Whole-Body Cryotherapy in Athletes: From Therapy to Stimulation. An Updated Review of the Literature," *Frontiers in Physiology* 8 (2017): 258–58, https://doi.org/10.3389/fphys.2017.00258; Chris M. Bleakley, François Bieuzen, Gareth W. Davison,

and Joseph T. Costello, "Whole-Body Cryotherapy: Empirical Evidence and Theoretical Perspectives," *Open Access Journal of Sports Medicine* 5 (2014): 25–36, https://doi.org/10.2147/OAJSM.S41655.

2. S. M. Cooper and R. P. R. Dawber, "The History of Cryosurgery," *Journal of the Royal Society of Medicine* 94, no. 4 (2001): 196–201, https://doi.org/10.1177/014107680109400416.

3. Hippocrates, "Aphorisms," in *The Genuine Works of Hippocrates*, ed. Charles Darwin Adams (New York: Dover, 1868), https://www.chlt.org/hippocrates/Adams/page.316.a.php.

4. M. H. Armstrong Davison, "The Evolution of Anaesthesia," *British Journal of Anaesthesia* 31, no. 3 (1959): 134, https://doi.org/10.1093/bja/31.3.134.

5. Davison, "The Evolution," 135.

6. Cooper and Dawber, "The History"; Lyman Weeks Crossman, Frederick M. Allen, Vincent Hurley, Wilfred Ruggiero, and Cyrus E. Warden, "Refrigeration Anesthesia," *Anesthesia and Analgesia* 21, no. 1 (1942): 241–54, https://doi.org/10.1213/00000539-194201000-00059.

7. Frederick M. Allen, "Refrigeration Anesthesia for Limb Operations," *Anesthesiology* 4, no. 1 (1943): 12–16, https://doi.org/10.1097/00000542-194301000-00003; Crossman et al., "Refrigeration Anesthesia"; Sergei S. Yudin, "Refrigeration Anesthesia for Amputations," *Anesthesia and Analgesia* 24, no. 5 (1945): 216–19, https://doi.org/10.1213/00000539-194509000-00005; Harry E. Mock and Harry E. Mock Jr., "Refrigeration Anesthesia in Amputations," *Journal of the American Medical Association* 123, no. 1 (1943): 81–89, https://doi.org/10.1001/jama.1943.02840360015003.

8. Yudin, "Refrigeration Anesthesia."

9. Crossman et al., "Refrigeration Anesthesia," 243.

10. Yudin, "Refrigeration Anesthesia."

11. Barclay Moon Newman, "Shockless Surgery," *Scientific American* 166, no. 4 (1942): 182, https://doi.org/10.1038/scientificamerican0442-182.

12. Michael E. DeBakey and Fiorindo Simeone, "Acute Battle-Incurred Arterial Injuries," in *Vascular Surgery in World War II*, ed. Daniel C. Elkin and Michael E. Debakey (Washington, DC: Office of the Surgeon General, Department of the Army, 1955), 74, 100.

13. Yudin, "Refrigeration Anesthesia"; E. S. R. Hughes, "Refrigeration Anaesthesia," *British Medical Journal* 1, no. 4508 (1947): 761–64, https://doi.org/10.1136/bmj.1.4508.761.

14. Jacques Bruneau and Peter Heinbecker, "Effects of Cooling on Experimentally Infected Tissues," *Annals of Surgery* 120, no. 5 (1944): 716–26; A. Large and P. Heinbecker, "The Effect of Cooling on Wound Healing," *Annals of Surgery* 120, no. 5 (1944): 727–41, https://doi.org/10.1097/00000658-194411000-00005.

15. Bruneau and Heinbecker, "Effects of Cooling"; Large and Heinbecker, "The Effect of Cooling"; A. Large and P. Heinbecker, "Refrigeration in Clinical Surgery," *Annals of Surgery* 120, no. 5 (1944): 707–15, https://doi.org/10.1097/00000658-194411000-00003; Mock and Mock Jr., "Refrigeration Anesthesia."

16. Ethel A. Noble, "Refrigeration Anaesthesia," *Anesthesiology* 10, no. 1 (1948): 121–22, https://doi.org/10.1097/00000542-194901000-00028; Dermont W. Melick, "Refrigeration Anesthesia," *American Journal of Surgery* 70, no. 3 (1945): 364–68, https://doi.org/10.1016/0002-9610(45)90184-X.

17. Allen, "Refrigeration Anesthesia." In the 1940s, a tourniquet was used to reduce blood circulation in limbs.

18. Jeffrey J. Ciolek, "Cryotherapy. Review of Physiological Effects and Clinical Application," *Cleveland Clinic Quarterly* 52, no. 2 (1985): 193–201, https://doi.org/10.3949/ccjm.52.2.193; Tricia J. Hubbard, Stephanie L. Aronson, and Craig R. Denegar, "Does Cryotherapy Hasten Return to Participation? A Systematic Review," *Journal of Athletic Training* 39, no. 1 (2004): 88–94; Mark A. Merrick, "Secondary Injury After Musculoskeletal Trauma: A Review and Update," *Journal of Athletic Training* 37, no. 2 (2002): 209–17.

19. Chris Bleakley, Suzanne McDonough, and Domhnall MacAuley, "The Use of Ice in the Treatment of Acute Soft-Tissue Injury: A Systematic Review of Randomized Controlled Trials," *American Journal of Sports Medicine* 32, no. 1 (2004): 251–61, https://doi.org/10.1177/0363546503260757.

20. Yudin, "Refrigeration Anesthesia."

21. Amin A. Algafly and Keith P. George, "The Effect of Cryotherapy on Nerve Conduction Velocity, Pain Threshold and Pain Tolerance," *British Journal of Sports Medicine* 41, no. 6 (2007): 365–69, https://doi.org/10.1136/bjsm .2006.031237.

22. Algafly and George, "The Effect," 366.

23. Joseph B. Scarcella and Bruce T. Cohn, "The Effect of Cold Therapy on the Postoperative Course of Total Hip and Knee Arthroplasty Patients," *American Journal of Orthopedics* 24, no. 11 (1995): 847.

24. N. C. Collins, "Is Ice Right? Does Cryotherapy Improve Outcome for Acute Soft Tissue Injury?," *Emergency Medicine Journal* 25, no. 2 (2008): 65–68, https://doi.org/10.1136/emj.2007.051664.

25. Lucy A. Lessard, Roger A. Scudds, Annunziato Amendola, and Margaret D. Vaz, "The Efficacy of Cryotherapy Following Arthroscopic Knee Surgery," *Journal of Orthopaedic and Sports Physical Therapy* 26, no. 1 (1997): 14–22, https://doi.org/10.2519/jospt.1997.26.1.14. The group that iced their knees took less pain medication and scored lower on the affective dimension of pain, which measures people's feelings and emotions associated with pain.

26. Dale M. Daniel, Mary Lou Stone, and Diana L. Arendt, "The Effect of Cold Therapy on Pain, Swelling, and Range of Motion after Anterior Cruciate Ligament Reconstructive Surgery," *Arthroscopy* 10, no. 5 (1994): 530–33. https://doi.org/10.1016/S0749-8063(05)80008-8.

27. Lucas Ogura Dantas, Carolina Carreira Breda, Paula Regina Mendes da Silva Serrao, Francisco Aburquerque-Sendín, Ana Elisa Serafim Jorge, Jonathan Emanuel Cunha, Germanna Medeiros Barbosa, Joao Luiz Quagliotti Durigan, and Tania de Fatima Salvini, "Short-Term Cryotherapy Did Not Substantially Reduce Pain and Had Unclear Effects on Physical Function and Quality of Life

in People with Knee Osteoarthritis: A Randomised Trial," *Journal of Physiotherapy* 65, no. 4 (2019): 215–21, https://doi.org/10.1016/j.jphys.2019.08.004.

28. Dantas et al., "Short-Term Cryotherapy," 219.

29. Merih Yurtkuran and Tuncer Kocagil, "TENS, Electroacupuncture and Ice Massage: Comparison of Treatment for Osteoarthritis of the Knee," *American Journal of Acupuncture* 27, nos. 3–4 (1999): 133–40; Olavi V. Airaksinen, Nils Kyrklund, Kyösti Latvala, Jukka P. Kouri, Mats Grönblad, and Pertti Kolari, "Efficacy of Cold Gel for Soft Tissue Injuries: A Prospective Randomized Double-Blinded Trial," *American Journal of Sports Medicine* 31, no. 5 (2003): 680–84, https://doi.org/10.1177/03635465030310050801.

30. Lucas Ogura Dantas, Roberta de Fátima Carreira Moreira, Flavia Maintinguer Norde, Paula Regina Mendes Silva Serrao, Francisco Alburquerque-Sendín, and Tania Fatima Salvini, "The Effects of Cryotherapy on Pain and Function in Individuals with Knee Osteoarthritis: A Systematic Review of Randomized Controlled Trials," *Clinical Rehabilitation* 33, no. 8 (2019): 1310–19, https://doi.org/10.1177/0269215519840406; Collins, "Is Ice Right?"

31. Airaksinen et al., "Efficacy of Cold Gel."

32. Gregory A. Konrath, Terrence Lock, Henry T. Goitz, and Jeb Scheidler, "The Use of Cold Therapy After Anterior Cruciate Ligament Reconstruction: A Prospective, Randomized Study and Literature Review," *American Journal of Sports Medicine* 24, no. 5 (1996): 629–33, https://doi.org/10.1177/036354659602400511. A fourth group received no cold treatment, real or fake.

33. James R. Broatch, Aaron Petersen, and David J. Bishop, "Postexercise Cold Water Immersion Benefits Are Not Greater Than the Placebo Effect," *Medicine and Science in Sports and Exercise* 46, no. 11 (2014): 2139–47, https://doi.org/10.1249/MSS.0000000000000348.

34. Llion A. Roberts, Truls Raastad, James F. Markworth, Vandre C. Figueiredo, Ingrid M. Egner, Anthony Shield, David Cameron-Smith, Jeff S. Coombes, and Jonathan M. Peake, "Post-Exercise Cold Water Immersion Attenuates Acute Anabolic Signalling and Long-Term Adaptations in Muscle to Strength Training," *Journal of Physiology* 593, no. 18 (2015): 4285–301, https://doi.org/10.1113/JP270570.

35. Ching-Yu Tseng, Jo-Ping Lee, Yung-Shen Tsai, Shin-Da Lee, Chung-Lan Kao, Te-Chih Liu, Cheng Hsiu Lai, M. Brennan Harris, and Chia-Hua Kuo, "Topical Cooling (Icing) Delays Recovery from Eccentric Exercise–Induced Muscle Damage," *Journal of Strength and Conditioning Research* 27, no. 5 (2013): 1354–61, https://doi.org/10.1519/JSC.0b013e318267a22c.

36. Haiyan Lu, Danping Huang, Noah Saederup, Israel F. Charo, Richard M. Ransohoff, and Lan Zhou, "Macrophages Recruited via CCR2 Produce Insulin-Like Growth Factor-1 to Repair Acute Skeletal Muscle Injury," *FASEB Journal* 25, no. 1 (2011): 358–69, https://doi.org/10.1096/fj.10-171579.

37. Domhnall C. Mac Auley, "Ice Therapy: How Good Is the Evidence?," *International Journal of Sports Medicine* 22, no. 5 (2001): 379–84, https://doi.org/10.1055/s-2001-15656; Bleakley, McDonough, and MacAuley, "The Use of Ice."

38. Linda S. Chesterton, Nadine E. Foster, and Lesley Ross, "Skin Temperature Response to Cryotherapy," *Archives of Physical Medicine and Rehabilitation* 83, no. 4 (2002): 543–49, https://doi.org/10.1053/apmr.2002.30926.
39. Ronald Bugaj, "The Cooling, Analgesic, and Rewarming Effects of Ice Massage on Localized Skin," *Physical Therapy* 55, no. 1 (1975): 11–19, https://doi.org/10.1093/ptj/55.1.11; Bleakley, McDonough, and MacAuley, "The Use of Ice."
40. Chesterton, Foster, and Ross, "Skin Temperature."
41. Bleakley, McDonough, and MacAuley, "The Use of Ice"; Chesterton, Foster, and Ross, "Skin Temperature."
42. Beth Elchek LaVelle and Mariah Snyder, "Differential Conduction of Cold Through Barriers," *Journal of Advanced Nursing* 10, no. 1 (1985): 55–61, https://doi.org/10.1111/j.1365-2648.1985.tb00492.x.
43. William J. Myrer, Kimberly A. Myrer, Gary J. Measom, Gilbert W. Fellingham, and Stacey L. Evers, "Muscle Temperature Is Affected by Overlying Adipose When Cryotherapy Is Administered," *Journal of Athletic Training* 36, no. 1 (2001): 32–36.
44. Shijia Chen, "體寒百病生？ [Cold Qi Begets All Ailments?]," Yahoo Health, February 19, 2019, https://www.edh.tw/article/21075.
45. Yijun Lin, "伤后不用冰敷？ [No Icing After Injury?]," June 22, 2019, https://www.epochtimes.com/gb/19/6/20/n11335054.htm.
46. David O. Draper, Shane Schulthies, Pasi Sorvisto, and Anna-Mari Hautala, "Temperature Changes in Deep Muscles of Humans During Ice and Ultrasound Therapies: An In Vivo Study," *Journal of Orthopaedic and Sports Physical Therapy* 21, no. 3 (1995): 153–57, https://doi.org/10.2519/jospt.1995.21.3.153. Ultrasound may have nonthermal effects on processes such as collagen synthesis. See Levent Özgönenel, Ebru Aytekin, and Gulis Durmuşog-lu, "A Double-Blind Trial of Clinical Effects of Therapeutic Ultrasound in Knee Osteoarthritis," *Ultrasound in Medicine and Biology* 35, no. 1 (2008): 44–49, https://doi.org/10.1016/j.ultrasmedbio.2008.07.009.
47. Noriko Ichinoseki-Sekine, Hisashi Naito, Norio Saga, Yuji Ogura, Minoru Shiraishi, Arrigo Giombini, Valentina Giovannini, and Shizuo Katamoto, "Changes in Muscle Temperature Induced by 434 MHz Microwave Hyperthermia," *British Journal of Sports Medicine* 41, no. 7 (2007): 425–29, https://doi.org/10.1136/bjsm.2006.032540. Diathermy may have nonthermal effects such as acceleration of cell growth. See David O. Draper, "Comparison of Shortwave Diathermy and Microwave Diathermy," *International Journal of Athletic Therapy and Training* 18, no. 6 (2013): 13–17, https://doi.org/10.1123/ijatt.18.6.13.
48. Chang W. Song, "Effect of Local Hyperthermia on Blood Flow and Microenvironment: A Review," Supplement, *Cancer Research* 44, no. 10 (1984): 4721s-30s.
49. Mac Auley, "Ice Therapy."
50. Jerrold S. Petrofsky, Michael S. Laymon, Faris S. Alshammari, and Haneul Lee, "Use of Low Level of Continuous Heat as an Adjunct to Physical Therapy Improves Knee Pain Recovery and the Compliance for Home Exercise in

Patients with Chronic Knee Pain: A Randomized Controlled Trial," *Journal of Strength and Conditioning Research* 30, no. 11 (2016): 3107–15, https://doi .org/10.1519/JSC.0000000000001409.

51. Petrofsky et al., "Use of Low Level."
52. Anne Chandler, Joanne Preece, and Sara Lister, "Using Heat Therapy for Pain Management," *Nursing Standard* 17, no. 9 (2002): 40–42, https://doi.org /10.7748/ns2002.11.17.9.40.c3297.
53. Chandler, Preece, and Lister, "Using Heat."
54. W. McCarberg, G. Erasala, M. Goodale, J. Grender, D. Hengehold, and L. Donikyan, "Therapeutic Benefits of Continuous Low-Level Heat Wrap Therapy (CLHT) for Osteoarthritis (OA) of the Knee," *Journal of Pain* 6, no. 3 (2005): S53, https://doi.org/10.1016/j.jpain.2005.01.208.
55. Arrigo Giombini, Annalisa Di Cesare, Mariachiara Di Cesare, Maurizio Ripani, and Nicola Maffulli, "Localized Hyperthermia Induced by Microwave Diathermy in Osteoarthritis of the Knee: A Randomized Placebo-Controlled Double-Blind Clinical Trial," *Knee Surgery, Sports Traumatology, Arthroscopy* 19, no. 6 (2011): 980–87, https://doi.org/10.1007/s00167-010-1350-7.
56. Özgönenel, Aytekin, and Durmuşog-lu, "A Double-Blind Trial."
57. Hiroaki Seto, Hiroshi Ikeda, Hidehiko Hisaoka, and Hisashi Kurosawa, "Effect of Heat- and Steam-Generating Sheet on Daily Activities of Living in Patients with Osteoarthritis of the Knee: Randomized Prospective Study," *Journal of Orthopaedic Science: Official Journal of the Japanese Orthopaedic Association* 13, no. 3 (2008): 187–91, https://doi.org/10.1007/s00776-008-1214-x.
58. Mariko Usuba, Yutaka Miyanaga, Shumpei Miyakawa, Toru Maeshima, and Yoshio Shirasaki, "Effect of Heat in Increasing the Range of Knee Motion After the Development of a Joint Contracture: An Experiment with an Animal Model," *Archives of Physical Medicine and Rehabilitation* 87, no. 2 (2006): 247–53, https://doi.org/10.1016/j.apmr.2005.10.015; Draper, "Comparison."
59. Usuba et al., "Effect of Heat."
60. Chris M. Bleakley and Joseph T. Costello, "Do Thermal Agents Affect Range of Movement and Mechanical Properties in Soft Tissues? A Systematic Review," *Archives of Physical Medicine and Rehabilitation* 94, no. 1 (2013): 149–63, https://doi.org/10.1016/j.apmr.2012.07.023.
61. Jerrold Scott Petrofsky, Michael Laymon, and Haneul Lee, "Effect of Heat and Cold on Tendon Flexibility and Force to Flex the Human Knee," *Medical Science Monitor* 19 (2013): 661–67, https://doi.org/10.12659/MSM.889145.
62. Hamish McGorm, Llion A. Roberts, Jeff S. Coombes, and Jonathan M. Peake, "Turning Up the Heat: An Evaluation of the Evidence for Heating to Promote Exercise Recovery, Muscle Rehabilitation and Adaptation," *Sports Medicine* 48, no. 6 (2018): 1311–28, https://doi.org/10.1007/s40279-018-0876-6.
63. McGorm et al., "Turning Up."
64. Kousuke Takeuchi, Takuya Hatade, Soushi Wakamiya, Naoto Fujita, Takamitsu Arakawa, and Akinori Miki, "Heat Stress Promotes Skeletal Muscle Regeneration

After Crush Injury in Rats," *Acta Histochemica* 116, no. 2 (2014): 327–34, https://doi.org/10.1016/j.acthis.2013.08.010.

65. McGorm et al., "Turning Up."
66. J. T. Viitasalo, K. Niemelä, R. Kaappola, T. Korjus, M. Levola, H. V. Mononen, H. K. Rusko, and T. E. S. Takala, "Warm Underwater Water-Jet Massage Improves Recovery from Intense Physical Exercise," *European Journal of Applied Physiology* 71, no. 5 (1995): 431–38, https://doi.org/10.1007/BF00635877.
67. Joanna Vaile, Shona Halson, Nicholas Gill, and Brian Dawson, "Effect of Hydrotherapy on the Signs and Symptoms of Delayed Onset Muscle Soreness," *European Journal of Applied Physiology* 103, no. 1 (2008): 121–22, https://doi.org/10.1007/s00421-007-0653-y.
68. Albertas Skurvydas, Sigitas Kamandulis, Aleksas Stanislovaitis, Vytautas Streckis, Gediminas Mamkus, and Adomas Drazdauskas, "Leg Immersion in Warm Water, Stretch-Shortening Exercise, and Exercise-Induced Muscle Damage," *Journal of Athletic Training* 43, no. 6 (2008): 592–99, https://doi.org/10.4085/1062-6050-43.6.592.
69. McGorm et al., "Turning Up"; Bleakley and Costello, "Do Thermal Agents"; Yu Wu, Shibo Zhu, Zenghui Lv, Shunli Kan, Qiuli Wu, Wenye Song, Guangzhi Ning, and Shiqing Feng, "Effects of Therapeutic Ultrasound for Knee Osteoarthritis: A Systematic Review and Meta-Analysis," *Clinical Rehabilitation* 33, no. 12 (2019): 1863–75, https://doi.org/10.1177/0269215519866494.
70. Bleakley and Costello, "Do Thermal Agents"; Anne W. S. Rutjes, Eveline Nüesch, Rebekka Sterchi, and Peter Jüni, "Therapeutic Ultrasound for Osteoarthritis of the Knee or Hip," *Cochrane Library* 2010, no. 1 (2010): CD003132-CD32, https://doi.org/10.1002/14651858.CD003132.pub2.
71. Craig R. Denegar, Devon R. Dougherty, Jacob E. Friedman, Maureen E. Schimizzi, James E. Clark, Brett A. Comstock, and William J. Kraemer, "Preferences for Heat, Cold, or Contrast in Patients with Knee Osteoarthritis Affect Treatment Response," *Clinical Interventions in Aging* 5 (2010): 199–206, https://doi.org/10.2147/CIA.S11431.
72. Sam Borden, "Colleges Swear by Football Knee Braces. Not All Players and Experts Do," *New York Times*, January 8, 2017, https://www.nytimes.com/2017/01/08/sports/ncaafootball/college-football-playoff-alabama-clemson-knee-braces.html.
73. Michael J. Salata, Aimee E. Gibbs, and Jon K. Sekiya, "The Effectiveness of Prophylactic Knee Bracing in American Football: A Systematic Review," *Sports Health: A Multidisciplinary Approach* 2, no. 5 (2010): 375–79, https://doi.org/10.1177/1941738110378986.
74. Borden, "Colleges Swear."
75. Salata, Gibbs, and Sekiya, "The Effectiveness."
76. Borden, "Colleges Swear."
77. George Anderson, Stuart C. Zeman, and Robert T. Rosenfeld, "The Anderson Knee Stabler," *Physician and Sportsmedicine* 7, no. 6 (1979): 125–27, https://doi.org/10.1080/00913847.1979.11710882.

78. Anderson, Zeman, and Rosenfeld, "The Anderson Knee Stabler."
79. John P. Albright, John W. Powell, Walter Smith, Al Martindale, Edward Crowley, Jeff Monroe, Russ Miller, et al., "Medial Collateral Ligament Knee Sprains in College Football: Brace Wear Preferences and Injury Risk," *American Journal of Sports Medicine* 22, no. 1 (1994): 2–11, https://doi.org/10.1177/036354659402200102.
80. Frank Randall, Harold Miller, and Donald Shurr, "The Use of Prophylactic Knee Orthoses at Iowa State University," *Orthotics and Prosthetics* 37, no. 4 (1983): 54–57; Byron L. Hansen, Jack C. Ward, and Richard C. Diehl Jr., "The Preventive Use of the Anderson Knee Stabler in Football," *Physician and Sportsmedicine* 13, no. 9 (1985): 75–81, https://doi.org/10.1080/00913847.1985.11708879.
81. Thomas G. Grace, Betty J. Skipper, James C. Newberry, Michael A. Nelson, Edward R. Sweetser, and Michael L. Rothman, "Prophylactic Knee Braces and Injury to the Lower Extremity," *Journal of Bone and Joint Surgery* 70, no. 3 (1988): 422–27, https://doi.org/10.2106/00004623-198870030-00015; Carol C. Teitz, Bonnie K. Hermanson, Richard A. Kronmal, and Paula H. Diehr, "Evaluation of the Use of Braces to Prevent Injury to the Knee in Collegiate Football Players," *Journal of Bone and Joint Surgery* 69, no. 1 (1987): 2–9, https://doi.org/10.2106/00004623-198769010-00002.
82. American Academy of Pediatrics Committee on Sports Medicine, "Knee Brace Use by Athletes," *Pediatrics* 85, no. 2 (1990): 228.
83. Soheil Najibi and John P. Albright, "The Use of Knee Braces, Part 1: Prophylactic Knee Braces in Contact Sports," *American Journal of Sports Medicine* 33, no. 4 (2005): 602–11, https://doi.org/10.1177/0363546505275128; Michael Sitler, Jack Ryan, William Hopkinson, James Wheeler, James Santomier, Rickey Kolb, and David Polley, "The Efficacy of a Prophylactic Knee Brace to Reduce Knee Injuries in Football: A Prospective, Randomized Study at West Point," *American Journal of Sports Medicine* 18, no. 3 (1990): 310–15, https://doi.org/10.1177/036354659001800315; Salata, Gibbs, and Sekiya, "The Effectiveness"; John P. Albright, John W. Powell, Walter Smith, Al Martindale, Edward Crowley, Jeff Monroe, Russ Miller, et al., "Medial Collateral Ligament Knee Sprains in College Football: Effectiveness of Preventive Braces," *American Journal of Sports Medicine* 22, no. 1 (1994): 12–18, https://doi.org/10.1177/036354659402200103.
84. Sitler et al., "The Efficacy of a Prophylactic."
85. Najibi and Albright, "The Use of Knee Braces."
86. E. Paul France and Lonnie E. Paulos, "Knee Bracing," *Journal of the American Academy of Orthopaedic Surgeons* 2, no. 5 (1994): 281–87, https://doi.org/10.5435/00124635-199409000-00006.
87. B. D. Beynnon, M. H. Pope, C. M. Wertheimer, R. J. Johnson, B. C. Fleming, C. E. Nichols, and J. G. Howe, "The Effect of Functional Knee-Braces on Strain on the Anterior Cruciate Ligament in Vivo," *Journal of Bone and Joint Surgery*, 74, no. 9 (1992): 1298–312, https://doi.org/10.2106/00004623-199274090-00003;

Charles Beck, David Drez Jr., John Young, W. Dilworth Cannon Jr., and Mary Lou Stone, "Instrumented Testing of Functional Knee Braces," *American Journal of Sports Medicine* 14, no. 4 (1986): 253–56, https://doi.org/10.1177 /036354658601400401; Thomas Branch, Robert Hunter, and Peter Reynolds, "Controlling Anterior Tibial Displacement Under Static Load: A Comparison of Two Braces," *Orthopedics* 11, no. 9 (1988): 1249–52.

88. John F. Kramer, Tracy Dubowitz, Peter Fowler, Candice Schachter, and Trevor Birmingham, "Functional Knee Braces and Dynamic Performance: A Review," *Clinical Journal of Sport Medicine* 7, no. 1 (1997): 32–39, https://doi .org/10.1097/00042752-199701000-00007; Edward M. Wojtys and Laura J. Huston, " 'Custom-Fit' Versus 'Off-the-Shelf' ACL Functional Braces," *American Journal of Knee Surgery* 14, no. 3 (2001): 157–62; Gloria K. H. Wu, Gabriel Y. F. Ng, and Arthur F. T. Mak, "Effects of Knee Bracing on the Functional Performance of Patients with Anterior Cruciate Ligament Reconstruction," *Archives of Physical Medicine and Rehabilitation* 82, no. 2 (2001): 282–85, https://doi.org/10.1053/apmr.2001.19020.

89. Xiong-gang Yang, Jiang-tao Feng, Xin He, Feng Wang, and Yong-cheng Hu, "The Effect of Knee Bracing on the Knee Function and Stability Following Anterior Cruciate Ligament Reconstruction: A Systematic Review and Meta-Analysis of Randomized Controlled Trials," *Orthopaedics and Traumatology, Surgery and Research* 105, no. 6 (2019): 1107–14, https://doi.org/10.1016/j.otsr .2019.04.015; May Arna Risberg, Inger Holm, Harald Steen, Jan Eriksson, and Arne Ekeland, "The Effect of Knee Bracing After Anterior Cruciate Ligament Reconstruction: A Prospective, Randomized Study with Two Years' Follow-Up," *American Journal of Sports Medicine* 27, no. 1 (1999): 76–83.

90. P. Bordes, E. Laboute, A. Bertolotti, J. F. Dalmay, P. Puig, P. Trouve, E. Verhaegue, et al., "No Beneficial Effect of Bracing After Anterior Cruciate Ligament Reconstruction in a Cohort of 969 Athletes Followed in Rehabilitation," *Annals of Physical and Rehabilitation Medicine* 60, no. 4 (2017): 230–36, https://doi.org /10.1016/j.rehab.2017.02.001.

91. Najibi and Albright, "The Use of Knee Braces"; Edward M. Wojtys, Sandip U. Kothari, and Laura J. Huston, "Anterior Cruciate Ligament Functional Brace Use in Sports," *American Journal of Sports Medicine* 24, no. 4 (1996): 539–46, https://doi.org/10.1177/036354659602400421; Carl L. Highgenboten, Allen Jackson, Neil Meske, and Jimmy Smith, "The Effects of Knee Brace Wear on Perceptual and Metabolic Variables During Horizontal Treadmill Running," *American Journal of Sports Medicine* 19, no. 6 (1991): 639–43, https://doi .org/10.1177/036354659101900615.

92. Risberg et al., "The Effect of Knee Bracing."

93. Wu, Ng, and Mak, "Effects of Knee Bracing."

94. Kramer et al., "Functional Knee Braces."

95. Kramer et al., "Functional Knee Braces."

96. Chetan Gohal, Ajaykumar Shanmugaraj, Patrick Tate, Nolan S. Horner, Asheesh Bedi, Anthony Adili, and Moin Khan, "Effectiveness of Valgus Offloading Knee

Braces in the Treatment of Medial Compartment Knee Osteoarthritis: A Systematic Review," *Sports Health* 10, no. 6 (2018): 500–14, https://doi.org /10.1177/1941738118763913; Richard K. Jones, Christopher J. Nester, Jim D. Richards, Winston Y. Kim, David S. Johnson, Sanjiv Jari, Philip Laxton, and Sarah F. Tyson, "A Comparison of the Biomechanical Effects of Valgus Knee Braces and Lateral Wedged Insoles in Patients with Knee Osteoarthritis," *Gait and Posture* 37, no. 3 (2012): 368–72, https://doi.org/10.1016/j.gaitpost .2012.08.002.

97. Jones et al., "A Comparison."
98. Gohal et al., "Effectiveness of Valgus Offloading"; Tijs Duivenvoorden, Reinoud W. Brouwer, Tom M. van Raaij, Arianne P. Verhagen, Jan A. N. Verhaar, and Sita M. A. Bierma-Zeinstra, "Braces and Orthoses for Treating Osteoarthritis of the Knee," *Cochrane Library* 2015, no. 3 (2015): CD004020-CD20, https:// doi.org/10.1002/14651858.CD004020.pub3.
99. Gohal et al., "Effectiveness of Valgus Offloading."
100. Richard D. Komistek, Douglas A. Dennis, Eric J. Northcut, Adam Wood, Andrew W. Parker, and Steve M. Traina, "An in Vivo Analysis of the Effectiveness of the Osteoarthritic Knee Brace During Heel-Strike of Gait," *Journal of Arthroplasty* 14, no. 6 (1999): 738–42, https://doi.org/10.1016/S0883-5403(99)90230-9; Jeffrey A. Haladik, William K. Vasileff, Cathryn D. Peltz, Terrence R. Lock, and Michael J. Bey, "Bracing Improves Clinical Outcomes but Does Not Affect the Medial Knee Joint Space in Osteoarthritic Patients During Gait," *Knee Surgery, Sports Traumatology, Arthroscopy* 22, no. 11 (2014): 2715–20, https://doi .org/10.1007/s00167-013-2596-7; Matthew C. Nadaud, Richard D. Komistek, Mohamed R. Mahfouz, Douglas A. Dennis, and Matthew R. Anderle, "In Vivo Three-Dimensional Determination of the Effectiveness of the Osteoarthritic Knee Brace: A Multiple Brace Analysis," *Journal of Bone and Joint Surgery* 87A, Suppl. 2 (2005): 114–19, https://doi.org/10.2106/00004623-200511002-00013.
101. Tomasz Cudejko, Martin van der Esch, Marike van der Leeden, Josien C. van den Noort, Leo D. Roorda, Willem Lems, Jos Twisk, et al., "The Immediate Effect of a Soft Knee Brace on Pain, Activity Limitations, Self-Reported Knee Instability, and Self-Reported Knee Confidence in Patients with Knee Osteoarthritis," *Arthritis Research and Therapy* 19, no. 1 (2017): 260–60, https:// doi.org/10.1186/s13075-017-1456-0; Raphael Schween, Dominic Gehring, and Albert Gollhofer, "Immediate Effects of an Elastic Knee Sleeve on Frontal Plane Gait Biomechanics in Knee Osteoarthritis," *PLoS One* 10, no. 1 (2015): e0115782-e82, https://doi.org/10.1371/journal.pone.0115782; Kyue-nam Park and Si-hyun Kim, "Effects of Knee Taping During Functional Activities in Older People with Knee Osteoarthritis: A Randomized Controlled Clinical Trial: Effects of Taping on Knee Osteoarthritis," *Geriatrics and Gerontology International* 18, no. 8 (2018): 1206–10, https://doi.org/10.1111/ggi.13448; David W. Edmonds, Jenny McConnell, Jay R. Ebert, Tim R. Ackland, and Cyril J. Donnelly, "Biomechanical, Neuromuscular and Knee Pain Effects Following Therapeutic Knee Taping Among Patients with Knee Osteoarthritis

During Walking Gait," *Clinical Biomechanics* 39 (2016): 38–43, https://doi
.org/10.1016/j.clinbiomech.2016.09.003.

102. B. S. Hassan, S. Mockett, and M. Doherty, "Influence of Elastic Bandage on
Knee Pain, Proprioception, and Postural Sway in Subjects with Knee Osteoarthritis," *Annals of the Rheumatic Diseases* 61, no. 1 (2002): 24–28, https://
doi.org/10.1136/ard.61.1.24; Kai-Yu Ho, Ryan Epstein, Ron Garcia, Nicole
Riley, and Szu-Ping Lee, "Effects of Patellofemoral Taping on Patellofemoral Joint Alignment and Contact Area During Weight Bearing," *Journal of
Orthopaedic and Sports Physical Therapy* 47, no. 2 (2017): 115–23, https://
doi.org/10.2519/jospt.2017.6936; Park and Kim, "Effects."

103. D. S. Barrett, A. G. Cobb, and G. Bentley, "Joint Proprioception in Normal,
Osteoarthritic and Replaced Knees," *Journal of Bone and Joint Surgery* 73, no. 1
(1991): 53–56, https://doi.org/10.1302/0301-620X.73B1.1991775; Damien
Van Tiggelen, Pascal Coorevits, and Erik Witvrouw, "The Effects of a Neoprene Knee Sleeve on Subjects with a Poor Versus Good Joint Position Sense
Subjected to an Isokinetic Fatigue Protocol," *Clinical Journal of Sport Medicine*
18, no. 3 (2008): 259–65, https://doi.org/10.1097/JSM.0b013e31816d78c1.

104. Barrett, Cobb, and Bentley, "Joint Proprioception"; Shih-Hung Chuang, Mao-
Hsiung Huang, Tien-Wen Chen, Ming-Chang Weng, Chin-Wei Liu, and Chia-
Hsin Chen, "Effect of Knee Sleeve on Static and Dynamic Balance in Patients
with Knee Osteoarthritis," *Kaohsiung Journal of Medical Sciences* 23, no. 8
(2007): 405–11, https://doi.org/10.1016/S0257-5655(07)70004-4; Amber T.
Collins, J. Troy Blackburn, Chris W. Olcott, Jodie Miles, Joanne Jordan, Douglas
R. Dirschl, and Paul S. Weinhold, "Stochastic Resonance Electrical Stimulation
to Improve Proprioception in Knee Osteoarthritis," *The Knee* 18, no. 5 (2011):
317–22, https://doi.org/10.1016/j.knee.2010.07.001; Van Tiggelen, Coorevits, and
Witvrouw, "The Effects of a Neoprene Knee Sleeve"; Michael J. Callaghan, James
Selfe, Pam J. Bagley, and Jacqueline A. Oldham, "The Effects of Patellar Taping on
Knee Joint Proprioception," *Journal of Athletic Training* 37, no. 1 (2002): 19–24.

105. Barrett, Cobb, and Bentley, "Joint Proprioception"; Van Tiggelen, Coorevits,
and Witvrouw, "The Effects of a Neoprene Knee Sleeve."

106. Callaghan et al., "The Effects of Patellar Taping."

107. Barrett, Cobb, and Bentley, "Joint Proprioception," 54.

108. Hassan, Mockett, and Doherty, "Influence of Elastic Bandage."

109. Kramer et al., "Functional Knee Braces"; Najibi and Albright, "The Use of Knee
Braces."

7. THE HURTFUL KNEE

The quote in the first chapter epigraph comes from Kaweewit Kaewjinda,
"Death of Young Thai Kickboxer Brings Focus on Dangers," *Seattle Times*,
November 18, 2018, https://www.seattletimes.com/nation-world/death-of
-young-thai-kickboxer-brings-focus-on-dangers/.

The quote in the second chapter epigraph comes from Maggie Koerth, "The Two Autopsies of George Floyd Aren't as Different as They Seem," FiveThirtyEight, June 8, 2020, https://fivethirtyeight.com/features/the-two -autopsies-of-george-floyd-arent-as-different-as-they-seem/. The duration of time Chauvin knelt on Floyd was later determined to be nine minutes and twenty-nine seconds.

1. Jeremy DeSilva, *First Steps: How Walking Upright Made Us Human* (New York, NY: HarperCollins, 2021), 53.

2. DeSilva, *First Steps*; Deepika Babu and Bruno Bordoni, "Anatomy, Bony Pelvis and Lower Limb, Medial Longitudinal Arch of the Foot," in *StatPearls* (Treasure Island, FL: StatPearls, 2021), 32965960; James T. Webber and David A. Raichlen, "The Role of Plantigrady and Heel-Strike in the Mechanics and Energetics of Human Walking with Implications for the Evolution of the Human Foot," *Journal of Experimental Biology* 219 (2016): 3729–37, https://doi.org/10.1242 /jeb.138610.

3. Alberto Corbí and Olga Santos, "Myshikko: Modelling Knee Walking in Aikido Practice," in *Proceedings for the 26th Conference on User Modeling, Adaptation and Personalization* (New York: Association for Computing Machinery, 2018), 217–18.

4. William Scott Wilson, trans. *Ideals of the Samurai: Writings of Japanese Warriors* (Burbank, CA: Ohara, 1982).

5. Walther G. von Krenner, Damon Apodaca, and Ken Jeremiah, *Aikido Ground Fighting: Grappling and Submission Techniques* (Berkeley, CA: Blue Snake Books, 2013).

6. von Krenner, Apodaca, and Jeremiah, *Aikido Ground Fighting*.

7. von Krenner, Apodaca, and Jeremiah, *Aikido Ground Fighting*.

8. von Krenner, Apodaca, and Jeremiah, *Aikido Ground Fighting*. Another idea suggests that knee techniques help to cultivate internal power.

9. Gaku Homma, "No Suwariwaza (Kneeling Techniques) at Nippon Kan," Nippon Kan Kancho, June 15, 2007, http://www.nippon-kan.org/no-suwariwaza -kneeling-techniques-at-nippon-kan/.

10. Md. Ayub Ali, Teruo Uetake, and Fumio Ohtsuki, "Secular Changes in Relative Leg Length in Post-War Japan," *American Journal of Human Biology* 12, no. 3 (2000): 405–16.

11. P. Grasgruber, M. Sebera, E. Hrazdíra, J. Cacek, and T. Kalina, "Major Correlates of Male Height: A Study of 105 Countries," *Economics and Human Biology* 21 (2016): 172–95, https://doi.org/10.1016/j.ehb.2016.01.005. Genetics and other environmental factors may also play a role.

12. Homma, "No Suwariwaza."

13. Homma, "No Suwariwaza."

14. Homma, "No Suwariwaza."

15. R. B. Birrer and S. P. Halbrook, "Martial Arts Injuries: The Results of a Five Year National Survey," *American Journal of Sports Medicine* 16, no. 4 (1988): 408–10, https://doi.org/10.1177/036354658801600418; Mark McPherson

and William Pickett, "Characteristics of Martial Art Injuries in a Defined Canadian Population: A Descriptive Epidemiological Study," *BMC Public Health* 10, no. 1 (2010): 795–95, https://doi.org/10.1186/1471-2458-10-795; M. N. Zetaruk, M. A. Violán, D. Zurakowski, and L. J. Micheli, "Injuries in Martial Arts: A Comparison of Five Styles," *British Journal of Sports Medicine* 39, no. 1 (2005): 29–33, https://doi.org/10.1136/bjsm.2003.010322.

16. Harvey Kurland, "A Comparison of Judo and Aikido Injuries," *Physician and Sportsmedicine* 8, no. 6 (1980): 71–74, https://doi.org/10.1080/00913847.198 0.11948618; McPherson and Pickett, "Characteristics of Martial Art Injuries"; Zetaruk, "Injuries in Martial Arts."

17. E. R. Serina and D. K. Lieu, "Thoracic Injury Potential of Basic Competition Taekwondo Kicks," *Journal of Biomechanics* 24, no. 10 (1991): 951–60, https://doi.org/10.1016/0021-9290(91)90173-K; F. Pieter and W. Pieter, "Speed and Force in Selected Taekwondo Techniques," *Biology of Sport* 12, no. 4 (1995): 257–66.

18. Pieter and Pieter, "Speed and Force"; Jacek Wąsik, "Kinematics and Kinetics of Taekwon-Do Side Kick," *Journal of Human Kinetics* 30 (2011): 13–20, https://doi.org/10.2478/v10078-011-0068-z; Serina and Lieu, "Thoracic Injury."

19. "Stealth Fighters," *Fight Science*, Season 1, Episode 5, Created by Michael Stern, National Geographic, 2008.

20. Sutima Thibordee and Orawan Prasartwuth, "Effectiveness of Roundhouse Kick in Elite Taekwondo Athletes," *Journal of Electromyography and Kinesiology* 24, no. 3 (2014): 353–58, https://doi.org/10.1016/j.jelekin.2014.02.002; Emanuel Preuschl, Michaela Hassmann, and Arnold Baca, "A Kinematic Analysis of the Jumping Front-Leg Axe-Kick in Taekwondo," *Journal of Sports Science and Medicine* 15, no. 1 (2016): 92–101; Young Kwan Kim, Yoon Hyuk Kim, and Shin Ja Im, "Inter-Joint Coordination in Producing Kicking Velocity of Taekwondo Kicks," *Journal of Sports Science and Medicine* 10, no. 1 (2011): 31–38.

21. Preuschl, Hassmann, and Baca, "A Kinematic Analysis"; David Burke, Samir al-Adawi, Daniel Burke, Paolo Bonato, and Casey Leong, "The Kicking Process in Tae Kwon Do: A Biomechanical Analysis," *International Physical Medicine and Rehabilitation Journal* 1, no. 1 (2017), https://doi.org/10.15406/ipmrj .2017.01.00002.

22. The Different Types of Knees in Muay Thai," *Evolve Mixed Martial Arts* (blog), 2021, https://evolve-vacation.com/blog/the-different-types-of-knees-in-muay -thai/.

23. P. Rachnavy, T. Khaothin, and W. Rittiwat, "Kinematics Analysis of Muay Thai Knee Techniques," International Conference of Sport Science-AESA, 2018, https://journal.aesasport.com/index.php/AESA-Conf/article/view/84.

24. "Muay Thai," *Human Weapon*, Season 1 Episode 1, created by Terry Bullman, History Channel, 2007.

25. Rachnavy, Khaothin, and Rittiwat, "Kinematics Analysis."

26. Guilherme Cruz, "Photos: 'Cyborg' Santos Before and After Surgery," MMA Fighting, July 28, 2016, https://www.mmafighting.com/2016/7/28/12319768 /photos-cyborg-santos-before-and-after-surgery.

27. Alan Dawson, "Yes, You Can Make Millions as a Professional Fighter—but Only 19 out of 21,000 Successfully Manage It," Business Insider, June 9, 2017, https://www.businessinsider.com/mcgregor-mayweather-fight-economics-ufc-mma-bellator-boxing-paul-daley-2017-6; Jeffrey Hays, "Muay Thai (Thai Kick Boxing) and Olympic and Pro Boxing in Thailand," Facts and Details, updated May 2014, https://factsanddetails.com/southeast-asia/Thailand/sub5_8e/entry-3270.html.

28. Ben Solomon, "'Destroying Our Children for Sport': Thailand May Limit Underage Boxing," New York Times, December 23, 2018, https://www.nytimes.com/2018/12/23/world/asia/thailand-children-muay-thai.html; Athit Perawongmetha and Jiraporn Kuhakan, "Punching out of Poverty: Despite Risks, 9-Year-Old Thai Fighter Eager to Return to Ring," Reuters, April 6, 2021, https://www.reuters.com/article/us-thailand-muaythai-widerimage/punching-out-of-poverty-despite-risks-9-year-old-thai-fighter-eager-to-return-to-ring-idUSKBN2BT31S.

29. "Muay Thai," Human Weapon.

30. "Muay Thai," Human Weapon; Solomon, "Destroying Our Children."

31. "Muay Thai, Human Weapon."

32. Jiraporn Laothamatas, Adisak Plitponkarnpim, Onousa Sangfai, Thirawat Suparatpriyakon, Mattana Pongsopon, Daochompu Nakawiro, Chakrit Sukying, Anannit Visudtibhan, and Witaya Sungkarat, "Child Muaythai Boxing: Conflict of Health and Culture," Injury Prevention 24, no. Suppl. 1 (2018): A126, https://doi.org/10.1136/injuryprevention-2018-safety.349.

33. Julie Power and Kate Geraghty, "Inside Muay Thai: Where Culture and Children's Well-Being Collide," Sydney Morning Herald, November 24, 2018, https://www.smh.com.au/sport/boxing/inside-muay-thai-where-culture-and-children-s-well-being-collide-20181123-p50htq.html.

34. Power and Geraghty, "Inside Muay Thai."

35. "Muay Thai Children Fighting for Cash," Channel 4, British Public Broadcast Service, 2014, https://www.youtube.com/watch?v=u2ueOF7tm1k; "Thailand's Child Fighters," Al Jazeera, October 3, 2019, https://www.youtube.com/watch?v=bh2pviMAB9I.

36. "Muay Thai Children."

37. Solomon, "Destroying Our Children."

38. "Muay Thai Children."

39. U.S. Department of Labor, "Child Labor and Forced Labor Reports: Thailand" (Bureau of International Labor Affairs, 2020), https://www.dol.gov/sites/dolgov/files/ILAB/child_labor_reports/tda2020/Thailand.pdf.

40. Dean Cornish, "'We Will Never Have Champions If Fighters Can't Start Young.' Getting to Know Thailand's Child Fighters," Dateline, May 21, 2019, https://www.sbs.com.au/news/dateline/we-will-never-have-champions-if-fighters-can-t-start-young-getting-to-know-thailand-s-child-fighters.

41. U.S. Department of Labor, "Child Labor and Forced Labor Reports: Thailand" (Bureau of International Labor Affairs, 2018), https://www.dol.gov/sites/dolgov/files/ILAB/child_labor_reports/tda2018/Thailand.pdf.

42. Associated Press, "13-Year-Old Dies After Getting KO'd in Thai Kickboxing Match," *New York Post*, November 13, 2018, https://nypost.com/2018/11/13/13-year-old-dies-after-getting-kod-in-thai-kickboxing-match/.
43. "Thailand's Child Fighters."
44. "Thailand's Child Fighters."
45. James Surowiecki, "Beautiful. Violent. American. The N.F.L. At 100," *New York Times*, December 19, 2019, https://www.nytimes.com/2019/12/19/sports/football/nfl-100-violence-american-culture.html.
46. Surowiecki, "Beautiful. Violent. American."
47. Bob Carter, "The Violent World," ESPN Classic, n.d., https://www.espn.com/classic/biography/s/Huff_Sam.html.
48. Surowiecki, "Beautiful. Violent. American."
49. Thomas P. Dompier, Zachary Y. Kerr, Stephen W. Marshall, Brian Hainline, Erin M. Snook, Ross Hayden, and Janet E. Simon, "Incidence of Concussion During Practice and Games in Youth, High School, and Collegiate American Football Players," *JAMA Pediatrics* 169, no. 7 (2015): 659–65, https://doi.org/10.1001/jamapediatrics.2015.0210; Ray W. Daniel, Steven Rowson, and Stefan M. Duma, "Head Impact Exposure in Youth Football," *Annals of Biomedical Engineering* 40, no. 4 (2012): 976–81, https://doi.org/10.1007/s10439-012-0530-7.
50. Jay G. Ingram, Sarah K. Fields, Ellen E. Yard, and R. Dawn Comstock, "Epidemiology of Knee Injuries Among Boys and Girls in US High School Athletics," *American Journal of Sports Medicine* 36, no. 6 (2008): 1116–22, https://doi.org/10.1177/0363546508314400.
51. Daniel R. Clifton, James A. Onate, Eric Schussler, Aristarque Djoko, Thomas P. Dompier, and Zachary Y. Kerr, "Epidemiology of Knee Sprains in Youth, High School, and Collegiate American Football Players," *Journal of Athletic Training* 52, no. 5 (2017): 464–73, https://doi.org/10.4085/1062-6050-52.3.09; Zachary Y. Kerr, Gary B. Wilkerson, Shane V. Caswell, Dustin W. Currie, Lauren A. Pierpoint, Erin B. Wasserman, Sarah B. Knowles, et al., "The First Decade of Web-Based Sports Injury Surveillance: Descriptive Epidemiology of Injuries in United States High School Football (2005–2006 Through 2013–2014) and National Collegiate Athletic Association Football (2004–2005 Through 2013–2014)," *Journal of Athletic Training* 53, no. 8 (2018): 738–51, https://doi.org/10.4085/1062-6050-144-17.
52. Dompier et al., "Incidence of Concussion."
53. Julie M. Stamm, Alexandra P. Bourlas, Christine M. Baugh, Nathan G. Fritts, Daniel H. Daneshvar, Brett M. Martin, Michael D. McClean, Yorghos Tripodis, and Robert A. Stern, "Age of First Exposure to Football and Later-Life Cognitive Impairment in Former NFL Players," *Neurology* 84, no. 11 (2015): 1114–20, https://doi.org/10.1212/WNL.0000000000001358.
54. Michael L. Alosco, Jesse Mez, Yorghos Tripodis, Patrick T. Kiernan, Bobak Abdolmohammadi, Lauren Murphy, Neil W. Kowall, et al., "Age of First

Exposure to Tackle Football and Chronic Traumatic Encephalopathy," *Annals of Neurology* 83, no. 5 (2018): 886–901, https://doi.org/10.1002/ana.25245.

55. Shaila Dewan and Sheri Fink, "Does It Matter Whether Chauvin Knelt on Floyd's Neck Versus His Shoulder?," *New York Times*, April 8, 2021, https://www.nytimes.com/2021/04/08/us/does-it-matter-whether-chauvin-knelt-on-floyds-neck-versus-his-shoulder.html; Nicholas Bogel-Burroughs, "George Floyd Showed Signs of a Brain Injury 4 Minutes Before Derek Chauvin Lifted His Knee, a Doctor Testifies," *New York Times*, April 8, 2021, https://www.nytimes.com/2021/04/08/us/george-floyd-knee-on-neck.html.

56. *State of Minnesota v. Derek Michael Chauvin, Tou Thao, J. Alexander Kueng, Thomas Kiernan Lane*, Hennepin County Court File No. 27-CR-12646, 27-CR-20-12949, 27-CR-20-12953, 27-CR-20-12951," Minnesota Judicial Branch, 2020, https://www.mncourts.gov/mncourtsgov/media/High-Profile-Cases/27-CR-20-12646/ExhibitMtD08282020.pdf.

57. The City of Minneapolis banned neck restraints and chokeholds after Floyd's death.

58. Casey Tolan, "Two-Thirds of People Put in Neck Restraints by Minneapolis Police Were Black, Department Data Shows," CNN, June 2, 2020, https://www.cnn.com/2020/06/02/us/mn-minneapolis-police-neck-restraints-george-floyd-invs/index.html.

59. Michael Schlosser, "Unlocking the Confusion Around Chokeholds," *Police* 43, no. 3 (2019): 36.

60. Donald T. Reay and John W. Eisele, "Death from Law Enforcement Neck Holds," *American Journal of Forensic Medicine and Pathology* 3, no. 3 (1982): 253–58, https://doi.org/10.1097/00000433-198209000-00012.

61. Reay and Eisele, "Death from Law Enforcement."

62. Bill Chappell, "Chauvin's Restraint on Floyd's Neck Isn't Taught by Police, Use-of-Force Trainer Says," NPR, April 6, 2021, https://www.npr.org/sections/trial-over-killing-of-george-floyd/2021/04/06/984717386/watch-live-minneapolis-police-crisis-intervention-trainer-testifies-in-chauvin-t.

63. Lou Raguse, "MPD Training Materials Show Knee-to-Neck Restraint Similar to the One Used on Floyd," KARE11, July 8, 2020, https://www.kare11.com/article/news/local/george-floyd/minneapolis-police-training-materials-show-knee-to-neck-restraint-similar-to-used-on-george-floyd/89-9f002e3f-972a-4410-86cb-50a1237fc496.

64. Raguse, "MPD Training."

65. Amy Forliti, Steve Karnowski, and Tammy Webber, "Police Chief: Kneeling on Floyd's Neck Violated Policy," Associated Press News, April 5, 2021, https://apnews.com/article/derek-chauvin-trial-live-updates-c3e3fe08773cd2f-012654e782e326f6e.; Raguse, "MPD Training."

66. Deena Winter, "Dr. Andrew Baker, Key Witness, Stands by Homicide Determination in Chauvin Trial," *Florida Phoenix*, April 10, 2021.

67. *State of Minnesota v. Derek Michael Chauvin.*

68. *State of Minnesota v. Derek Michael Chauvin.*
69. Winter, "Dr. Andrew Baker."
70. Amy Forliti, Steve Karnowski, and Tammy Webber, "Medical Examiner Blames Police Pressure for Floyd's Death," Associated Press News, April 9, 2021; Chao Xiong, Paul Walsh, and Rochelle Olson, "Cardiac Arrest and Drugs, Not Low Oxygen, Caused Floyd's Death, Defense Expert Says," *Star Tribune*, April 15, 2021, https://www.startribune.com/cardiac-arrest-and-drugs-not -low-oxygen-caused-floyd-s-death-defense-expert-says/600046038/; Winter, "Dr. Andrew Baker."
71. Winter, "Dr. Andrew Baker."
72. Forliti, Karnowski, and Webber, "Medical Examiner Blames Police."
73. Bogel-Burroughs, "George Floyd Showed Signs."
74. Bogel-Burroughs, "George Floyd Showed Signs."
75. Steven B. Karch, "The Problem of Police-Related Cardiac Arrest," *Journal of Forensic and Legal Medicine* 41 (2016): 36–41, https://doi.org/10.1016/j.jflm .2016.04.008.
76. Xiong, Walsh, and Olson, "Cardiac Arrest"; Laurel Wamsley and Vanessa Romo, "Defense Medical Expert: Floyd's Manner of Death 'Undetermined,' Not 'Homicide," NPR, April 14, 2021, https://www.npr.org/sections/trial-over -killing-of-george-floyd/2021/04/14/987134841/watch-live-defense-testimony -resumes-in-derek-chauvins-trial.
77. Mark W. Kroll, Michael A. Brave, Scott R. Kleist, Mollie B. Ritter, Darrell L. Ross, and Steven B. Karch, "Applied Force During Prone Restraint: Is Officer Weight a Factor?," *American Journal of Forensic Medicine and Pathology* 40, no. 1 (2019): 1–7, https://doi.org/10.1097/PAF.0000000000000457.
78. Bogel-Burroughs, "George Floyd Showed Signs."
79. Wamsley and Romo, "Defense Medical Expert."
80. Wamsley and Romo, "Defense Medical Expert."
81. Marc Rowe and Lee Wedlake, "The Carotid Choke: To Sleep, Perchance to Die?," *Journal of Asian Martial Arts* 18, no. 3 (2009): 50–69.
82. Darrell L. Ross and Michael H. Hazlett, "A Prospective Analysis of the Outcomes of Violent Prone Restraint Incidents in Policing," *Forensic Research and Criminology International Journal* 2, no. 1 (2016): 16–24, https://doi.org /10.15406/frcij.2016.02.00040.
83. Ross and Hazlett, "A Prospective Analysis."
84. Donald T. Reay, Corinne L. Fligner, Allan D. Stilwell, and Judy Arnold, "Positional Asphyxia During Law-Enforcement Transport," *American Journal of Forensic Medicine and Pathology* 13, no. 2 (1992): 90–97, https://doi.org/10.1097 /00000433-199206000-00002.
85. B. Paterson, P. Bradley, C. Stark, D. Saddler, D. Leadbetter, and D. Allen, "Deaths Associated with Restraint Use in Health and Social Care in the UK: The Results of a Preliminary Survey," *Journal of Psychiatric and Mental Health Nursing* 10, no. 1 (2003): 3–15, https://doi.org/10.1046/j.1365-2850.2003 .00523.x.

86. Paterson et al., "Deaths Associated with Restraint"; Michael S. Pollanen, David A. Chiasson, James T. Cairns, and James G. Young, "Unexpected Death Related to Restraint for Excited Delirium: A Retrospective Study of Deaths in Police Custody and in the Community," *Canadian Medical Association Journal* 158, no. 12 (1998): 1603–7; Ronald L. O'Halloran and Janice G. Frank, "Asphyxial Death During Prone Restraint Revisited: A Report of 21 Cases," *American Journal of Forensic Medicine and Pathology* 21, no. 1 (2000): 39–52, https://doi.org/10.1097/00000433-200003000-00007; Samuel J. Stratton, Christopher Rogers, Karen Brickett, and Ginger Grunzinski, "Factors Associated with Sudden Death of Individuals Requiring Restraint for Excited Delirium," *American Journal of Emergency Medicine* 19, no. 3 (2001): 187–91, https://doi.org/10.1053/ajem.2001.22665.

87. Stratton et al., "Factors Associated with Sudden Death."

88. Bogel-Burroughs, "George Floyd Showed Signs."

89. Holly Bailey, "George Floyd Died of Low Level of Oxygen, Medical Expert Testifies; Derek Chauvin Kept Knee on His Neck 'Majority of the Time,'" *Washington Post*, April 8, 2021, https://www.washingtonpost.com/nation/2021/04/08/derek-chauvin-trial-2/.

90. Ross and Hazlett, "A Prospective Analysis."

91. Theodore C. Chan, Gary M. Vilke, Tom Neuman, and Jack L. Clausen, "Restraint Position and Positional Asphyxia," *Annals of Emergency Medicine* 30, no. 5 (1997): 578–86, https://doi.org/10.1016/S0196-0644(97)70072-6.

92. Betty A. Michalewicz, Theodore C. Chan, Gary M. Vilke, Susan S. Levy, Tom S. Neuman, and Fred W. Kolkhorst, "Ventilatory and Metabolic Demands During Aggressive Physical Restraint in Healthy Adults," *Journal of Forensic Sciences* 52, no. 1 (2007): 171–75, https://doi.org/10.1111/j.1556-4029.2006.00296.x.

93. Kroll et al., "Applied Force."

94. Mark W. Kroll, G. Keith Still, Tom S. Neuman, Michael A. Graham, and Lanny V. Griffin, "Acute Forces Required for Fatal Compression Asphyxia: A Biomechanical Model and Historical Comparisons," *Medicine, Science, and the Law* 57, no. 2 (2017): 61–68, https://doi.org/10.1177/0025802417695711.

95. Similarly, in these studies, it is impossible to simulate the kind of psychological stress that arrestees may face in real life.

96. O'Halloran and Frank, "Asphyxial Death." In a study with obese participants, however, prone restraint did not cause clinically significant changes in heart function or breathing. See Christian Sloane, Theodore C. Chan, Fred Kolkhorst, Tom Neuman, Edward M. Castillo, and Gary M. Vilke, "Evaluation of the Ventilatory Effects of the Prone Maximum Restraint (PMR) Position on Obese Human Subjects," *Forensic Science International* 237 (2014): 86–89, https://doi.org/10.1016/j.forsciint.2014.01.017.

97. Pollanen et al., "Unexpected Death"; Stratton et al., "Factors Associated with Sudden Death."

98. Pollanen et al., "Unexpected Death"; O'Halloran and Frank, "Asphyxial Death"; Stratton et al., "Factors Associated with Sudden Death."

99. Pollanen et al., "Unexpected Death"; Deborah C. Mash, "Excited Delirium and Sudden Death: A Syndromal Disorder at the Extreme End of the Neuropsychiatric Continuum," *Frontiers in Physiology* 7 (2016): 435–35, https://doi.org/10.3389/fphys.2016.00435.

100. John L. Hick, Stephen W. Smith, and Michael T. Lynch, "Metabolic Acidosis in Restraint-Associated Cardiac Arrest: A Case Series," *Academic Emergency Medicine* 6, no. 3 (1999): 239–43; Alon Steinberg, "Prone Restraint Cardiac Arrest: A Comprehensive Review of the Scientific Literature and an Explanation of the Physiology," *Medicine, Science and the Law* 61, no. 3 (2021): 215–26, https://doi.org/10.1177/0025802420988370.

101. Winter, "Dr. Andrew Baker."

102. Forliti, Karnowski, and Webber, "Medical Examiner Blames Police."

103. BJJ World, "BJJ Knee on Belly," October 4, 2020, https://bjj-world.com/bjj-knee-on-belly/.

104. Rebecca Schroll, Alison Smith, Norman E. McSwain, Jr., John Myers, Kristin Rocchi, Kenji Inaba, Stefano Siboni, et al., "A Multi-Institutional Analysis of Prehospital Tourniquet Use," *Journal of Trauma and Acute Care Surgery* 79, no. 1 (2015): 10–14, https://doi.org/10.1097/TA.0000000000000689.

105. John P. Slevin, Cierra Harrison, Eric Da Silva, and Nathan J. White, "Martial Arts Technique for Control of Severe External Bleeding," *Emergency Medicine Journal* 36, no. 3 (2019): 154–58, https://doi.org/10.1136/emermed-2018-207966.

106. Slevin et al., "Martial Arts."

107. Slevin et al., "Martial Arts."

8. RACE AND MONEY

The quote in the first chapter epigraph comes from Theodore Dalrymple, "Torn Jeans: The Politics of a Fashion Statement," *City Journal*, Autumn 2004, https://www.city-journal.org/html/torn-jeans-12831.html.

The quote in the second chapter epigraph comes from Nishi Chaturvedi and Yoav Ben-Shlomo, "From the Surgery to the Surgeon: Does Deprivation Influence Consultation and Operation Rates?," *British Journal of General Practice* 45, no. 392 (1995): 129–30.

1. Ian J. Wallace, Steven Worthington, David T. Felson, Robert D. Jurmain, Kimberly T. Wren, Heli Maijanen, Robert J. Woods, and Daniel E. Lieberman, "Knee Osteoarthritis Has Doubled in Prevalence Since the Mid-20th Century," *Proceedings of the National Academy of Sciences* 114, no. 35 (2017): 9332–36, https://doi.org/10.1073/pnas.1703856114; Center for Disease Control and Prevention, "Osteoarthritis," July 27, 2020, https://www.cdc.gov/arthritis/basics/osteoarthritis.htm.

2. Reva C. Lawrence, David T. Felson, Charles G. Helmick, Lesley M. Arnold, Hyon Choi, Richard A. Deyo, Sherine Gabriel, et al., "Estimates of the Prevalence

of Arthritis and Other Rheumatic Conditions in the United States: Part II," *Arthritis and Rheumatism* 58, no. 1 (2008): 26–35, https://doi.org/10.1002/art.23176.

3. G. Peat, R. McCarney, and P. Croft, "Knee Pain and Osteoarthritis in Older Adults: A Review of Community Burden and Current Use of Primary Health Care," *Annals of the Rheumatic Diseases* 60, no. 2 (2001): 91–97, https://doi.org/10.1136/ard.60.2.91.

4. R. L. Neame, K. Muir, S. Doherty, and M. Doherty, "Genetic Risk of Knee Osteoarthritis: A Sibling Study," *Annals of the Rheumatic Diseases* 63, no. 9 (2004): 1022–27, https://doi.org/10.1136/ard.2003.014498.

5. Lawrence et al., "Estimates of the Prevalence."

6. Nicole C. Wright, Gail Kershner Riggs, Jeffrey R. Lisse, and Zhao Chen, "Self-Reported Osteoarthritis, Ethnicity, Body Mass Index, and Other Associated Risk Factors in Postmenopausal Women—Results from the Women's Health Initiative," *Journal of the American Geriatrics Society (JAGS)* 56, no. 9 (2008): 1736–43, https://doi.org/10.1111/j.1532-5415.2008.01812.x; Wallace et al., "Knee Osteoarthritis."

7. Neame et al., "Genetic Risk."

8. Tim D. Spector, Flavia Cicuttini, Juliet Baker, John Loughlin, and Deborah Hart, "Genetic Influences on Osteoarthritis in Women: A Twin Study," *BMJ* 312, no. 7036 (1996): 940–43, https://doi.org/10.1136/bmj.312.7036.940.

9. Wei Yu, Melinda Clyne, Muin J. Khoury, and Marta Gwinn, "Phenopedia and Genopedia: Disease-Centered and Gene-Centered Views of the Evolving Knowledge of Human Genetic Associations," *Bioinformatics* (2021), https://phgkb.cdc.gov/PHGKB/phenoPedia.action?firstQuery=Osteoarthritis&cuiID=C0029408&typeSubmit=GO&check=y&which=2&pubOrderType=pubD.

10. Noriyuki Tsumaki, Kazuhiro Tanaka, Eri Arikawa-Hirasawa, Takanobu Nakase, Tomoatsu Kimura, J. Terrig Thomas, Takahiro Ochi, Frank P. Luyten, and Yoshihiko Yamada, "Role of CDMP-1 in Skeletal Morphogenesis: Promotion of Mesenchymal Cell Recruitment and Chondrocyte Differentiation," *Journal of Cell Biology* 144, no. 1 (1999): 161–73, https://doi.org/10.1083/jcb.144.1.161.

11. Paul Buxton, Christopher Edwards, Charles W. Archer, and Philippa Francis-West, "Growth/Differentiation Factor-5 (GDF-5) and Skeletal Development," *Journal of Bone and Joint Surgery* 83-A, Suppl. 1 (2001): S23–S30.

12. Evangelos Evangelou, Kay Chapman, Ingrid Meulenbelt, Fotini B. Karassa, John Loughlin, Andrew Carr, Michael Doherty, et al., "Large-Scale Analysis of Association Between *GDF5* and *FRZB* Variants and Osteoarthritis of the Hip, Knee, and Hand," *Arthritis and Rheumatism* 60, no. 6 (2009): 1710–21, https://doi.org/10.1002/art.24524.

13. Eleni Zengini, Konstantinos Hatzikotoulas, Ioanna Tachmazidou, Julia Steinberg, Fernando P. Hartwig, Lorraine Southam, Sophie Hackinger, et al., "Genome-Wide Analyses Using UK Biobank Data Provide Insights Into the Genetic Architecture of Osteoarthritis," *Nature Genetics* 50, no. 4 (2018): 549–58, https://doi.org/10.1038/s41588-018-0079-y.

14. Ana M. Valdes and Tim D. Spector, "Genetic Epidemiology of Hip and Knee Osteoarthritis," *Nature Reviews: Rheumatology* 7, no. 1 (2011): 23–32, https://doi.org/10.1038/nrrheum.2010.191.
15. Giorgio Sirugo, Scott M. Williams, and Sarah A. Tishkoff, "The Missing Diversity in Human Genetic Studies," *Cell* 177, no. 1 (2019): 26–31, https://doi.org/10.1016/j.cell.2019.02.048.
16. Susanne B. Haga, "Impact of Limited Population Diversity of Genome-Wide Association Studies," *Genetics in Medicine* 12, no. 2 (2010): 81–84, https://doi.org/10.1097/GIM.0b013e3181ca2bbf.
17. Alice B. Popejoy and Stephanie M. Fullerton, "Genomics Is Failing on Diversity," *Nature* 538, no. 7624 (2016): 161–64, https://doi.org/10.1038/538161a.
18. Haga, "Impact of Limited Population Diversity."
19. Popejoy and Fullerton, "Genomics Is Failing."
20. Darcell P. Scharff, Katherine J. Mathews, Pamela Jackson, Jonathan Hoffsuemmer, Emeobong Martin, and Dorothy Edwards, "More Than Tuskegee: Understanding Mistrust About Research Participation," *Journal of Health Care for the Poor and Underserved* 21, no. 3 (2010): 879–97, https://doi.org/10.1353/hpu.0.0323.
21. Popejoy and Fullerton, "Genomics Is Failing."
22. Maulana Bachtiar and Caroline G. L. Lee, "Genetics of Population Differences in Drug Response," *Current Genetic Medicine Reports* 1, no. 3 (2013): 162–70, https://doi.org/10.1007/s40142-013-0017-3.
23. Evangelou et al., "Large-Scale Analysis."
24. Yoshinari Miyamoto, Akihiko Mabuchi, Dongquan Shi, Toshikazu Kubo, Yoshio Takatori, Susumu Saito, Mikihiro Fujioka, et al., "A Functional Polymorphism in the 5` UTR of GDF5 Is Associated with Susceptibility to Osteoarthritis," *Nature Genetics* 39, no. 4 (2007): 529–33, https://doi.org/10.1038/2005.
25. Miyamoto et al., "A Functional Polymorphism."
26. Hideki Kizawa, Ikuyo Kou, Aritoshi Iida, Akihiro Sudo, Yoshinari Miyamoto, Akira Fukuda, Akihiko Mabuchi, et al., "An Aspartic Acid Repeat Polymorphism in Asporin Inhibits Chondrogenesis and Increases Susceptibility to Osteoarthritis," *Nature Genetics* 37, no. 2 (2005): 138–44, https://doi.org/10.1038/ng1496.
27. Kizawa et al., "An Aspartic Acid."
28. Qing Jiang, Dongquan Shi, Long Yi, Shiro Ikegawa, Yong Wang, Takahiro Nakamura, Di Qiao, Cheng Liu, and Jin Dai, "Replication of the Association of the Aspartic Acid Repeat Polymorphism in the Asporin Gene with Knee-Osteoarthritis Susceptibility in Han Chinese," *Journal of Human Genetics* 51, no. 12 (2006): 1068–72, https://doi.org/10.1007/s10038-006-0065-6.
29. U. Atif, A. Philip, J. Aponte, E. M. Woldu, S. Brady, V. B. Kraus, J. M. Jordan, et al., "Absence of Association of Asporin Polymorphisms and Osteoarthritis Susceptibility in US Caucasians," *Osteoarthritis and Cartilage* 16, no. 10 (2008): 1174–77, https://doi.org/10.1016/j.joca.2008.03.007; Zehra Mustafa,

Barbara Dowling, Kay Chapman, Janet S. Sinsheimer, Andrew Carr, and John Loughlin, "Investigating the Aspartic Acid (D) Repeat of Asporin as a Risk Factor for Osteoarthritis in a UK Caucasian Population," *Arthritis and Rheumatism* 52, no. 11 (2005): 3502–6, https://doi.org/10.1002/art.21399.

30. Sirugo, Williams, and Tishkoff, "The Missing Diversity."

31. Sirugo, Williams, and Tishkoff, "The Missing Diversity."

32. Sirugo, Williams, and Tishkoff, "The Missing Diversity"; David R. Williams, Selina A. Mohammed, Jacinta Leavell, and Chiquita Collins, "Race, Socioeconomic Status, and Health: Complexities, Ongoing Challenges, and Research Opportunities," *Annals of the New York Academy of Sciences* 1186, no. 1 (2010): 69–101, https://doi.org/10.1111/j.1749-6632.2009.05339.x.

33. Popejoy and Fullerton, "Genomics Is Failing."

34. Popejoy and Fullerton, "Genomics Is Failing."

35. Lawrence et al., "Estimates of the Prevalence"; Joanne M. Jordan, Charles G. Helmick, Jordan B. Renner, Gheorghe Luta, Anca D. Dragomir, Janice Woodard, Fang Fang, et al., "Prevalence of Knee Symptoms and Radiographic and Symptomatic Knee Osteoarthritis in African Americans and Caucasians: The Johnston County Osteoarthritis Project," *Journal of Rheumatology* 34, no. 1 (2007): 172–80.

36. Jordan et al., "Prevalence of Knee Symptoms."

37. Zengini et al., "Genome-Wide Analyses."

38. Kathryn A. Thompson, Ellen L. Terry, Kimberly T. Sibille, Ethan W. Gossett, Erin N. Ross, Emily J. Bartley, Toni L. Glover, et al., "At the Intersection of Ethnicity/Race and Poverty: Knee Pain and Physical Function," *Journal of Racial and Ethnic Health Disparities* 6, no. 6 (2019): 1131–43, https://doi.org/10.1007/s40615-019-00615-7.

39. Thompson et al., "At the Intersection."

40. For people who have osteoarthritis in only some parts of the knee, a partial knee replacement surgery may be performed to resurface and replace those parts that are damaged.

41. Michael Murphy, Simon Journeaux, and Trevor Russell, "High-Flexion Total Knee Arthroplasty: A Systematic Review," *International Orthopaedics* 33, no. 4 (2009): 887–93, https://doi.org/10.1007/s00264-009-0774-5.

42. Murphy, Journeaux, and Russell, "High-Flexion Total Knee."

43. Jonathan Skinner, Weiping Zhou, and James Weinstein, "The Influence of Income and Race on Total Knee Arthroplasty in the United States," *Journal of Bone and Joint Surgery* 88, no. 10 (2006): 2159–66, https://doi.org/10.2106/JBJS.E.00271; A. M. Cavanaugh, M. J. Rauh, C. A. Thompson, J. Alcaraz, W. M. Mihalko, C. E. Bird, C. B. Eaton, et al., "Racial and Ethnic Disparities in Utilization of Total Knee Arthroplasty Among Older Women," *Osteoarthritis and Cartilage* 27, no. 12 (2019): 1746–54, https://doi.org/10.1016/j.joca.2019.07.015; Dorothy D. Dunlop, Larry M. Manheim, Jing Song, Min-Woong Sohn, Joseph M. Feinglass, Huan J. Chang, and Rowland W. Chang, "Age and

Racial/Ethnic Disparities in Arthritis-Related Hip and Knee Surgeries," *Medical Care* 46, no. 2 (2008): 200–8, https://doi.org/10.1097/MLR.0b013e31815cecd8; Nicholas Steel, Allan Clark, Iain A. Lang, Robert B. Wallace, and David Melzer, "Racial Disparities in Receipt of Hip and Knee Joint Replacements Are Not Explained by Need: The Health and Retirement Study 1998–2004," *Journals of Gerontology. Series A, Biological Sciences and Medical Sciences* 63, no. 6 (2008): 629–34, https://doi.org/10.1093/gerona/63.6.629.

44. Skinner, Zhou, and Weinstein, "The Influence of Income."

45. Steel et al., "Racial Disparities."

46. Dunlop et al., "Age and Racial/Ethnic Disparities."

47. Cavanaugh et al., "Racial and Ethnic Disparities."

48. Candace H. Feldman, Yan Dong, Jeffrey N. Katz, Laurel A. Donnell-Fink, and Elena Losina, "Association Between Socioeconomic Status and Pain, Function and Pain Catastrophizing at Presentation for Total Knee Arthroplasty," *BMC Musculoskeletal Disorders* 16, no. 1 (2015), https://doi.org/10.1186/s12891-015 -0475-8; Somnath Saha, Jose J. Arbelaez, and Lisa A. Cooper, "Patient-Physician Relationships and Racial Disparities in the Quality of Health Care," *American Journal of Public Health* 93, no. 10 (2003): 1713–19, https://doi.org/10.2105 /AJPH.93.10.1713.

49. Said A. Ibrahim, Laura A. Siminoff, Christopher J. Burant, and C. Kent Kwoh, "Variation in Perceptions of Treatment and Self-Care Practices in Elderly with Osteoarthritis: A Comparison Between African American and White Patients," *Arthritis and Rheumatism* 45, no. 4 (2001): 340–45.

50. Susan M. Goodman, Lisa A. Mandl, Michael L. Parks, Meng Zhang, Kelly R. McHugh, Yuo-Yu Lee, Joseph T. Nguyen, et al., "Disparities in TKA Outcomes: Census Tract Data Show Interactions Between Race and Poverty," *Clinical Orthopaedics and Related Research* 474, no. 9 (2016): 1986–95, https:// doi.org/10.1007/s11999-016-4919-8.

51. Anne R. Bass, Kelly McHugh, Kara Fields, Rie Goto, Michael L. Parks, and Susan M. Goodman, "Higher Total Knee Arthroplasty Revision Rates Among United States Blacks Than Whites: A Systematic Literature Review and Meta-Analysis," *Journal of Bone and Joint Surgery* 98, no. 24 (2016): 2103–8, https:// doi.org/10.2106/JBJS.15.00976.

52. Bass et al., "Higher Total Knee Arthroplasty."

53. Karl Kronebusch, Bradford H. Gray, and Mark Schlesinger, "Explaining Racial/Ethnic Disparities in Use of High-Volume Hospitals: Decision-Making Complexity and Local Hospital Environments," *Inquiry: Journal of Health Care Organization, Provision, and Financing* 51, no. 1 (2014): 1–21, https:// doi.org/10.1177/0046958014545575.

54. Janet K. Freburger, George M. Holmes, Li-Jung E. Ku, Malcolm P. Cutchin, Kendra Heatwole-Shank, and Lloyd J. Edwards, "Disparities in Post–Acute Rehabilitation Care for Joint Replacement," *Arthritis Care and Research (2010)* 63, no. 7 (2011): 1020–30, https://doi.org/10.1002/acr.20477.

55. Feldman et al., "Association Between Socioeconomic Status."
56. Said A. Ibrahim, Laura A. Siminoff, Christopher J. Burant, and C. Kent Kwoh, "Differences in Expectations of Outcome Mediate African American/White Patient Differences in 'Willingness' to Consider Joint Replacement," *Arthritis and Rheumatism* 46, no. 9 (2002): 2429–35, https://doi.org/10.1002/art.10494.
57. Maria E. Suarez-Almazor, Julianne Souchek, P. Adam Kelly, Kimberly O'Malley, Margaret Byrne, Marsha Richardson, and Chong Pak, "Ethnic Variation in Knee Replacement: Patient Preferences or Uninformed Disparity?," *Archives of Internal Medicine* 165, no. 10 (2005): 1117–24, https://doi.org/10.1001/archinte.165.10.1117.
58. Boonsin Tangtrakulwanich, Virasakdi Chongsuvivatwong, and Alan F. Geater, "Habitual Floor Activities Increase Risk of Knee Osteoarthritis," *Clinical Orthopaedics and Related Research* 454 (2007): 147–54, https://doi.org/10.1097/01.blo.0000238808.72164.1d; Yuqing Zhang, Ling Xu, Michael C. Nevitt, Piran Aliabadi, Wei Yu, Mingwei Qin, Li-Yung Lui, and David T. Felson, "Comparison of the Prevalence of Knee Osteoarthritis Between the Elderly Chinese Population in Beijing and Whites in the United States: The Beijing Osteoarthritis Study," *Arthritis and Rheumatism* 44, no. 9 (2001): 2065–71.
59. Sumaiyah Mat, Mohamad Hasif Jaafar, Chin Teck Ng, Sargunan Sockalingam, Jasmin Raja, Shahrul Bahyah Kamaruzzaman, Ai-Vyrn Chin, et al., "Ethnic Differences in the Prevalence, Socioeconomic and Health Related Risk Factors of Knee Pain and Osteoarthritis Symptoms in Older Malaysians," *PLoS One* 14, no. 11 (2019): e0225075. Methods for diagnosis differ across studies. Some studies use clinical symptoms; others use radiographic evidence.
60. Hyung Joon Cho, Vivek Morey, Jong Yeal Kang, Ki Woong Kim, and Tae Kyun Kim, "Prevalence and Risk Factors of Spine, Shoulder, Hand, Hip, and Knee Osteoarthritis in Community-Dwelling Koreans Older Than Age 65 Years," *Clinical Orthopaedics and Related Research* 473, no. 10 (2015): 3307–14, https://doi.org/10.1007/s11999-015-4450-3.
61. Tangtrakulwanich, Chongsuvivatwong, and Geater, "Habitual Floor Activities."
62. Tangtrakulwanich, Chongsuvivatwong, and Geater, "Habitual Floor Activities."
63. Yuqing Zhang, David J. Hunter, Michael C. Nevitt, Ling Xu, Jingbo Niu, Li-Yung Lui, Wei Yu, Piran Aliabadi, and David T. Felson, "Association of Squatting with Increased Prevalence of Radiographic Tibiofemoral Knee Osteoarthritis: The Beijing Osteoarthritis Study," *Arthritis and Rheumatism* 50, no. 4 (2004): 1187–92, https://doi.org/10.1002/art.20127; Young Hoo Kim, Shuichi Matsuda, and Tae Kyun Kim, "Clinical Faceoff: Do We Need Special Strategies for Asian Patients with TKA?," *Clinical Orthopaedics and Related Research* 474, no. 5 (2016): 1102–7, https://doi.org/10.1007/s11999-016-4716-4.
64. Zhang et al., "Association of Squatting"; Michihiko Fukunaga and Kentaro Morimoto, "Calculation of the Knee Joint Force at Deep Squatting and Kneeling," *Journal of Biomechanical Science and Engineering* 10, no. 4 (2015): 15-00452-15-52, https://doi.org/10.1299/jbse.15-00452.

65. Mohamed Samir Hefzy, Brian P. Kelly, T. Derek V. Cooke, Abdel Mohsen Al-Baddah, and Laurie Harrison, "Knee Kinematics in-Vivo of Kneeling in Deep Flexion Examined by Bi-Planar Radiographs," *Biomedical Sciences Instrumentation* 33 (1997): 453–58.

66. Shinichi Demura and Masanobu Uchiyama, "Effect of Japanese Sitting Style (Seiza) on the Center of Foot Pressure after Standing," *Journal of Physiological Anthropology and Applied Human Science* 24, no. 2 (2005): 167–73, https://doi.org/10.2114/jpa.24.167; Fukunaga and Morimoto, "Calculation of the Knee Joint."

67. Susan J. Mulholland and URS P. Wyss, "Activities of Daily Living in Non-Western Cultures: Range of Motion Requirements for Hip and Knee Joint Implants," *International Journal of Rehabilitation Research* 24, no. 3 (2001): 191–98, https://doi.org/10.1097/00004356-200109000-00004.

68. Shreyasee Amin, Joyce Goggins, Jingbo Niu, Ali Guermazi, Mikayel Grigoryan, David J. Hunter, Harry K. Genant, and David T. Felson, "Occupation-Related Squatting, Kneeling, and Heavy Lifting and the Knee Joint: A Magnetic Resonance Imaging-Based Study in Men," *Journal of Rheumatology* 35, no. 8 (2008): 1645–49.

69. Warapat Virayavanich, Hamza Alizai, Thomas Baum, Lorenzo Nardo, Michael C. Nevitt, John A. Lynch, Charles E. McCulloch, and Thomas M. Link, "Association of Frequent Knee Bending Activity with Focal Knee Lesions Detected with 3T Magnetic Resonance Imaging: Data from the Osteoarthritis Initiative," *Arthritis Care and Research (2010)* 65, no. 9 (2013): 1441–48, https://doi.org/10.1002/acr.22017; Amin et al., "Occupation-Related Squatting"; David Coggon, Peter Croft, Samantha Kellingray, David Barrett, Magnus McLaren, and Cyrus Cooper, "Occupational Physical Activities and Osteoarthritis of the Knee," *Arthritis and Rheumatism* 43, no. 7 (2000): 1443–49.

70. W. Andrew Hodge, Melinda K. Harman, and Scott A. Banks, "Patterns of Knee Osteoarthritis in Arabian and American Knees," *Journal of Arthroplasty* 24, no. 3 (2009): 448–53, https://doi.org/10.1016/j.arth.2007.12.012.

71. Tangtrakulwanich, Chongsuvivatwong, and Geater, "Habitual Floor Activities."

72. Zhang et al., "Association of Squatting."

73. David A. Raichlen, Herman Pontzer, Theodore W. Zderic, Jacob A. Harris, Audax Z. P. Mabulla, Marc T. Hamilton, and Brian M. Wood, "Sitting, Squatting, and the Evolutionary Biology of Human Inactivity," *Proceedings of the National Academy of Sciences* 117, no. 13 (2020): 7115–21, https://doi.org/10.1073/pnas.1911868117.

74. Zimmer Biomet, "NexGen®CR-Flex and LPS-Flex Knees: Design Rationale," 2016, https://www.zimmerbiomet.com/content/dam/zimmer-biomet/medical-professionals/knee/nexgen-complete-knee-solution-legacy-knee-posterior-stabilized/nexgen-cr-flex-and-lps-flex-knees-design-rationale.pdf.

75. Zimmer Biomet, "NexGen®CR-Flex."

76. Zimmer Biomet, "NexGen®CR-Flex."

77. Zimmer Biomet, "NexGen®CR-Flex."

78. Hsuan-Ti Huang, Jiing Yuan Su, and Gwo-Jaw Wang, "The Early Results of High-Flex Total Knee Arthroplasty: A Minimum of 2 Years of Follow-Up," *Journal of Arthroplasty* 20, no. 5 (2005): 674–79, https://doi.org/10.1016/j.arth.2004.09.053.

79. Aree Tanavalee, Srihatach Ngarmukos, Saran Tantavisut, and Arak Limtrakul, "High-Flexion TKA in Patients with a Minimum of 120 Degrees of Pre-Operative Knee Flexion: Outcomes at Six Years of Follow-Up," *International Orthopaedics* 35, no. 9 (2010): 1321–26, https://doi.org/10.1007/s00264-010-1140-3.

80. Hyuk-Soo Han and Seung-Baik Kang, "Does High-Flexion Total Knee Arthroplasty Allow Deep Flexion Safely in Asian Patients?," *Clinical Orthopaedics and Related Research* 471, no. 5 (2013): 1492–97, https://doi.org/10.1007/s11999-012-2628-5.

81. Zimmer Biomet, "NexGen®CR-Flex"; Masahiro Kurosaka, Shinichi Yoshiya, Kiyonori Mizuno, and Tetsuji Yamamoto, "Maximizing Flexion After Total Knee Arthroplasty: The Need and the Pitfalls," *Journal of Arthroplasty* 17, no. 4 (2002): 59–62, https://doi.org/10.1054/arth.2002.32688.

82. Han and Kang, "Does High-Flexion."

83. T. H. Kim, D. H. Lee, and S. I. Bin, "The NexGen LPS-Flex to the Knee Prosthesis at a Minimum of Three Years," *Journal of Bone and Joint Surgery* 90, no. 10 (2008): 1304–10, https://doi.org/10.1302/0301-620X.90B10.21050.

84. Bum-Sik Lee, Jong-Won Chung, Jong-Min Kim, Kyung-Ah Kim, and Seong-Il Bin, "High-Flexion Prosthesis Improves Function of TKA in Asian Patients Without Decreasing Early Survivorship," *Clinical Orthopaedics and Related Research* 471, no. 5 (2012): 1504–11, https://doi.org/10.1007/s11999-012-2661-4; Tanavalee et al., "High-Flexion TKA."

85. Kim et al., "The NexGen LPS-Flex."

86. Howard Hall, "'Distressed' Look Not Well-Dressed but Distressing," *Bradenton Herald* (Bradenton, FL), September 11, 1987,

87. Jake Woolf, "Everything You Need to Know Before Buying Ripped Jeans," *GQ*, March 24, 2016, https://www.gq.com/story/ripped-jeans-guide-wearing-gq.

88. Woolf, "Everything You Need."

89. James Sullivan, *Jeans: A Cultural History of an American Icon* (New York: Gotham Books, 2006).

90. Graham Marsh and Paul Trynka, *Denim: From Cowboys to Catwalks: A Visual History of the World's Most Legendary Fabric* (London: Aurum, 2002).

91. Marsh and Trynka, *Denim: From Cowboys.*

92. Sullivan, *Jeans: A Cultural History.*

93. James B. Salazar, "Fashioning the Historical Body: The Political Economy of Denim," *Social Semiotics* 20, no. 3 (2010): 293–308, https://doi.org/10.1080/10350331003722851.

94. Sullivan, *Jeans: A Cultural History,* 204.

95. Sarah Rainey, "Why Is Everyone Wearing Ripped Jeans . . . And Why Do They Cost More Than Ones with No Holes?," *Daily Mail,* July 24, 2017, https://www.dailymail.co.uk/femail/article-4726702/Why-wearing-ripped-jeans.html.

96. Catrin Rosquist, *Still Fashion Victims? Monitoring a Ban on Sandblasted Denim*, Fair Trade Center, 2012, https://fairaction.se/wp-content/uploads/2015/06/English-version_Still-fashion-victims-Monitoring-a-ban-on-sandblasted-denim-2012_0.pdf.

97. Clean Clothes Campaign, *Deadly Denim: Sandblasting in the Bangladesh Garment Industry*, 2012, https://archive.cleanclothes.org/resources/publications/ccc-deadly-denim.pdf/view.

98. Christopher Riddselius, *Fashion Victims: A Report on Sand Blasted Denim*, Swedish Fair Trade Centre and the Clean Clothes Campaign, 2010, https://archive.cleanclothes.org/resources/national-cccs/fashion-victims-a-report-on-sandblasted-denim/view.

99. The following account of Turkish workers is based on M. Akgun, O. Araz, I. Akkurt, A. Eroglu, F. Alper, L. Saglam, A. Mirici, M. Gorguner, and B. Nemery, "An Epidemic of Silicosis Among Former Denim Sandblasters," *European Respiratory Journal* 32, no. 5 (2008): 1295–303, https://doi.org/10.1183/09031936.00093507.

100. See Metin Akgun, Metin Gorguner, Mehmet Meral, Atila Turkyilmaz, Fazli Erdogan, Leyla Saglam, and Arzu Mirici, "Silicosis Caused by Sandblasting of Jeans in Turkey: A Report of Two Concomitant Cases," *Journal of Occupational Health* 47, no. 4 (2005): 346–49, https://doi.org/10.1539/joh.47.346; Metin Akgun, Arzu Mirici, Elif Yilmazel Ucar, Mecit Kantarci, Omer Araz, and Metin Gorguner, "Silicosis in Turkish Denim Sandblasters," *Occupational Medicine* 56, no. 8 (2006): 554–58, https://doi.org/10.1093/occmed/kql094; Akgun et al., "An Epidemic of Silicosis."

101. Rosquist, *Still Fashion Victims?*

102. Akgun et al., "Silicosis in Turkish."

103. Akgun et al., "Silicosis in Turkish"; Akgun et al., "An Epidemic of Silicosis."

104. Akgun et al., "An Epidemic of Silicosis."

105. Riddselius, *Fashion Victims.*

106. Clean Clothes Campaign, *Deadly Denim.*

107. Heather Somerville, "Retailer Sandblasting Bans Have Changed Little in the Garment Industry," *Mercury News*, October 28, 2013, https://www.mercurynews.com/2013/10/28/retailer-sandblasting-bans-have-changed-little-in-the-garment-industry-2/; Clean Cloth Campaign, *Deadly Denim.*

108. Clean Cloth Campaign, *Deadly Denim.*

109. Somerville, "Retailer Sandblasting Bans."

110. Andrew Brooks, *Clothing Poverty: The Hidden World of Fast Fashion and Second-Hand Clothes* (London: Zed Books, 2015).

111. Brooks, *Clothing Poverty.*

112. Brooks, *Clothing Poverty.*

113. David Heiden, "Clothes, Poverty, and the Global Economy," *Western Journal of Medicine* 175, no. 1 (2001): 72–72, https://doi.org/10.1136/ewjm.175.1.72; Brooks, *Clothing Poverty.*

114. Brooks, *Clothing Poverty.*

115. Brooks, *Clothing Poverty.*

Bibliography

1. KNEES BEFORE THE BRAIN

Begun, David R. "The Earliest Hominins: Is Less More?" *Science* 303, no. 5663 (2004): 1478–80.

Bennett, M. R., S. C. Reynolds, S. A. Morse, and M. Budka. "Laetoli's Lost Tracks: 3D Generated Mean Shape and Missing Footprints." *Scientific Reports* 6, no. 1 (2016): 21916. https://doi.org/10.1038/srep21916.

Berger, Lee R., Darryl J. De Ruiter, Steven E. Churchill, Peter Schmid, Kristian J. Carlson, Paul H. G. M. Dirks, and Job M. Kibii. "*Australopithecus sediba*: A New Species of *Homo*-Like Australopith from South Africa." *Science* 328, no. 5975 (2010): 195–204. https://doi.org/10.1126/science.1184944.

Berger, Lee R., John Hawks, Darryl J. de Ruiter, Steven E. Churchill, Peter Schmid, Lucas K. Delezene, Tracy L. Kivell, et al. "*Homo naledi*, A New Species of the Genus *Homo* from the Dinaledi Chamber, South Africa." *eLife* 4 (2015). https://doi.org/10.7554/eLife.09560.

Bertelsman, Tim, and Brandon Steele. "Foot Hyperpronation." Illinois Chiropractic Society. Accessed November 11, 2021. https://ilchiro.org/foot-hyperpronation/.

Cerling, Thure E., Yang Wang, and Jay Quade. "Expansion of C4 Ecosystems as an Indicator of Global Ecological Change in the Late Miocene." *Nature* 361, no. 6410 (1993): 344–45. https://doi.org/10.1038/361344a0.

Crompton, Robin H. "Making the Case for Possible Hominin Footprints from the Late Miocene (c. 5.7 Ma) of Crete?" *Proceedings of the Geologists' Association* 128, nos. 5–6 (2017): 692–93. https://doi.org/10.1016/j.pgeola.2017.09.001.

Crompton, Robin H., Todd C. Pataky, Russell Savage, Kristiaan D'Août, Matthew R. Bennett, Michael H. Day, Karl Bates, Sarita Morse, and William I. Sellers. "Human-Like External Function of the Foot, and Fully Upright Gait, Confirmed in the 3.66 Million Year Old Laetoli Hominin Footprints by Topographic Statistics, Experimental Footprint-Formation and Computer Simulation." *Journal of the Royal Society, Interface* 9, no. 69 (2012): 707–19. https://doi.org/10.1098/rsif.2011.0258.

Deino, Alan L. "40ar/39ar Dating of Laetoli, Tanzania." In *Paleontology and Geology of Laetoli: Human Evolution in Context.* Vol. 1, *Geology, Geochronology,*

Paleoecology and Paleoenvironment, ed. Terry Harrison, 77–97. Dordrecht: Springer Netherlands, 2011.

DeSilva, Jeremy. *First Steps: How Walking Upright Made Us Human*. New York: HarperCollins, 2021.

DeSilva, Jeremy M., Kenneth G. Holt, Steven E. Churchill, Kristian J. Carlson, Christopher S. Walker, Bernhard Zipfel, and Lee R. Berger. "The Lower Limb and Mechanics of Walking in *Australopithecus sediba*." *Science* 340, no. 6129 (2013): 1232999. https://doi.org/10.1126/science.1232999.

Détroit, Florent, Armand Salvador Mijares, Julien Corny, Guillaume Daver, Clément Zanolli, Eusebio Dizon, Emil Robles, Rainer Grün, and Philip J. Piper. "A New Species of *Homo* from the Late Pleistocene of the Philippines." *Nature* 568, no. 7751 (2019): 181–86. https://doi.org/10.1038/s41586-019-1067-9.

Gibbens, Sarah. "Why This Gorilla Prefers to Walk Upright." *National Geographic*, March 19, 2018. https://www.nationalgeographic.com/news/2018/03/gorilla -walking-upright-bipedal-philadelphia-zoo-spd/.

Gibbons, Ann. "A New Kind of Ancestor: Ardipithecus Unveiled." *Science* 326, no. 5949 (2009): 36–40. https://doi.org/10.1126/science.326.5949.36.

"Gibbons." *National Geographic*. Accessed November 11, 2021. https://www .nationalgeographic.com/animals/mammals/group/gibbons/.

Gierliński, Gerard D., Grzegorz Niedźwiedzki, Martin G. Lockley, Athanassios Athanassiou, Charalampos Fassoulas, Zofia Dubicka, Andrzej Boczarowski, Matthew R. Bennett, and Per Erik Ahlberg. "Possible Hominin Footprints from the Late Miocene (c. 5.7 Ma) of Crete?" *Proceedings of the Geologists' Association* 128, nos. 5–6 (2017): 697–710. https://doi.org/10.1016/j.pgeola.2017.07.006.

Gordon, Adam D., David J. Green, and Brian G. Richmond. "Strong Postcranial Size Dimorphism in *Australopithecus afarensis*: Results from Two New Resampling Methods for Multivariate Data Sets with Missing Data." *American Journal of Physical Anthropology* 135, no. 3 (2008): 311–28. https://doi.org/10.1002 /ajpa.20745.

"Gorilla Strolls on Hind Legs." NBC News, January 27, 2011. http://www.nbcnews .com/id/41292533/ns/technology_and_science-science/t/gorilla-strolls-hind -legs/#.XrBe1mXQguX.

Harmand, Sonia, Jason E. Lewis, Craig S. Feibel, Christopher J. Lepre, Sandrine Prat, Arnaud Lenoble, Xavier Boës, et al. "3.3-Million-Year-Old Stone Tools from Lomekwi 3, West Turkana, Kenya." *Nature* 521, no. 7552 (2015): 310–15. https://doi.org/10.1038/nature14464.

Hart, Donna, and Robert Sussman. *Man the Hunted: Primates, Predators, and Human Evolution*. Boulder, CO: Westview Press, 2009.

Hayama, S., M. Nakatsukasa, and Y. Kunimatsu. "Monkey Performance: The Development of Bipedalism in Trained Japanese Monkeys." *Acta Anat Nippon* 67, no. 3 (1992): 169–85.

Hewes, Gordon W. "Food Transport and the Origin of Hominid Bipedalism." *American Anthropologist* 63, no. 4 (1961): 687–710. https://doi.org/10.1525/aa .1961.63.4.02a00020.

——. "Hominid Bipedalism: Independent Evidence for the Food-Carrying Theory." *Science* 146, no. 3642 (1964): 416–18. https://doi.org/10.1126/science.146 .3642.416.

Hunt, Kevin D. "The Postural Feeding Hypothesis: An Ecological Model for the Evolution of Bipedalism." *South African Journal of Science* 92, no. 2 (1996): 77–90.

Isaac, Glynn L. "Models of Human Evolution." *Science* 217, no. 4557 (1982): 295.

Jablonski, Nina G., and George Chaplin. "Origin of Habitual Terrestrial Bipedalism in the Ancestor of the Hominidae." *Journal of Human Evolution* 24, no. 4 (1993): 259–80. https://doi.org/10.1006/jhev.1993.1021.

Janković, Ivor. "Certain Medical Problems Resulting from Evolutionary Processes: Bipedalism as an Example." *Periodicum Biologorum* 117, no. 1 (2015): 17–26.

Johanson, Donald C., and Maitland Edey. *Lucy: The Beginnings of Humankind.* New York: Simon & Schuster, 1981.

Johanson, Donald. C., Lovejoy C. Owen, A. H. Burstein, and K. G. Heiple. "Functional Implications of the Afar Knee Joint." *American Journal of Physical Anthropology* 44, no. 1 (1976): 188.

Johanson, Donald C., and Kate Wong. *Lucy's Legacy: The Quest for Human Origins.* New York: Three Rivers Press, 2009.

Kimbel, William H., and Lucas K. Delezene. "'Lucy' Redux: A Review of Research on *Australopithecus afarensis.*" *American Journal of Physical Anthropology* 140, no. S49 (2009): 2–48. https://doi.org/10.1002/ajpa.21183.

Latimer, Bruce. "The Perils of Being Bipedal." *Annals of Biomedical Engineering* 33, no. 1 (2005): 3–6. https://doi.org/10.1007/s10439-005-8957-8.

Leigh, Steven R., and Brian T. Shea. "Ontogeny and the Evolution of Adult Body Size Dimorphism in Apes." *American Journal of Primatology* 36, no. 1 (1995): 37–60. https://doi.org/10.1002/ajp.1350360104.

Lockwood, Charles A., Brian G. Richmond, William L. Jungers, and William H. Kimbel. "Randomization Procedures and Sexual Dimorphism in *Australopithecus afarensis.*" *Journal of Human Evolution* 31, no. 6 (1996): 537–48. https://doi .org/10.1006/jhev.1996.0078.

Lovejoy, C. Owen. "Evolution of Human Walking." *Scientific American* 259, no. 5 (1988): 118–25. https://doi.org/10.1038/scientificamerican1188-118.

——. "The Natural History of Human Gait and Posture." *Gait and Posture* 25, no. 3 (2007): 325–41. https://doi.org/10.1016/j.gaitpost.2006.05.001.

——. "The Origin of Man." *Science* 211, no. 4480 (1981): 341–50.

——. "Reexamining Human Origins in Light of *Ardipithecus ramidus.*" *Science* 326, no. 5949 (2009): 74, 74e1–e8. https://doi.org/10.1126/science.1175834.

Niemitz, Carsten. "The Evolution of the Upright Posture and Gait—a Review and a New Synthesis." *Die Naturwissenschaften* 97, no. 3 (2010): 241–63. https://doi .org/10.1007/s00114-009-0637-3.

Philadelphia Zoo. "Snacking on the Run." March 5, 2018. https://www.facebook .com/watch/?v=10155414572887934

Raichlen, David A., Adam D. Gordon, William E. H. Harcourt-Smith, Adam D. Foster, and Wm. Randall Haas Jr. "Laetoli Footprints Preserve Earliest Direct

Evidence of Human-Like Bipedal Biomechanics." *PLoS One* 5, no. 3 (2010): e9769. https://doi.org/10.1371/journal.pone.0009769.

Rak, Yoel. "Lucy's Pelvic Anatomy: Its Role in Bipedal Gait." *Journal of Human Evolution* 20, no. 4 (1991): 283–90. https://doi.org/10.1016/0047-2484(91)90011-J.

Reno, Philip L., Richard S. Meindl, Melanie A. McCollum, and Owen Lovejoy. "Sexual Dimorphism in *Australopithecus afarensis* Was Similar to That of Modern Humans." *Proceedings of the National Academy of Sciences of the United States* 100, no. 16 (2003): 9404–409. https://doi.org/10.1073/pnas.1133180100.

Richmond, Brian G., and William L. Jungers. "*Orrorin tugenensis* Femoral Morphology and the Evolution of Hominin Bipedalism." *Science* 319, no. 5870 (2008): 1662–65. https://doi.org/10.1126/science.1154197.

Schacht, Ryan, and Karen L. Kramer. "Are We Monogamous? A Review of the Evolution of Pair-Bonding in Humans and Its Contemporary Variation Cross-Culturally." *Frontiers in Ecology and Evolution* 7 (2019). https://doi.org/10.3389/fevo.2019.00230.

Semaw, S., P. Renne, J. W. K. Harris, C. S. Feibel, R. L. Bernor, N. Fesseha, and K. Mowbray. "2.5-Million-Year-Old Stone Tools from Gona, Ethiopia." *Nature* 385, no. 6614 (1997): 333–36. https://doi.org/10.1038/385333a0.

Sockol, Michael D., David A. Raichlen, and Herman Pontzer. "Chimpanzee Locomotor Energetics and the Origin of Human Bipedalism." *Proceedings of the National Academy of Sciences of the United States of America* 104, no. 30 (2007): 12265–69. https://doi.org/10.1073/pnas.0703267104.

Smithsonian National Museum of Natural History. "Laetoli Footprint Trails." Updated December 17, 2020. http://humanorigins.si.edu/evidence/behavior/footprints/laetoli-footprint-trails.

Stern, Jack T. "Climbing to the Top: A Personal Memoir of *Australopithecus afarensis*." *Evolutionary Anthropology: Issues, News, and Reviews* 9, no. 3 (2000): 113–33. https://doi.org/10.1002/1520-6505(2000)9:3<113::AID-EVAN2>3.0.CO;2-W.

Tardieu, C., Y. Glard, E. Garron, C. Boulay, J. L. Jouve, O. Dutour, G. Boetsch, and G. Bollini. "Relationship Between Formation of the Femoral Bicondylar Angle and Trochlear Shape: Independence of Diaphyseal and Epiphyseal Growth." *American Journal of Physical Anthropology* 130, no. 4 (2006): 491–500. https://doi.org/10.1002/ajpa.20373.

Tardieu, Christine, and Erik Trinkaus. "Early Ontogeny of the Human Femoral Bicondylar Angle." *American Journal of Physical Anthropology* 95, no. 2 (1994): 183–95. https://doi.org/10.1002/ajpa.1330950206.

Venkateswaran, T. V. "Did Ardi Walk for Sex? Gender, Science and World Views." *Economic and Political Weekly* 46, no. 3 (2011): 19–23.

Walter, Robert C. "Age of Lucy and the First Family: Single-Crystal ^{40}Ar/^{39}Ar Dating of the Denen Dora and Lower Kada Hadar Members of the Hadar Formation, Ethiopia." *Geology* 22, no. 1 (1994): 6–10. https://doi.org/10.1130/0091-7613(1994)022<0006:AOLATF>2.3.CO;2.

Ward, Carol V. "Interpreting the Posture and Locomotion of *Australopithecus afarensis*: Where Do We Stand?" *Yearbook of Physical Anthropology* 45 (2002): 185–215.

Washburn, S. L. "Speculations on the Interrelations of the History of Tools and Biological Evolution." *Human Biology* 31, no. 1 (1959): 21–31.

White, Tim D., C. Owen Lovejoy, Berhane Asfaw, Joshua P. Carlson, and Gen Suwa. "Neither Chimpanzee nor Human, Ardipithecus Reveals the Surprising Ancestry of Both." *Proceedings of the National Academy of Sciences of the United States of America* 112, no. 16 (2015): 4877–84. https://doi.org/10.1073/pnas .1403659111.

Zaitsev, Anatoly N., Thomas Wenzel, John Spratt, Terry C. Williams, Stanislav Strekopytov, Victor V. Sharygin, Sergey V. Petrov, Tamara A. Golovina, Elena O. Zaitseva, and Gregor Markl. "Was Sadiman Volcano a Source for the Laetoli Footprint Tuff?" *Journal of Human Evolution* 61, no. 1 (2011): 121–24. https://doi .org/10.1016/j.jhevol.2011.02.004.

2. CONFUSED ANATOMY

Abulhasan, Jawad, and Michael Grey. "Anatomy and Physiology of Knee Stability." *Journal of Functional Morphology and Kinesiology* 2, no. 4 (2017): 34. https:// doi.org/10.3390/jfmk2040034.

Arnoczky, Steven P., and Russell F. Warren. "Microvasculature of the Human Meniscus." *American Journal of Sports Medicine* 10, no. 2 (1982): 90–95. https:// doi.org/10.1177/036354658201000205.

Beers, Amanda, Michael Ryan, Zenya Kasubuchi, Scott Fraser, and Jack E. Taunton. "Effects of Multi-Modal Physiotherapy, Including Hip Abductor Strengthening, in Patients with Iliotibial Band Friction Syndrome." *Physiotherapy Canada* 60, no. 2 (2008): 180–88. https://doi.org/10.3138/physio.60.2.180.

Behm, David G., and Jan Wilke. "Do Self-Myofascial Release Devices Release Myofascia? Rolling Mechanisms: A Narrative Review." *Sports Medicine* (Auckland) 49, no. 8 (2019): 1173–81. https://doi.org/10.1007/s40279-019-01149-y.

Bradbury-Squires, David J., Jennifer C. Noftall, Kathleen M. Sullivan, David G. Behm, Kevin E. Power, and Duane C. Button. "Roller-Massager Application to the Quadriceps and Knee-Joint Range of Motion and Neuromuscular Efficiency During a Lunge." *Journal of Athletic Training* 50, no. 2 (2015): 133–40. https:// doi.org/10.4085/1062-6050-49.5.03.

Bull, Anthony M. J., and Andrew A. Amis. "The Pivot-Shift Phenomenon: A Clinical and Biomechanical Perspective." *The Knee* 5, no. 3 (1998): 141–58. https:// doi.org/10.1016/s0968-0160(97)10027-8.

Cartner, Jacob L., Zane M. Hartsell, William M. Ricci, and Paul Tornetta III. "Can We Trust Ex Vivo Mechanical Testing of Fresh-Frozen Cadaveric Specimens? The Effect of Postfreezing Delays." *Journal of Orthopaedic Trauma* 25, no. 8 (2011): 459–61. https://doi.org/10.1097/BOT.0b013e318225b875.

Chahla, Jorge, Gilbert Moatshe, Chase S. Dean, and Robert F. LaPrade. "Posterolateral Corner of the Knee: Current Concepts." *Archives of Bone and Joint Surgery* 4, no. 2 (2016): 97–103.

Claes, Steven, Evie Vereecke, Michael Maes, Jan Victor, Peter Verdonk, and Johan Bellemans. "Anatomy of the Anterolateral Ligament of the Knee." *Journal of Anatomy* 223, no. 4 (2013): 321–28. https://doi.org/10.1111/joa.12087.

Conte, Melissa Nicol, and Peter Kessler. "Method and device for therapeutic treatment of iliotibial band syndrome, myofascial and musculoskeletal dysfunctions." US Patent 20,150,313,788, filed November 5, 2015. http://appft.uspto.gov/netacgi /nph-Parser?Sect1=PTO1&Sect2=HITOFF&p=1&u=/netahtml/PTO/srchnum .html&r=1&f=G&l=50&d=PG01&s1=20150313788.

Cox, C. F., M. A. Sinkler, and J. B. Hubbard. "Anatomy, Bony Pelvis and Lower Limb, Knee Patella." In *StatPearls*, 30137819. Treasure Island, FL: StatPearls, 2020.

Cox, Chandler F., and John B. Hubbard. "Anatomy, Bony Pelvis and Lower Limb, Knee Lateral Meniscus." In StatPearls, 30137778. Treasure Island, FL: StatPearls, 2020.

Daggett, Matt, Steven Claes, Camilo P. Helito, Pierre Imbert, Edoardo Monaco, Christian Lutz, and Bertrand Sonnery-Cottet. "The Role of the Anterolateral Structures and the ACL in Controlling Laxity of the Intact and ACL-Deficient Knee: Letter to the Editor." *American Journal of Sports Medicine* 44, no. 4 (2016): NP14–NP15. https://doi.org/10.1177/0363546516638069.

Diamantopoulos, Andreas, Anastasios Tokis, Matheus Tzurbakis, Iraklis Patsopoulos, and Anastasios Georgoulis. "The Posterolateral Corner of the Knee: Evaluation Under Microsurgical Dissection." *Arthroscopy* 21, no. 7 (2005): 826–33. https://doi.org/10.1016/j.arthro.2005.03.021.

Dombrowski, Malcolm E., Joanna M. Costello, Bruno Ohashi, Christopher D. Murawski, Benjamin B. Rothrauff, Fabio V. Arilla, Nicole A. Friel, et al. "Macroscopic Anatomical, Histological and Magnetic Resonance Imaging Correlation of the Lateral Capsule of the Knee." *Knee Surgery, Sports Traumatology, Arthroscopy* 24, no. 9 (2016): 2854–60. https://doi.org/10.1007/s00167-015-3517-8.

Ellis, Richard, Wayne Hing, and Duncan Reid. "Iliotibial Band Friction Syndrome— A Systematic Review." *Manual Therapy* 12, no. 3 (2007): 200–208. https://doi.org /10.1016/j.math.2006.08.004.

Eng, Carolyn M., Allison S. Arnold, Andrew A. Biewener, and Daniel E. Lieberman. "The Human Iliotibial Band Is Specialized for Elastic Energy Storage Compared with the Chimp Fascia Lata." *Journal of Experimental Biology* 218, no. 15 (2015): 2382–93. https://doi.org/10.1242/jeb.117952.

Eng, Carolyn M., Allison S. Arnold, Daniel E. Lieberman, and Andrew A. Biewener. "The Capacity of the Human Iliotibial Band to Store Elastic Energy During Running." *Journal of Biomechanics* 48, no. 12 (2015): 3341–48. https://doi.org /10.1016/j.jbiomech.2015.06.017.

Evans, Jennifer, and Jeffery I. Nielson. "Anterior Cruciate Ligament Knee Injuries." In *StatPearls*, 29763023. Treasure Island, FL: StatPearls, 2020.

Fairbank, T. J. "Knee Joint Changes After Meniscectomy." *Journal of Bone and Joint Surgery* 30-B, no. 4 (1948): 664–70. https://doi.org/10.1302/0301-620x.30b4.664.

Fairclough, John, Koji Hayashi, Hechmi Toumi, Kathleen Lyons, Graeme Bydder, Nicola Phillips, Thomas M. Best, and Mike Benjamin. "The Functional Anatomy of the Iliotibial Band During Flexion and Extension of the Knee: Implications for Understanding Iliotibial Band Syndrome." *Journal of Anatomy* 208, no. 3 (2006): 309–16. https://doi.org/10.1111/j.1469-7580.2006.00531.x.

——. "Is Iliotibial Band Syndrome Really a Friction Syndrome?" *Journal of Science and Medicine in Sport* 10, no. 2 (2007): 74–76. https://doi.org/10.1016/j.jsams .2006.05.017.

Falvey, E. C., R. A. Clark, A. Franklyn-Miller, A. L. Bryant, C. Briggs, and P. R. McCrory. "Iliotibial Band Syndrome: An Examination of the Evidence Behind a Number of Treatment Options." *Scandinavian Journal of Medicine and Science in Sports* 20, no. 4 (2010): 580–87. https://doi.org/10.1111/j.1600-0838.2009.00968.x.

Farrell, Connor, Alan G. Shamrock, and John Kiel. "Anatomy, Bony Pelvis and Lower Limb, Medial Meniscus." In *StatPearls*, 30725961. Treasure Island, FL: StatPearls, 2020.

Flato, Russell, Giovanni J. Passanante, Matthew R. Skalski, Dakshesh B. Patel, Eric A. White, and George R. Matcuk. "The Iliotibial Tract: Imaging, Anatomy, Injuries, and Other Pathology." *Skeletal Radiology* 46, no. 5 (2017): 605–22. https:// doi.org/10.1007/s00256-017-2604-y.

Fox, Alice J. S., Asheesh Bedi, and Scott A. Rodeo. "The Basic Science of Human Knee Menisci: Structure, Composition, and Function." *Sports Health* 4, no. 4 (2012): 340–51. https://doi.org/10.1177/1941738111429419.

Fredericson, Michael, Curtis L. Cookingham, Ajit M. Chaudhari, Brian C. Dowdell, Nina Oestreicher, and Shirley A. Sahrmann. "Hip Abductor Weakness in Distance Runners with Iliotibial Band Syndrome." *Clinical Journal of Sport Medicine* 10, no. 3 (2000): 169–75. https://doi.org/10.1097/00042752-200007000-00004.

Fredericson, Michael, Jeremy J. White, John M. MacMahon, and Thomas P. Andriacchi. "Quantitative Analysis of the Relative Effectiveness of 3 Iliotibial Band Stretches." *Archives of Physical Medicine and Rehabilitation* 83, no. 5 (2002): 589–92. https://doi.org/10.1053/apmr.2002.31606.

Gallacher, P. D., R. E. Gilbert, G. Kanes, S. N. J. Roberts, and D. Rees. "White on White Meniscal Tears to Fix or Not to Fix?" *The Knee* 17, no. 4 (2010): 270–73. https://doi.org/10.1016/j.knee.2010.02.016.

Geiger, Daniel, Eric Y. Chang, Mini N. Pathria, and Christine B. Chung. "Posterolateral and Posteromedial Corner Injuries of the Knee." *Magnetic Resonance Imaging Clinics of North America* 22, no. 4 (2014): 581–99. https://doi.org/10.1016 /j.mric.2014.08.001.

Godin, Jonathan A., Jorge Chahla, Gilbert Moatshe, Bradley M. Kruckeberg, Kyle J. Muckenhirn, Alexander R. Vap, Andrew G. Geeslin, and Robert F. LaPrade. "A Comprehensive Reanalysis of the Distal Iliotibial Band: Quantitative Anatomy, Radiographic Markers, and Biomechanical Properties." *American Journal of Sports Medicine* 45, no. 11 (2017): 2595–603. https://doi.org/10.1177/0363546517707961.

Grau, S., I. Krauss, C. Maiwald, R. Best, and T. Horstmann. "Hip Abductor Weakness Is Not the Cause for Iliotibial Band Syndrome." *International Journal of Sports Medicine* 29, no. 7 (2008): 579–83. https://doi.org/10.1055/s-2007-989323.

Guenther, Daniel, Amir A. Rahnemai-Azar, Kevin M. Bell, Sebastián Irarrázaval, Freddie H. Fu, Volker Musahl, and Richard E. Debski. "The Anterolateral Capsule of the Knee Behaves Like a Sheet of Fibrous Tissue." *American Journal of Sports Medicine* 45, no. 4 (2016): 849–55. https://doi.org/10.1177/0363546516674477.

Gunter, P., and M. P. Schwellnus. "Local Corticosteroid Injection in Iliotibial Band Friction Syndrome in Runners: A Randomised Controlled Trial." *British Journal of Sports Medicine* 38, no. 3 (2004): 269–72. https://doi.org/10.1136/bjsm.2003.000283.

Herbst, Elmar, Marcio Albers, Andreas Imhoff, Freddie Fu, and Volker Musahl. "The Anterolateral Complex of the Knee." *Orthopaedic Journal of Sports Medicine* 5, no. 10 (2017): 2325967117730805. https://doi.org/10.1177/2325967118S00031.

Hirschmann, Michael T., and Werner Müller. "Complex Function of the Knee Joint: The Current Understanding of the Knee." *Knee Surgery, Sports Traumatology, Arthroscopy* 23, no. 10 (2015): 2780–88. https://doi.org/10.1007/s00167-015-3619-3.

Hughston, Jack C. "A Simple Meniscectomy." *Journal of Sports Medicine* 3, no. 4 (1975): 179–87. https://doi.org/10.1177/036354657500300406.

Jelsing, Elena J., Jonathan T. Finnoff, Andrea L. Cheville, Bruce A. Levy, and Jay Smith. "Sonographic Evaluation of the Iliotibial Band at the Lateral Femoral Epicondyle: Does the Iliotibial Band Move?" *Journal of Ultrasound in Medicine* 32, no. 7 (2013): 1199–206. https://doi.org/10.7863/ultra.32.7.1199.

Jones, Jennifer C., Robert Burks, Brett D. Owens, Rodney X. Sturdivant, Steven J. Svoboda, and Kenneth L. Cameron. "Incidence and Risk Factors Associated with Meniscal Injuries Among Active-Duty US Military Service Members." *Journal of Athletic Training* 47, no. 1 (2012): 67–73.

Kittl, Christoph, Hadi El-Daou, Kiron K. Athwal, Chinmay M. Gupte, Andreas Weiler, Andy Williams, and Andrew A. Amis. "The Role of the Anterolateral Structures and the ACL in Controlling Laxity of the Intact and ACL-Deficient Knee: Response." *American Journal of Sports Medicine* 44, no. 4 (2016): NP15–NP18. https://doi.org/10.1177/0363546516638070.

Louw, Maryke, and Clare Deary. "The Biomechanical Variables Involved in the Aetiology of Iliotibial Band Syndrome in Distance Runners—A Systematic Review of the Literature." *Physical Therapy in Sport* 15, no. 1 (2014): 64–75. https://doi.org/10.1016/j.ptsp.2013.07.002.

Lovejoy, C. Owen. "The Natural History of Human Gait and Posture." *Gait and Posture* 25, no. 3 (2007): 325–41. https://doi.org/10.1016/j.gaitpost.2006.05.001.

MacDonald, Graham Z., Michael D. H. Penney, Michelle E. Mullaley, Amanda L. Cuconato, Corey D. J. Drake, David G. Behm, and Duane C. Button. "An Acute Bout of Self-Myofascial Release Increases Range of Motion Without a Subsequent Decrease in Muscle Activation or Force." *Journal of Strength and Conditioning Research* 27, no. 3 (2013): 812–21. https://doi.org/10.1519/JSC.0b013e31825c2bc1.

Matsushita, Takehiko, Shinya Oka, Kouki Nagamune, Tomoyuki Matsumoto, Yuichiro Nishizawa, Yuichi Hoshino, Seiji Kubo, Masahiro Kurosaka, and Ryosuke Kuroda. "Differences in Knee Kinematics Between Awake and Anesthetized Patients During the Lachman and Pivot-Shift Tests for Anterior Cruciate

Ligament Deficiency." *Orthopaedic Journal of Sports Medicine* 1, no. 1 (2013): 2325967113487855–55. https://doi.org/10.1177/2325967113487855.

McMurray, T. P. "The Semilunar Cartilages." *British Journal of Surgery* 29, no. 116 (1942): 407–14. https://doi.org/10.1002/bjs.18002911612.

Messier, Stephen P., David G. Edwards, David F. Martin, Robert B. Lowery, D. Wayne Cannon, Margaret K. James, Walton W. Curl, Hank M. Read Jr., and D. Monte Hunter. "Etiology of Iliotibial Band Friction Syndrome in Distance Runners." *Medicine and Science in Sports and Exercise* 27, no. 7 (1995): 951–60. https://doi.org/10.1249/00005768-199507000-00002.

Monaco, E., A. Ferretti, L. Labianca, B. Maestri, A. Speranza, M. J. Kelly, and C. D'Arrigo. "Navigated Knee Kinematics After Cutting of the ACL and Its Secondary Restraint." *Knee Surgery, Sports Traumatology, Arthroscopy* 20, no. 5 (2011): 870–77. https://doi.org/10.1007/s00167-011-1640-8.

Mordecai, Simon C., Nawfal Al-Hadithy, Howard E. Ware, and Chinmay M. Gupte. "Treatment of Meniscal Tears: An Evidence Based Approach." *World Journal of Orthopedics* 5, no. 3 (2014): 233–41. https://doi.org/10.5312/wjo.v5.i3.233.

Naqvi, Usker, and Andrew I. Sherman. "Medial Collateral Ligament Knee Injuries." In *StatPearls*, 28613747. Treasure Island, FL: StatPearls, 2020.

Nickinson, Richard, Clare Darrah, and Simon Donell. "Accuracy of Clinical Diagnosis in Patients Undergoing Knee Arthroscopy." *International Orthopaedics* 34, no. 1 (2010): 39–44. https://doi.org/10.1007/s00264-009-0760-y.

Niemuth, Paul E., Robert J. Johnson, Marcella J. Myers, and Thomas J. Thieman. "Hip Muscle Weakness and Overuse Injuries in Recreational Runners." *Clinical Journal of Sport Medicine* 15, no. 1 (2005): 14–21. https://doi.org/10.1097/00042752-200501000-00004.

Nitri, Marco, Matthew T. Rasmussen, Brady T. Williams, Samuel G. Moulton, Raphael Serra Cruz, Grant J. Dornan, Mary T. Goldsmith, and Robert F. LaPrade. "An in Vitro Robotic Assessment of the Anterolateral Ligament, Part 2: Anterolateral Ligament Reconstruction Combined with Anterior Cruciate Ligament Reconstruction." *American Journal of Sports Medicine* 44, no. 3 (2016): 593–601. https://doi.org/10.1177/0363546515620183.

Noehren, Brian, Anne Schmitz, Ross Hempel, Carolyn Westlake, and William Black. "Assessment of Strength, Flexibility, and Running Mechanics in Men with Iliotibial Band Syndrome." *Journal of Orthopaedic and Sports Physical Therapy* 44, no. 3 (2014): 217–22. https://doi.org/10.2519/jospt.2014.4991.

Noyes, Frank R., Ryan C. Chen, Sue D. Barber-Westin, and Hollis G. Potter. "Greater Than 10-Year Results of Red-White Longitudinal Meniscal Repairs in Patients 20 Years of Age or Younger." *American Journal of Sports Medicine* 39, no. 5 (2011): 1008–17. https://doi.org/10.1177/0363546510392014.

Nyland, John, Narusha Lachman, Yavuz Kocabey, Joseph Brosky, Remziye Altun, and David Caborn. "Anatomy, Function, and Rehabilitation of the Popliteus Musculotendinous Complex." *Journal of Orthopaedic and Sports Physical Therapy* 35, no. 3 (2005): 165–79. https://doi.org/10.2519/jospt.2005.35.3.165.

Orchard, John W., Peter A. Fricker, Anna T. Abud, and Bruce R. Mason. "Biomechanics of Iliotibial Band Friction Syndrome in Runners." *American Journal of Sports Medicine* 24, no. 3 (1996): 375–79. https://doi.org/10.1177/036354659602400321.

Patil, Shantanu Sudhakar, Anshu Shekhar, and Sachin Ramchandra Tapasvi. "Meniscal Preservation Is Important for the Knee Joint." *Indian Journal of Orthopaedics* 51, no. 5 (2017): 576–87. https://doi.org/10.4103/ortho.ijortho_247_17.

Pepper, Talin M., Jean-Michel Brismée, Phillip S. Sizer Jr., Jeegisha Kapila, Gesine H. Seeber, Christopher A. Huggins, and Troy L. Hooper. "The Immediate Effects of Foam Rolling and Stretching on Iliotibial Band Stiffness: A Randomized Controlled Trial." *International Journal of Sports Physical Therapy* 16, no. 3 (2021): 651–61. https://doi.org/10.26603/001c.23606.

Phillips, Jake, David Diggin, Deborah L. King, and Gary A. Sforzo. "Effect of Varying Self-Myofascial Release Duration on Subsequent Athletic Performance." *Journal of Strength and Conditioning Research* 35, no. 3 (2018): 746–53. https://doi.org/10.1519/JSC.0000000000002751.

Raj, Marc A., Ahmed Mabrouk, and Matthew Varacallo. "Posterior Cruciate Ligament Knee Injuries." In *StatPearls*, 28613477. Treasure Island, FL: StatPearls, 2020.

Rasmussen, Matthew T., Marco Nitri, Brady T. Williams, Samuel G. Moulton, Raphael Serra Cruz, Grant J. Dornan, Mary T. Goldsmith, and Robert F. LaPrade. "An in Vitro Robotic Assessment of the Anterolateral Ligament, Part 1: Secondary Role of the Anterolateral Ligament in the Setting of an Anterior Cruciate Ligament Injury." *American Journal of Sports Medicine* 44, no. 3 (2015): 585–92. https://doi.org/10.1177/0363546515618387.

Renne, J. W. "The Iliotibial Band Friction Syndrome." *Journal of Bone and Joint Surgery* 57, no. 8 (1975): 1110–11. https://doi.org/10.2106/00004623-197557080-00014.

Rosas, Humberto G. "Unraveling the Posterolateral Corner of the Knee." *Radiographics* 36, no. 6 (2016): 1776–91. https://doi.org/10.1148/rg.2016160027.

Runer, Armin, Stephan Birkmaier, Mathias Pamminger, Simon Reider, Elmar Herbst, Karl-Heinz Künzel, Erich Brenner, and Christian Fink. "The Anterolateral Ligament of the Knee: A Dissection Study." *The Knee* 23, no. 1 (2016): 8–12. https://doi.org/10.1016/j.knee.2015.09.014.

Scholten, Rob J. P. M., Wim Opstelten, Cees G. van der Plas, Dick Bijl, Walter L. J. M. Deville, and Lex M. Bouter. "Accuracy of Physical Diagnostic Tests for Assessing Ruptures of the Anterior Cruciate Ligament: A Meta-Analysis." *Journal of Family Practice* 52, no. 9 (2003): 689–94.

Schwellnus, M. P., L. Theunissen, T. D. Noakes, and S. G. Reinach. "Anti-Inflammatory and Combined Anti-Inflammatory/Analgesic Medication in the Early Management of Iliotibial Band Friction Syndrome. A Clinical Trial." *South African Medical Journal* 79, no. 10 (1991): 602–6.

Shea, Kevin G., Matthew D. Milewski, Peter C. Cannamela, Theodore J. Ganley, Peter D. Fabricant, Elizabeth B. Terhune, Alexandra C. Styhl, Allen F. Anderson, and John D. Polousky. "Anterolateral Ligament of the Knee Shows Variable Anatomy in Pediatric Specimens." *Clinical Orthopaedics and Related Research* 475, no. 6 (2017): 1583–91. https://doi.org/10.1007/s11999-016-5123-6.

Sher, Irene, Hilary Umans, Sherry A. Downie, Keith Tobin, Ritika Arora, and Todd R. Olson. "Proximal Iliotibial Band Syndrome: What Is It and Where Is It?" *Skeletal Radiology* 40, no. 12 (2011): 1553–56. https://doi.org/10.1007/s00256-011-1168-5.

Sonnery-Cottet, Bertrand, Mathieu Thaunat, Benjamin Freychet, Barbara H. B. Pupim, Colin G. Murphy, and Steven Claes. "Outcome of a Combined Anterior Cruciate Ligament and Anterolateral Ligament Reconstruction Technique with a Minimum 2-Year Follow-Up." *American Journal of Sports Medicine* 43, no. 7 (2015): 1598–605. https://doi.org/10.1177/0363546515571571.

Taunton, J. E., M. B. Ryan, D. B. Clement, D. C. McKenzie, D. R. Lloyd-Smith, and B. D. Zumbo. "A Retrospective Case-Control Analysis of 2002 Running Injuries." *British Journal of Sports Medicine* 36, no. 2 (2002): 95–101. https://doi.org/10.1136/bjsm.36.2.95.

Thein, Ran, James Boorman-Padgett, Kyle Stone, Thomas L. Wickiewicz, Carl W. Imhauser, and Andrew D. Pearle. "Biomechanical Assessment of the Anterolateral Ligament of the Knee: A Secondary Restraint in Simulated Tests of the Pivot Shift and of Anterior Stability." *Journal of Bone and Joint Surgery* 98, no. 11 (2016): 937–43. https://doi.org/10.2106/JBJS.15.00344.

Tria, Alfred J., Christopher D. Johnson, and Joseph P. Zawadsky. "The Popliteus Tendon." *Journal of Bone and Joint Surgery* 71, no. 5 (1989): 714–16. https://doi.org/10.2106/00004623-198971050-00011.

van der Worp, Maarten P., Nick van der Horst, Anton de Wijer, Frank J. G. Backx, and Maria W. G. Nijhuis-van der Sanden. "Iliotibial Band Syndrome in Runners: A Systematic Review." *Sports Medicine* 42, no. 11 (2012): 969–92. https://doi.org/10.1007/BF03262306.

van Trommel, Michiel F., Peter T. Simonian, Hollis G. Potter, and Thoma L. Wickiewicz. "Arthroscopic Meniscal Repair with Fibrin Clot of Complete Radial Tears of the Lateral Meniscus in the Avascular Zone." *Arthroscopy* 14, no. 4 (1998): 360–65. https://doi.org/10.1016/S0749-8063(98)70002-7.

Vaudreuil, Nicholas J., Benjamin B. Rothrauff, Darren de Sa, and Volker Musahl. "The Pivot Shift: Current Experimental Methodology and Clinical Utility for Anterior Cruciate Ligament Rupture and Associated Injury." *Current Reviews in Musculoskeletal Medicine* 12, no. 1 (2019): 41–49. https://doi.org/10.1007/s12178-019-09529-7.

Vincent, Jean-Philippe, Robert A. Magnussen, Ferittu Gezmez, Arnaud Uguen, Matthias Jacobi, Florent Weppe, Ma'ad F. Al-Saati, et al. "The Anterolateral Ligament of the Human Knee: An Anatomic and Histologic Study." *Knee Surgery, Sports Traumatology, Arthroscopy* 20, no. 1 (2012): 147–52. https://doi.org/10.1007/s00167-011-1580-3.

Walbron, Paul, Adrien Jacquot, Jean-Marc Geoffroy, François Sirveaux, and Daniel Molé. "Iliotibial Band Friction Syndrome: An Original Technique of Digastric Release of the Iliotibial Band from Gerdy's Tubercle." *Orthopaedics and Traumatology, Surgery and Research* 104, no. 8 (2018): 1209–13. https://doi.org/10.1016/j.otsr.2018.08.013.

Wang, Hsing-Kuo, Tiffany Ting-Fang Shih, Kwan-Hwa Lin, and Tyng-Guey Wang. "Real-Time Morphologic Changes of the Iliotibial Band During Therapeutic Stretching; An Ultrasonographic Study." *Manual Therapy* 13, no. 4 (2007): 334–40. https://doi.org/10.1016/j.math.2007.03.002.

Wang, Tyng-Guey, Mei-Hwa Jan, Kwan-Hwa Lin, and Hsing-Kuo Wang. "Assessment of Stretching of the Iliotibial Tract with Ober and Modified Ober Tests: An Ultrasonographic Study." *Archives of Physical Medicine and Rehabilitation* 87, no. 10 (2006): 1407–11. https://doi.org/10.1016/j.apmr.2006.06.007.

Wilhelm, Mark, Omer Matthijs, Kevin Browne, Gesine Seeber, Anja Matthijs, Phillip S. Sizer, Jean-Michel Brismée, C. Roger James, and Kerry K. Gilbert. "Deformation Response of the Iliotibial Band-Tensor Fascia Lata Complex to Clinical-Grade Longitudinal Tension Loading in-Vitro." *International Journal of Sports Physical Therapy* 12, no. 1 (2017): 16–24.

Wilke, Jan, Anna-Lena Müller, Florian Giesche, Gerard Power, Hamid Ahmedi, and David G. Behm. "Acute Effects of Foam Rolling on Range of Motion in Healthy Adults: A Systematic Review with Multilevel Meta-Analysis." *Sports Medicine* 50, no. 2 (2020): 387–402. https://doi.org/10.1007/s40279-019-01205-7.

Yaras, Reed J., Nicholas O'Neill, and Amjad M. Yaish. "Lateral Collateral Ligament Knee Injuries." In *StatPearls*, 32809682. Treasure Island, FL: StatPearls, 2020.

Yoo, Hannah, and Raghavendra Marappa-Ganeshan. "Anatomy, Bony Pelvis and Lower Limb, Knee Anterior Cruciate Ligament." In *StatPearls*, 32644659. Treasure Island, FL: StatPearls, 2020.

3. BARE KNEES, DICEY POWER

"Against Immodest Fashions." *Pittsburgh Press* (Pittsburgh, PA). March 20, 1921.

"The American and French Fashions Contrasted." *Water-Cure Journal* 12, no. 4 (1851): 96.

"Announcing Flapper Beauty Contest." *The Flapper*. June 1922.

"Are Bloomers Ugly?" *Hanford Semi-Weekly Journal* (Hanford, CA). November 5, 1895.

"Arrest for Wearing 'Male Attire.'" *Times-Picayune* (New Orleans, LA). June 23, 1866.

Associated Press. "Flappers in Verse Defend Near Nudity." *Morning News* (Wilmington, DE). August 25, 1923.

——. "Flappers-Puritans Declare Open War in Pennsylvania." *Daily Tribune* (Wisconsin Rapids, WI). August 24, 1923.

Awasthi, Bhuvanesh. "From Attire to Assault: Clothing, Objectification, and De-Humanization—a Possible Prelude to Sexual Violence?" *Frontiers in Psychology* 8 (2017): 338. https://doi.org/10.3389/fpsyg.2017.00338.

Barber, Nigel. "Women's Dress Fashions as a Function of Reproductive Strategy." *Sex Roles* 40, no. 5 (1999): 459–71. https://doi.org/10.1023/A:1018823727012.

Barmash, Isadore. "Furor Over the Mini-Skirt." *Des Moines Register* (Des Moines, IA). December 7, 1966.

Becker, Annette. "The Body." In *A Cultural History of Dress and Fashion in the Age of Empire*, ed. Denise Amy Baxter, 59–80. New York: Bloomsbury, 2017.

Beery, Zoë. "Flappers Didn't Really Wear Fringed Dresses." Racked. May 19, 2017, https://www.racked.com/2017/5/19/15612000/flappers-fringe-myth.

"Big Business Banishes the Flapper." *Morning Tulsa Daily World* (Tulsa, OK). July 16, 1922.

Bliven, Bruce. "Flapper Jane." *New Republic*. September 9, 1925.

"Bloomers in Paris." *Philadelphia Inquirer* (Philadelphia, Pennsylvania). March 15, 1896.

"Chicago Pastor Does Not Like Flappers." *Springfield News-Leader* (Springfield, MO). May 12, 1922.

Cooper, Ella H. "Bicycle skirt." US Patent 555,211. Filed July 18, 1895. Issued February 25, 1896.

Dempsey, Mary V. *The Occupational Progress of Women, 1910 to 1930*. Washington, DC: U.S. Government Printing Office, 1933.

"Design Painted on Knee of Stocking Newest in London." *St. Louis Star and Times* (St. Louis, MO). April 20, 1928.

"Discuss the Bloomer." *Chicago Tribune* (Chicago, IL). September 8, 1895.

"Does It Pay to Visit Yo Semite?" *Leavenworth Times* (Leavenworth, KS). October 4, 1870.

"Flapper." Accessed November 27, 2021. http://www.tomandrodna.com/Flapper .jpg. Contributed by Dave Bumgardner.

"Flapper Suits by Lady Duff-Gordon." *Ogden Standard-Examiner* (Ogden, UT). March 26, 1922.

"For the Knickerbocker." *San Francisco Examiner* (San Francisco, CA). June 23, 1895.

Gilbert, Theodosia. "An Eye Sore." *Water-Cure Journal* 11, no. 5 (1851): 116–17.

"Girls' Painted Knees Bothering Professor." *San Francisco Examiner* (San Francisco, CA). May 14, 1925.

Graham, Rubye. "Beauty Industry Goes out on a Limb." *Philadelphia Inquirer* (Philadelphia, PA). July 3, 1966.

Hall, Linda. "Fashion and Style in the Twenties: The Change." *Historian* 34, no. 3 (1972): 485–97. https://doi.org/10.1111/j.1540-6563.1972.tb00424.x.

"Hand-Painted Knees Latest Beauty Stunt." *Chattanooga News* (Chattanooga, OK). August 27, 1925.

Harlow, Mary. *A Cultural History of Dress and Fashion*. Vol. 1. New York: Bloomsbury, 2017.

Jewell, David A. "Exit the Mini-Skirts Enter the Micros." *News-Journal* (Mansfield, OH). May 14, 1967.

Kim, Soohyun, and Insook Ahn. "Impact of Macro-Economic Factors on the Hemline Cycles." In *ITAA Proceedings*, 72, 1–2. Santa Fe, NM: International Textile and Apparel Association, 2015. https://iastatedigitalpress.com/itaa/article/2476 /galley/2349/view/

Krauss, Bob. "Mini? Not Many Downtown." *Honolulu Advertiser* (Honolulu, HI). December 1, 1967.

Kriebl, Karen J. "From Bloomers to Flappers: The American Women's Dress Reform Movement, 1840–1920." PhD diss., Ohio State University, 1998. ProQuest.

Lennon, Theresa L., Sharron J. Lennon, and Kim K. P. Johnson. "Is Clothing Probative of Attitude or Intent? Implications for Rape and Sexual Harassment Cases."
Law and Inequality: A Journal of Theory and Practice 11, no. 2 (1993): 391–415.

Levy, Ariel. *Female Chauvinist Pigs: Women and the Rise of Raunch Culture*. New
York: Free Press, 2006.

Mabry, Mary Ann. "The Relationship Between Fluctuations in Hemlines and Stock
Market Averages from 1921 to 1971." Master's thesis, University of Tennessee,
1971. https://trace.tennessee.edu/utk_gradthes/1121.

Marks, Julie. "What Caused the Stock Market Crash of 1929?" History. Updated
April 27, 2021. https://www.history.com/news/what-caused-the-stock-market
-crash-of-1929.

"Mini-Skirt Comes to Mt. Carmel In 'Modified' Form, Study Reveals." *Daily
Republican-Register* (Mount Carmel, IL). September 30, 1966.

"Much Smartness in School Rainment." *Journal and Tribune* (Knoxville, TN).
January 17, 1915.

Nichols, M. S. Gove. "Woman the Physician." *Water-Cure Journal* 12, no. 4 (1851):
73–75.

"Painted Knee Fad Hits." *Nebraska State Journal* (Lincoln, NE). July 23, 1925.

Payne, Blanche. *History of Costume: From the Ancient Egyptians to the Twentieth
Century*. New York: Harper & Row, 1965.

"The Psychology of Knees." *The Flapper*. June 1922.

Rabinovitch-Fox, Einav. "Fabricating Black Modernity: Fashion and African American Womanhood During the First Great Migration." *International Journal of
Fashion Studies* 6, no. 2 (2019): 239–60.

——. "This Is What a Feminist Looks Like: The New Woman Image, American
Feminism, and the Politics of Women's Fashion 1890–1930." PhD diss., New
York University, 2014. ProQuest.

Richards, Lynne. "The Rise and Fall of It All: The Hemlines and Hiplines of the
1920s." *Clothing and Textiles Research Journal* 2, no. 1 (1983): 42–48.

Rose, Clare. *Making, Selling and Wearing Boys' Clothes in Late-Victorian England*.
Surrey, England: Ashgate, 2010.

"Rouged Knee Mode Interests Paris." *Boston Post* (Boston, MA). July 14, 1921.

"Rubber Roll Garters." *El Paso Herald* (El Paso, TX). August 3, 1928.

Serviss, Myrna. "Form a Flapper Flock of Your Own." *The Flapper*. June 1922, 20.

——. "News of the Flapper Flocks." *The Flapper*. November 1922, 32–33.

Smothers, David. "Leg-Islators Find Short Skirts No Mini-Controversy." *Wisconsin
State Journal* (Madison, WI). April 13, 1969.

Stanton, Elizabeth Cady, Susan B. Anthony, Emmeline Pankhurst, Anna Howard
Shaw, Millicent Garrett Fawcett, Jane Addams, Lucy Stone, Carrie Chapman
Catt, and Alice Paul. *The Women of the Suffrage Movement: Autobiographies and
Biographies of the Most Influential Suffragettes*. N.p.: Musaicum Books, 2018.

"Stock Market Crash of 1929." History. Updated April 27, 2021. https://www.history
.com/topics/great-depression/1929-stock-market-crash.

Studer, Brigitte. "'1968' and the Formation of the Feminist Subject." *Twentieth Century Communism*, no. 3 (2011): 38–69.

Tsui, Bonnie. *She Went to the Field: Women Soldiers of the Civil War*. Guilford, CT: TwoDot, 2006.

"Typical Flappers." *Weekly Journal-Miner* (Prescott, AZ). August 2, 1922.

"Underpinning the 1920s: Brassieres, Bandeaux, and Bust Flatteners." witness2fashion. April 27, 2014. https://witness2fashion.wordpress.com/tag/breast-binding-1920s/.

van Baardwijk, Marjolein, and Philip Hans Franses. "The Hemline and the Economy: Is There Any Match?" Report No. EI 2010-40. Econometric Institute, Erasmus University Rotterdam, 2010. https://repub.eur.nl/pub/20147.

"Wants School Girls to Hide Their Knees." *New York Times*. January 27, 1922.

"Weep Not, Girls, Breeze'll Make Them Red, Soon." *Capital Times* (Madison, WI). August 1, 1925.

"What Is a Flapper? The Critics Disagree." *Kansas City Times* (Kansas City, MO). March 18, 1922.

"What Shall the New Woman Wear, Skirts or Bloomers." *Los Angeles Herald* (Los Angeles, CA). September 15, 1895,

"Women Now Rouging Their Knees, Says a N.Y. Beauty Parlor Manager." *Birmingham News* (Birmingham, AL). May 26, 1921.

Yellis, Kenneth A. "Prosperity's Child: Some Thoughts on the Flapper." *American Quarterly* 21, no. 1 (1969): 44–64. https://doi.org/10.2307/2710772.

Zeitz, Joshua. *Flapper: A Madcap Story of Sex, Style, Celebrity, and the Women Who Made America Modern*. New York: Broadway Books, 2006.

4. THE WEAKER SEX?

Adachi, Noriko, Koji Nawata, Michio Maeta, and Youichi Kurozawa. "Relationship of the Menstrual Cycle Phase to Anterior Cruciate Ligament Injuries in Teen-aged Female Athletes." *Archives of Orthopaedic and Trauma Surgery* 128, no. 5 (2008): 473–78. https://doi.org/10.1007/s00402-007-0461-1.

Agel, Julie, Boris Bershadsky, and Elizabeth A. Arendt. "Hormonal Therapy: ACL and Ankle Injury." *Medicine and Science in Sports and Exercise* 38, no. 1 (2006): 7–12. https://doi.org/10.1249/01.mss.0000194072.13021.78.

Anderson, Allen F., David C. Dome, Shiva Gautam, Mark H. Awh, and Gregory W. Rennirt. "Correlation of Anthropometric Measurements, Strength, Anterior Cruciate Ligament Size, and Intercondylar Notch Characteristics to Sex Differences in Anterior Cruciate Ligament Tear Rates." *American Journal of Sports Medicine* 29, no. 1 (2001): 58–66. https://doi.org/10.1177/03635465010290011501.

Arendt, Elizabeth, and Randall Dick. "Knee Injury Patterns Among Men and Women in Collegiate Basketball and Soccer: NCAA Data and Review of Literature." *American Journal of Sports Medicine* 23, no. 6 (1995): 694–701. https://doi.org/10.1177/036354659502300611.

Asahina, Shintaro, Takeshi Muneta, and Yoichi Ezura. "Notchplasty in Anterior Cruciate Ligament Reconstruction: An Experimental Animal Study." *Arthroscopy* 16, no. 2 (2000): 165–72. https://doi.org/10.1016/s0749-8063(00)90031-8.

Aune, Arne K., Patrick W. Cawley, and Arne Ekeland. "Quadriceps Muscle Contraction Protects the Anterior Cruciate Ligament During Anterior Tibial Translation." *American Journal of Sports Medicine* 25, no. 2 (1997): 187–90. https://doi.org/10.1177/036354659702500208.

Bendjaballah, M. Z., A. Shirazi-Adl, and D. J. Zukor. "Finite Element Analysis of Human Knee Joint in Varus-Valgus." *Clinical Biomechanics* (Bristol) 12, no. 3 (1997): 139–48. https://doi.org/10.1016/S0268-0033(97)00072-7.

Betti, Lia, and Andrea Manica. "Human Variation in the Shape of the Birth Canal Is Significant and Geographically Structured." *Proceedings of the Royal Society B: Biological Sciences* 285, no. 1889 (2018): 20181807. https://doi.org/10.1098/rspb.2018.1807.

Boykoff, Jules. "Tokyo Olympics Head Yoshiro Mori Called out by Naomi Osaka and Others for Sexism." NBC News. February 10, 2021. https://www.nbcnews.com/think/opinion/tokyo-olympics-head-yoshiro-mori-called-out-naomi-osaka-others-ncna1257163.

Brooke-Marciniak, Beth A., and Donna de Varona. "Amazing Things Happen When You Give Female Athletes the Same Funding as Men." World Economic Forum. August 25, 2016. https://www.weforum.org/agenda/2016/08/sustaining-the-olympic-legacy-women-sports-and-public-policy/.

Cahill, Larry. "Denying the Neuroscience of Sex Differences." Quillette. March 29, 2019. https://quillette.com/2019/03/29/denying-the-neuroscience-of-sex-differences/.

Case, Mary Anne. "Heterosexuality as a Factor in the Long History of Women's Sports." *Law and Contemporary Problems* 80, no. 4 (2018): 25–46.

Chamberlain, Andrew, Daniel Zhao, and Amanda Stansell. *Progress on the Gender Pay Gap: 2019.* Glassdoor Economic Research. March 27, 2019. https://www.glassdoor.com/research/gender-pay-gap-2019/#.

Cooper, Danielle, and Heba Mahdy. "Oral Contraceptive Pills." In *StatPearls,* 28613632. Treasure Island, FL: StatPearls, 2020.

Dienst, Michael, Guenther Schneider, Katrin Altmeyer, Kristina Voelkering, Thomas Georg, Bernhard Kramann, and Dieter Kohn. "Correlation of Intercondylar Notch Cross Sections to the ACL Size: A High Resolution MR Tomographic in Vivo Analysis." *Archives of Orthopaedic and Trauma Surgery* 127, no. 4 (2007): 253–60. https://doi.org/10.1007/s00402-006-0177-7.

Dragoo, Jason L., Tiffany N. Castillo, Hillary J. Braun, Bethany A. Ridley, Ashleigh C. Kennedy, and S. Raymond Golish. "Prospective Correlation Between Serum Relaxin Concentration and Anterior Cruciate Ligament Tears Among Elite Collegiate Female Athletes." *American Journal of Sports Medicine* 39, no. 10 (2011): 2175–80. https://doi.org/10.1177/0363546511413378.

Dragoo, Jason L., Richard S. Lee, Prosper Benhaim, Gerald A. M. Finerman, and Sharon L. Hame. "Relaxin Receptors in the Human Female Anterior Cruciate

Ligament." *American Journal of Sports Medicine* 31, no. 4 (2003): 577–84. https://doi.org/10.1177/03635465030310041701.

Eliot, Lise. "Neurosexism: The Myth That Men and Women Have Different Brains." *Nature* 566 (2019): 453–54.

Emami, Mohammad-Jafar, Mohammad-Hossein Ghahramani, Farzad Abdinejad, and Hamid Namazi. "Q-Angle: An Invaluable Parameter for Evaluation of Anterior Knee Pain." *Archives of Iranian Medicine* 10, no. 1 (2007): 24–26.

Fernández-Jaén, Tomás, Juan Manuel López-Alcorocho, Elena Rodriguez-Iñigo, Fabián Castellán, Juan Carlos Hernández, and Pedro Guillén-García. "The Importance of the Intercondylar Notch in Anterior Cruciate Ligament Tears." *Orthopaedic Journal of Sports Medicine* 3, no. 8 (2015): 2325967115597882–82. https://doi.org/10.1177/2325967115597882.

Ferretti, Andrea, Paola Papandrea, Fabio Conteduca, and Pier Paolo Mariani. "Knee Ligament Injuries in Volleyball Players." *American Journal of Sports Medicine* 20, no. 2 (1992): 203–7. https://doi.org/10.1177/036354659202000219.

Fevold, H. L., Frederick L. Hisaw, and R. K. Meyer. "The Relaxative Hormone of the Corpus Luteum. Its Purification and Concentration." *Journal of the American Chemical Society* 52, no. 8 (1930): 3340–48. https://doi.org/10.1021/ja01371a051.

Ford, Kevin R., Gregory D. Myer, and Timothy E. Hewett. "Valgus Knee Motion During Landing in High School Female and Male Basketball Players." *Medicine and Science in Sports and Exercise* 35, no. 10 (2003): 1745–50. https://doi.org/10.1249/01.MSS.0000089346.85744.D9.

Ford, Kevin R., Gregory D. Myer, Laura C. Schmitt, Timothy L. Uhl, and Timothy E. Hewett. "Preferential Quadriceps Activation in Female Athletes with Incremental Increases in Landing Intensity." *Journal of Applied Biomechanics* 27, no. 3 (2011): 215–22. https://doi.org/10.1123/jab.27.3.215.

Fridén, Cecilia, Angelica Lindén Hirschberg, Tönu Saartok, and Per Renström. "Knee Joint Kinaesthesia and Neuromuscular Coordination During Three Phases of the Menstrual Cycle in Moderately Active Women." *Knee Surgery, Sports Traumatology, Arthroscopy* 14, no. 4 (2006): 383–89. https://doi.org/10.1007/s00167-005-0663-4.

Goldman, Bruce. "Two Minds: The Cognitive Differences Between Men and Women." *Stanford Medicine*. 2017. https://stanmed.stanford.edu/2017spring/how-mens-and-womens-brains-are-different.html.

Goldsmith, Laura T., and Gerson Weiss. "Relaxin in Human Pregnancy." *Annals of the New York Academy of Sciences* 1160, no. 1 (2009): 130–35. https://doi.org/10.1111/j.1749-6632.2008.03800.x.

Good, Lars, Magnus Odensten, and Jan Gillquist. "Intercondylar Notch Measurements with Special Reference to Anterior Cruciate Ligament Surgery." *Clinical Orthopaedics and Related Research*, no. 263 (1991): 185–89. https://doi.org/10.1097/00003086-199102000-00022.

Grelsamer, R. P., A. Dubey, and C. H. Weinstein. "Men and Women Have Similar Q Angles: A Clinical and Trigonometric Evaluation." *Journal of Bone and Joint Surgery* 87, no. 11 (2005): 1498–501. https://doi.org/10.1302/0301-620X.87B11.16485.

Haim, Amir, Moshe Yaniv, Samuel Dekel, and Hagay Amir. "Patellofemoral Pain Syndrome: Validity of Clinical and Radiological Features." *Clinical Orthopaedics and Related Research* 451 (2006): 223–28. https://doi.org/10.1097/01.blo.0000229284.45485.6c.

Halpern, Diane F., Camilla P. Benbow, David C. Geary, Ruben C. Gur, Janet Shibley Hyde, and Morton Ann Gernsbacher. "The Science of Sex Differences in Science and Mathematics." *Psychological Science in the Public Interest* 8, no. 1 (2007): 1–51. https://doi.org/10.1111/j.1529-1006.2007.00032.x.

Hanson, Ashley M., Darin A. Padua, J. Troy Blackburn, William E. Prentice, and Christopher J. Hirth. "Muscle Activation During Side-Step Cutting Maneuvers in Male and Female Soccer Athletes." *Journal of Athletic Training* 43, no. 2 (2008): 133–43. https://doi.org/10.4085/1062-6050-43.2.133.

Hertel, Jay, Nancy I. Williams, Lauren C. Olmsted-Kramer, Heather J. Leidy, and Margot Putukian. "Neuromuscular Performance and Knee Laxity Do Not Change Across the Menstrual Cycle in Female Athletes." *Knee Surgery, Sports Traumatology, Arthroscopy* 14, no. 9 (2006): 817–22. https://doi.org/10.1007/s00167-006-0047-4.

Herzberg, Simone D., Makalapua L. Motu'apuaka, William Lambert, Rongwei Fu, Jacqueline Brady, and Jeanne-Marie Guise. "The Effect of Menstrual Cycle and Contraceptives on ACL Injuries and Laxity: A Systematic Review and Meta-Analysis." *Orthopaedic Journal of Sports Medicine* 5, no. 7 (2017). https://doi.org/10.1177/2325967117718781.

Hewett, Timothy E., and Gregory D. Myer. "Reducing Knee and Anterior Cruciate Ligament Injuries Among Female Athletes: A Systematic Review of Neuromuscular Training Interventions." *Journal of Knee Surgery* 18, no. 1 (2005): 82–88. https://doi.org/10.1055/s-0030-1248163.

Hewett, Timothy E., Amanda L. Stroupe, Thomas A. Nance, and Frank R. Noyes. "Plyometric Training in Female Athletes: Decreased Impact Forces and Increased Hamstring Torques." *American Journal of Sports Medicine* 24, no. 6 (1996): 765–73. https://doi.org/10.1177/036354659602400611.

Hewett, T. E., J. S. Torg, and B. P. Boden. "Video Analysis of Trunk and Knee Motion During Non-Contact Anterior Cruciate Ligament Injury in Female Athletes: Lateral Trunk and Knee Abduction Motion Are Combined Components of the Injury Mechanism." *British Journal of Sports Medicine* 43, no. 6 (2009): 417–22. https://doi.org/10.1136/bjsm.2009.059162.

Hewett, Timothy E., Bohdanna T. Zazulak, and Gregory D. Myer. "Effects of the Menstrual Cycle on Anterior Cruciate Ligament Injury Risk: A Systematic Review." *American Journal of Sports Medicine* 35, no. 4 (2007): 659–68. https://doi.org/10.1177/0363546506295699.

Horton, Melissa G., and Terry L. Hall. "Quadriceps Femoris Muscle Angle: Normal Values and Relationships with Gender and Selected Skeletal Measures." *Physical Therapy* 69, no. 11 (1989): 897–901. https://doi.org/10.1093/ptj/69.11.897.

Huston, Laura J., and Edward M. Wojtys. "Neuromuscular Performance Characteristics in Elite Female Athletes." *American Journal of Sports Medicine* 24, no. 4 (1996): 427–36. https://doi.org/10.1177/036354659602400405.

Khasawneh, Ramada R., Mohammed Z. Allouh, and Ejlal Abu-El-Rub. "Measurement of the Quadriceps (Q) Angle with Respect to Various Body Parameters in Young Arab Population." *PLoS One* 14, no. 6 (2019): e0218387-e87. https://doi.org/10.1371/journal.pone.0218387.

LaPrade, Robert F., Glenn C. Terry, Ronald D. Montgomery, David Curd, and David J. Simmons. "The Effects of Aggressive Notchplasty on the Normal Knee in Dogs." *American Journal of Sports Medicine* 26, no. 2 (1998): 193–200. https://doi.org/10.1177/03635465980260020801.

Lefevre, N., Y. Bohu, S. Klouche, J. Lecocq, and S. Herman. "Anterior Cruciate Ligament Tear During the Menstrual Cycle in Female Recreational Skiers." *Orthopaedics and Traumatology: Surgery and Research* 99, no. 5 (Sep 2013): 571–75. https://doi.org/10.1016/j.otsr.2013.02.005.

Li, Zheng, Changshu Li, Li Li, and Ping Wang. "Correlation Between Notch Width Index Assessed Via Magnetic Resonance Imaging and Risk of Anterior Cruciate Ligament Injury: An Updated Meta-Analysis." *Surgical and Radiologic Anatomy* 42, no. 10 (2020): 1209–17. https://doi.org/10.1007/s00276-020-02496-6.

Liu, Stephen H., Raad Al-Shaikh, Vahé Panossian, Rong-Sen Yang, Scott D. Nelson, Neptune Soleiman, Gerald A. M. Finerman, and Joseph M. Lane. "Primary Immunolocalization of Estrogen and Progesterone Target Cells in the Human Anterior Cruciate Ligament." *Journal of Orthopaedic Research* 14, no. 4 (1996): 526–33. https://doi.org/10.1002/jor.1100140405.

Lombardo, Stephen, Paul M. Sethi, and Chad Starkey. "Intercondylar Notch Stenosis Is Not a Risk Factor for Anterior Cruciate Ligament Tears in Professional Male Basketball Players: An 11-Year Prospective Study." *American Journal of Sports Medicine* 33, no. 1 (2005): 29–34. https://doi.org/10.1177/0363546504266482.

Loudon, Janice K., Walter Jenkins, and Karen L. Loudon. "The Relationship Between Static Posture and ACL Injury in Female Athletes." *Journal of Orthopaedic and Sports Physical Therapy* 24, no. 2 (1996): 91–97. https://doi.org/10.2519/jospt.1996.24.2.91.

Lun, V., W. H. Meeuwisse, P. Stergiou, and D. Stefanyshyn. "Relation Between Running Injury and Static Lower Limb Alignment in Recreational Runners." *British Journal of Sports Medicine* 38, no. 5 (2004): 576–80. https://doi.org/10.1136/bjsm.2003.005488.

Lund-Hanssen, Hakon, James Gannon, Lars Engebretsen, Ketil J. Holen, Svein Anda, and Lars Vatten. "Intercondylar Notch Width and the Risk for Anterior Cruciate Ligament Rupture: A Case-Control Study in 46 Female Handball Players." *Acta Orthopaedica* 65, no. 5 (1994): 529–32. https://doi.org/10.3109/17453679409000907.

Malinzak, Robert A., Scott M. Colby, Donald T. Kirkendall, Bing Yu, and William E. Garrett. "A Comparison of Knee Joint Motion Patterns Between Men and Women in Selected Athletic Tasks." *Clinical Biomechanics* 16, no. 5 (2001): 438–45. https://doi.org/10.1016/S0268-0033(01)00019-5.

Mandelbaum, Bert R., Holly J. Silvers, Diane S. Watanabe, John F. Knarr, Stephen D. Thomas, Letha Y. Griffin, Donald T. Kirkendall, and William Garrett Jr. "Effectiveness of a Neuromuscular and Proprioceptive Training Program in Preventing Anterior Cruciate Ligament Injuries in Female Athletes." *American*

Journal of Sports Medicine 33, no. 7 (2005): 1003. https://doi.org/10.1177 /0363546504272261.

Medina McKeon, Jennifer M., and Jay Hertel. "Sex Differences and Representative Values for 6 Lower Extremity Alignment Measures." *Journal of Athletic Training* 44, no. 3 (2009): 249–55. https://doi.org/10.4085/1062-6050-44.3.249.

Mohamed, E. E., U. Useh, and B. F. Mtshali. "Q-Angle, Pelvic Width, and Inter-condylar Notch Width as Predictors of Knee Injuries in Women Soccer Players in South Africa." *African Health Sciences* 12, no. 2 (2012): 174–80. https://doi .org/10.4314/ahs.v12i2.15.

Muneta, Takeshi, Kazuo Takakuda, and Haruyasu Yamamoto. "Intercondylar Notch Width and Its Relation to the Configuration and Cross-Sectional Area of the Anterior Cruciate Ligament: A Cadaveric Knee Study." *American Journal of Sports Medicine* 25, no. 1 (1997): 69–72. https://doi.org/10.1177/036354659702500113.

Nagai, Takashi, Timothy C. Sell, John P. Abt, and Scott M. Lephart. "Reliability, Precision, and Gender Differences in Knee Internal/External Rotation Proprio-ception Measurements." *Physical Therapy in Sport* 13, no. 4 (2011): 233–37. https://doi.org/10.1016/j.ptsp.2011.11.004.

Nguyen, Anh-Dung, and Sandra J. Shultz. "Sex Differences in Clinical Measures of Lower Extremity Alignment." *Journal of Orthopaedic and Sports Physical Ther-apy* 37, no. 7 (2007): 389–98. https://doi.org/10.2519/jospt.2007.2487.

Park, Roberta. "Sport, Gender and Society in a Transatlantic Victorian Perspec-tive." In *From "Fair Sex" to Feminism: Sport and the Socialization of Women in the Industrial and Post-Industrial Eras*, ed. J. A. Mangan and Roberta Park, 58–96. Totowa, NJ: Frank Cass, 1987.

Pollard, Christine D., Barry Braun, and Joseph Hamill. "Influence of Gender, Estro-gen and Exercise on Anterior Knee Laxity." *Clinical Biomechanics* 21, no. 10 (2006): 1060–66. https://doi.org/10.1016/j.clinbiomech.2006.07.002.

Powell, John W., and Kim D. Barber-Foss. "Sex-Related Injury Patterns Among Selected High School Sports." *American Journal of Sports Medicine* 28, no. 3 (2000): 385–91. https://doi.org/10.1177/03635465000280031801.

Quatman, Carmen E., Kevin R. Ford, Gregory D. Myer, Mark V. Paterno, and Timothy E. Hewett. "The Effects of Gender and Pubertal Status on Generalized Joint Laxity in Young Athletes." *Journal of Science and Medicine in Sport* 11, no. 3 (2007): 257–63. https://doi.org/10.1016/j.jsams.2007.05.005.

Rahr-Wagner, Lene, Theis Muncholm Thillemann, Frank Mehnert, Alma Becic Ped-ersen, and Martin Lind. "Is the Use of Oral Contraceptives Associated with Opera-tively Treated Anterior Cruciate Ligament Injury? A Case-Control Study from the Danish Knee Ligament Reconstruction Registry." *American Journal of Sports Medi-cine* 42, no. 12 (2014): 2897–905. https://doi.org/10.1177/0363546514557240.

Ranuccio, Francesco, Filippo Familiari, Giuseppe Tedesco, Francesco La Camera, and Giorgio Gasparini. "Effects of Notchplasty on Anterior Cruciate Ligament Reconstruction: A Systematic Review." *Joints* 5, no. 3 (2017): 173–79. https:// doi.org/10.1055/s-0037-1605551.

Rauh, Mitchell J., Thomas D. Koepsell, Frederick P. Rivara, Stephen G. Rice, and Anthony J. Margherita. "Quadriceps Angle and Risk of Injury Among High School Cross-Country Runners." *Journal of Orthopaedic and Sports Physical Therapy* 37, no. 12 (2007): 725–33. https://doi.org/10.2519/jospt.2007.2453.

Raveendranath, Veeramani, Shankar Nachiket, Narayanan Sujatha, Ranganath Priya, and Devi Rema. "The Quadriceps Angle (Q Angle) in Indian Men and Women." *European Journal of Anatomy* 13, no. 3 (2009): 105–9.

Renström, P., S. W. Arms, T. S. Stanwyck, R. J. Johnson, and M. H. Pope. "Strain Within the Anterior Cruciate Ligament During Hamstring and Quadriceps Activity." *American Journal of Sports Medicine* 14, no. 1 (1986): 83–87. https://doi.org/10.1177/036354658601400114.

Rozzi, Susan L., Scott M. Lephart, William S. Gear, and Freddie H. Fu. "Knee Joint Laxity and Neuromuscular Characteristics of Male and Female Soccer and Basketball Players." *American Journal of Sports Medicine* 27, no. 3 (1999): 312–19. https://doi.org/10.1177/03635465990270030801.

Ruedl, Gerhard, Patrick Ploner, Ingrid Linortner, Alois Schranz, Christian Fink, Renate Sommersacher, Elena Pocecco, Werner Nachbauer, and Martin Burtscher. "Are Oral Contraceptive Use and Menstrual Cycle Phase Related to Anterior Cruciate Ligament Injury Risk in Female Recreational Skiers?" *Knee Surgery, Sports Traumatology, Arthroscopy* 17, no. 9 (2009): 1065–69. https://doi.org/10.1007/s00167-009-0786-0.

Russell, Kyla A., Riann M. Palmieri, Steven M. Zinder, and Christopher D. Ingersoll. "Sex Differences in Valgus Knee Angle During a Single-Leg Drop Jump." *Journal of Athletic Training* 41, no. 2 (2006): 166–71.

Schauberger, Charles W., Brenda L. Rooney, Laura Goldsmith, David Shenton, Paul D. Silva, and Ana Schaper. "Peripheral Joint Laxity Increases in Pregnancy but Does Not Correlate with Serum Relaxin Levels." *American Journal of Obstetrics and Gynecology* 174, no. 2 (1996): 667–71. https://doi.org/10.1016/S0002-9378(96)70447-7.

Seneviratne, Aruna, Erik Attia, Riley J. Williams, Scott A. Rodeo, and Jo A. Hannafin. "The Effect of Estrogen on Ovine Anterior Cruciate Ligament Fibroblasts: Cell Proliferation and Collagen Synthesis." *American Journal of Sports Medicine* 32, no. 7 (2004): 1613–18. https://doi.org/10.1177/0363546503262179.

Shambaugh, Philip J., Andrew Klein, and John H. Herbert. "Structural Measures as Predictors of Injury in Basketball Players." *Medicine and Science in Sports and Exercise* 23, no. 5 (1991): 522–27. https://doi.org/10.1249/00005768-199105000-00003.

Shelbourne, Donald K., Thorp J. Davis, and Thomas E. Klootwyk. "The Relationship Between Intercondylar Notch Width of the Femur and the Incidence of Anterior Cruciate Ligament Tears: A Prospective Study." *American Journal of Sports Medicine* 26, no. 3 (1998): 402–8.

Steiner, M. "Editorial Commentary: Size Does Matter—Anterior Cruciate Ligament Graft Diameter Affects Biomechanical and Clinical Outcomes." *Arthroscopy* 33, no. 5 (May 2017): 1014–15. https://doi.org/10.1016/j.arthro.2017.01.020.

Stevenson, J. Herbert, Chad S. Beattie, Jennifer B. Schwartz, and Brian D. Busconi. "Assessing the Effectiveness of Neuromuscular Training Programs in Reducing the Incidence of Anterior Cruciate Ligament Injuries in Female Athletes: A Systematic Review." *American Journal of Sports Medicine* 43, no. 2 (2015): 482–90. https://doi.org/10.1177/0363546514523388.

Stewart, Dennis R., Abbie C. Celniker, Clinton A. Taylor, Jeffrey R. Cragun, James W. Overstreet, and Bill L. Lasley. "Relaxin in the Peri-Implantation Period." *Journal of Clinical Endocrinology and Metabolism* 70, no. 6 (1990): 1771–73. https://doi.org/10.1210/jcem-70-6-1771.

Stijak, Lazar, Vidosava Radonjić, Valentina Nikolić, Zoran Blagojević, Milan Aksić, and Branislav Filipović. "Correlation Between the Morphometric Parameters of the Anterior Cruciate Ligament and the Intercondylar Width: Gender and Age Differences." *Knee Surgery, Sports Traumatology, Arthroscopy* 17, no. 7 (2009): 812–17. https://doi.org/10.1007/s00167-009-0807-z.

Strickland, Sabrina M., Thomas W. Belknap, Simon A. Turner, Timothy M. Wright, and Jo A. Hannafin. "Lack of Hormonal Influences on Mechanical Properties of Sheep Knee Ligaments." *American Journal of Sports Medicine* 31, no. 2 (2003): 210–15. https://doi.org/10.1177/03635465030310020901.

Sugimoto, Dai, Gregory D. Myer, Heather M. Bush, Maddie F. Klugman, Jennifer M. Medina McKeon, and Timothy E. Hewett. "Compliance with Neuromuscular Training and Anterior Cruciate Ligament Injury Risk Reduction in Female Athletes: A Meta-Analysis." *Journal of Athletic Training* 47, no. 6 (2012): 714–23. https://doi.org/10.4085/1062-6050-47.6.10.

Taunton, J. E., M. B. Ryan, D. B. Clement, D. C. McKenzie, D. R. Lloyd-Smith, and B. D. Zumbo. "A Retrospective Case-Control Analysis of 2002 Running Injuries." *British Journal of Sports Medicine* 36, no. 2 (2002): 95–101. https://doi.org/10.1136/bjsm.36.2.95.

Temesi, John, Pierrick J. Arnal, Thomas Rupp, Léonard Féasson, Régine Cartier, Laurent Gergelé, Samuel Verges, Vincent Martin, and Guillaume Y. Millet. "Are Females More Resistant to Extreme Neuromuscular Fatigue?" *Medicine and Science in Sports and Exercise* 47, no. 7 (2015): 1372–82. https://doi.org/10.1249/MSS.0000000000000540.

Unemori, Elaine N., L. Steven Beck, Wyne Pun Lee, Yvette Xu, Mark Siegel, Gilbert Keller, H. Denny Liggitt, Eugene A. Bauer, and Edward P. Amento. "Human Relaxin Decreases Collagen Accumulation in Vivo in Two Rodent Models of Fibrosis." *Journal of Investigative Dermatology* 101, no. 3 (1993): 280–85. https://doi.org/10.1111/1523-1747.ep12365206.

van Diek, Floor M., Megan R. Wolf, Christopher D. Murawski, Carola F. van Eck, and Freddie H. Fu. "Knee Morphology and Risk Factors for Developing an Anterior Cruciate Ligament Rupture: An MRI Comparison Between ACL-Ruptured and Non-Injured Knees." *Knee Surgery, Sports Traumatology, Arthroscopy* 22, no. 5 (2014): 987–94. https://doi.org/10.1007/s00167-013-2588-7.

Wall-Scheffler, Cara M., and Marcella J. Myers. "The Biomechanical and Energetic Advantages of a Mediolaterally Wide Pelvis in Women." *Anatomical Record* 300, no. 4 (2017): 764–75. https://doi.org/10.1002/ar.23553.

Warden, Stuart J., Leanne K. Saxon, Alesha B. Castillo, and Charles H. Turner. "Knee Ligament Mechanical Properties Are Not Influenced by Estrogen or Its Receptors." *American Journal of Physiology Endocrinology and Metabolism* 290, no. 5 (2006): 1034–40. https://doi.org/10.1152/ajpendo.00367.2005.

Wild, Catherine Y., Julie R. Steele, and Bridget J. Munro. "Why Do Girls Sustain More Anterior Cruciate Ligament Injuries Than Boys? A Review of the Changes in Estrogen and Musculoskeletal Structure and Function During Puberty." *Sports Medicine* 42, no. 9 (2012): 733–49. https://doi.org/10.1007/BF03262292.

Williams, Sophie. "Are Women Better Ultra-Endurance Athletes Than Men?" BBC News. August 11, 2019. https://www.bbc.com/news/world-49284389.

Wilson, Ross, and Alan A. Barhorst. "Intercondylar Notch Impingement of the Anterior Cruciate Ligament: A Cadaveric in Vitro Study Using Robots." *Journal of Healthcare Engineering* 2018 (2018): 8698167–27. https://doi.org/10.1155/2018/8698167.

Witvrouw, Erik, Roeland Lysens, Johan Bellemans, Dirk Cambier, and Guy Vanderstraeten. "Intrinsic Risk Factors for the Development of Anterior Knee Pain in an Athletic Population: A Two-Year Prospective Study." *American Journal of Sports Medicine* 28, no. 4 (2000): 480–89. https://doi.org/10.1177/03635465000280040701.

Wojtys, Edward M., Laura J. Huston, Melbourne D. Boynton, Kurt P. Spindler, and Thomas N. Lindenfeld. "The Effect of the Menstrual Cycle on Anterior Cruciate Ligament Injuries in Women as Determined by Hormone Levels." *American Journal of Sports Medicine* 30, no. 2 (2002): 182–88. https://doi.org/10.1177/03635465020300020601.

Woodhouse, Emma, Gregory A. Schmale, Peter Simonian, Allan Tencer, Phillipe Huber, and Kristy Seidel. "Reproductive Hormone Effects on Strength of the Rat Anterior Cruciate Ligament." *Knee Surgery, Sports Traumatology, Arthroscopy* 15, no. 4 (2007): 453–60. https://doi.org/10.1007/s00167-006-0237-0.

Wordeman, Samuel C., Carmen E. Quatman, Christopher C. Kaeding, and Timothy E. Hewett. "In Vivo Evidence for Tibial Plateau Slope as a Risk Factor for Anterior Cruciate Ligament Injury: A Systematic Review and Meta-Analysis." *American Journal of Sports Medicine* 40, no. 7 (2012): 1673–81. https://doi.org/10.1177/0363546512442307.

Yu, Warren D., Stephen H. Liu, Joshua D. Hatch, Vahé Panossian, and Gerald A. M. Finerman. "Effect of Estrogen on Cellular Metabolism of the Human Anterior Cruciate Ligament." *Clinical Orthopaedics and Related Research* 366 (1999): 229–38. https://doi.org/10.1097/00003086-199909000-00030.

Yu, Warren D., Vahé Panossian, Joshua D. Hatch, Stephen H. Liu, and Gerald A. M. Finerman. "Combined Effects of Estrogen and Progesterone on the Anterior Cruciate Ligament." *Clinical Orthopaedics and Related Research* 383 (2001): 268–81. https://doi.org/10.1097/00003086-200102000-00031.

Zazulak, Bohdanna T., Mark Paterno, Gregory D. Myer, William A. Romani, and Timothy E. Hewett. "The Effects of the Menstrual Cycle on Anterior Knee Laxity: A Systematic Review." *Sports Medicine* 36, no. 10 (2006): 847–62. https://doi.org/10.2165/00007256-200636100-00004.

Zeng, Chao, Shu-guang Gao, Jie Wei, Tu-bao Yang, Ling Cheng, Wei Luo, Min Tu, et al. "The Influence of the Intercondylar Notch Dimensions on Injury of the Anterior Cruciate Ligament: A Meta-Analysis." *Knee Surgery, Sports Traumatology, Arthroscopy* 21, no. 4 (2013): 804–15. https://doi.org/10.1007/s00167-012 -2166-4.

Zuckerman, Scott L., Adam M. Wegner, Karen G. Roos, Aristarque Djoko, Thomas P. Dompier, and Zachary Y. Kerr. "Injuries Sustained in National Collegiate Athletic Association Men's and Women's Basketball, 2009/2010–2014/2015." *British Journal of Sports Medicine* 52, no. 4 (2018): 261–68. https://doi.org/10.1136 /bjsports-2016-096005.

5. TO KNEEL, OR NOT TO KNEEL

Adnan, F. "Chinese Customers Furious after Samsung Executives Knelt to Apologize for the Galaxy Note 7." SamMobile. Updated November 2, 2016. https:// www.sammobile.com/2016/11/02/chinese-customers-furious-after-samsung -executives-knelt-to-apologize-for-the-galaxy-note-7/.

Allison, Keith. "Washington Redskins Teammates During the National Anthem Before a Game Against the Oakland Raiders at FedExField on September 24, 2017, in Landover, Maryland." Wikimedia Commons. September 24, 2017. https:// commons.wikimedia.org/wiki/File:Washington_Redskins_National_Anthem _Kneeling_(37301887651)_(cropped)_(cropped).jpg.

Around the NFL. "Roger Goodell: NFL 'Wrong' for Not Listening to Protesting Players Earlier." NFL. June 5, 2020. https://www.nfl.com/news/roger-goodell-nfl -wrong-for-not-listening-to-protesting-players-earlier.

Branch, John. "The Anthem Debate Is Back. But Now It's Standing That's Polarizing." *New York Times*. July 4, 2020. https://www.nytimes.com/2020/07/04 /sports/football/anthem-kneeling-sports.html.

Brinson, Will. "Here's How Nate Boyer Got Colin Kaepernick to Go from Sitting to Kneeling." CBS/NFL. September 27, 2016. https://www.cbssports.com/nfl/news /heres-how-nate-boyer-got-colin-kaepernick-to-go-from-sitting-to-kneeling/.

Burin, Eric. "Race, Dissent, and Patriotism in 21st Century America." In *Protesting on Bended Knee: Race, Dissent, and Patriotism in 21st Century America*, ed. Eric Burin, 1–83. Grand Forks, ND: Digital Press at the University of North Dakota, 2018.

Cahn, Dianna. "VFW, American Legion: NFL Protests Disrespectful to Vets; Others Disagree." Stripes, 2017. Accessed April 29, 2021. https://www.stripes.com/news /us/vfw-american-legion-nfl-protests-disrespectful-to-vets-others-disagree -1.489529.

Dallas, Kelsey. "The Religious Significance of Taking a Knee." Deseret News. August 14, 2020. https://www.deseret.com/indepth/2020/8/14/21362248/athletes-kneeling -national-anthem-colin-kaepernick-eric-reid-sam-coonrod-patriotism-religion.

Dan, Xingwu. "从马葛尔尼使华看国际体系之争 [A Study of Conflicting International Systems Through the Macartney Embassy to China]." 国际政治科学 2 (2006): 1–27.

Dann, Carrie. "NBC/WSJ Poll: Majority Say Kneeling During Anthem 'Not Appropriate'." NBC News. August 31, 2018. https://www.nbcnews.com/politics/first-read/nbc-wsj-poll-majority-say-kneeling-during-anthem-not-appropriate-n904891.

Eidson, Matt. "The Veteran View of Colin Kaepernick." In *Protesting on Bended Knee: Race, Dissent, and Patriotism in 21st Century America*, ed. Eric Burin, 253–56. Grand Forks, ND: Digital Press at The University of North Dakota, 2018.

Gao, Hao. "The 'Inner Kowtow Controversy' During the Amherst Embassy to China, 1816–1817." *Diplomacy and Statecraft* 27, no. 4 (2016): 595–614. https://doi.org/10.1080/09592296.2016.1238691.

"George Floyd: US Soccer Overturns Ban on Players Kneeling." BBC News. June 11, 2020. https://www.bbc.com/news/world-us-canada-53003816.

"George Floyd: What Happened in the Final Moments of His Life." BBC News. July 16, 2020. https://www.bbc.com/news/world-us-canada-52861726.

Gillingham, Paul. "The Macartney Embassy to China, 1792–94." *History Today*. 43, no. 11 (1993): 28–34.

Grovier, Kelly. "The Surprising Power of Kneeling." BBC Culture. September 29, 2017. https://www.bbc.com/culture/article/20170929-the-surprising-power-of-kneeling.

Hill, Evan, Ainara Tiefenthäler, Christiaan Triebert, Drew Jordan, Haley Willis, and Robin Stein. "How George Floyd Was Killed in Police Custody." *New York Times.* May 31, 2020. https://www.nytimes.com/2020/05/31/us/george-floyd-investigation.html.

Hill, Rachel (@r_hill3). "Unity." Twitter, June 30, 2020, 7:56 p.m. https://twitter.com/r_hill3/status/1278145465406627841.

Howard, Adam. "Colin Kaepernick National Anthem Protest Catches on in NFL." NBC News. September 12, 2016. https://www.nbcnews.com/news/us-news/colin-kaepernick-national-anthem-protest-catches-nfl-n646671.

Huang, Yinong. "印象与真相：清朝中英两国的观礼之争 [Impression and Reality: The Sino-British Ritual War During the Qing Dynasty]." 中央研究院历史语言研究所集刊 78, no. 1 (2007): 35–106.

"The Image of the Supplicant Slave: Advert or Advocate?" 1807 Commemorated. Institute for the Public Understanding of the Past, University of York. 2007. https://archives.history.ac.uk/1807commemorated/discussion/supplicant_slave.html.

JEFF (@jeffisrael25). "Thinking NFL players are 'protesting the flag' is like thinking Rosa Parks was protesting public transportation." Twitter, September 24, 2017, 3:24 p.m. https://twitter.com/jeffisrael25/status/912065134033539073?lang=en.

Kishi, Roudabeh, and Sam Jones. "Demonstrations and Political Violence in America: New Data for Summer 2020." Armed Conflict Location and Event Data Project (ACLED). September 2020. https://acleddata.com/2020/09/03/demonstrations-political-violence-in-america-new-data-for-summer-2020/.

Krugler, David F. "African American Patriotism During the World War I Era." In *Protesting on Bended Knee: Race, Dissent, and Patriotism in 21st Century America*, ed. Eric Burin, 187–93. Grand Forks, ND: Digital Press at The University of North Dakota, 2018.

Kyodo, Jiji. "Traditional Japanese Sitting Style to Be Recognized as Punishment Under New Law." *Japan Times*. December 4, 2019. https://www.japantimes.co.jp /news/2019/12/04/national/social-issues/japanese-sitting-style-recognized -punishment-new-law/.

Lewis, Aimee. "Colin Kaepernick: A Cultural Star Fast Turning Into a Global Icon." CNN. September 10, 2018. https://www.cnn.com/2018/09/07/sport/colin-kaepernick -protest-taking-the-knee-nate-boyer-spt-intl/index.html.

Maese, Rick, and Emily Guskin. "Most Americans Support Athletes Speaking Out, Say Anthem Protests Are Appropriate, Post Poll Finds." *Washington Post*. September 10, 2020. https://www.washingtonpost.com/sports/2020/09/10/poll -nfl-anthem-protests/.

Maske, Mark. "Americans Generally Disapprove of Players' Anthem Protests but Opinion Divided Along Racial Lines, Poll Finds." *Washington Post*. October 11, 2016. https://www.washingtonpost.com/news/sports/wp/2016/10/11/americans -generally-disapprove-of-players-anthem-protests-but-opinion-divided-along -racial-lines-poll-finds/.

Men's Health. "Survey: Men Bend Knee to Tradition." February 13, 2012. https:// www.menshealth.com/sex-women/a19515770/wedding-proposals/.

Nestel, M. L. "Trump Says Issue of NFL Players Kneeling 'Has Nothing to Do with Race'." ABC News. September 25, 2017. https://abcnews.go.com/US/trump -issue-nfl-players-kneeling-race/story?id=50074211.

NBC News. "NBC News Poll: June 1995." Cornell University, Ithaca, NY: Roper Center for Public Opinion Research. 1995. https://ropercenter.cornell.edu/ipoll /study/31106496.

Newsome, Bree (@BreeNewsome). "Don't allow racists to reframe #TakeAKnee as being a debate about anthem & flag. It's a protest of police brutality & racism." Twitter, September 23, 2017, 7:45 a.m. https://twitter.com/breenewsome/status /911587445296254982.

O'Kane, Caitlin. "Police Officers Kneel in Solidarity with Protesters in Several U.S. Cities." CBS News. June 1, 2020. https://www.cbsnews.com/news/protesters -police-kneel-solidarity-george-floyd/.

Perez, A. J. "Report: Colin Kaepernick, Eric Reid Got Less Than $10 Million in NFL Collusion Settlement." *USA Today*. March 21, 2019. https://www.usatoday .com/story/sports/nfl/2019/03/21/colin-kaepernick-eric-reid-nfl-collusion -settlement/3237678002/.

Pritchard, Earl H. "The Kotow in the Macartney Embassy to China in 1793." *Far Eastern Quarterly* 2, no. 2 (1943): 163–203. https://doi.org/10.2307/2049496.

"Protests Across the Globe After George Floyd's Death." CNN. June 13, 2020. https://www.cnn.com/2020/06/06/world/gallery/intl-george-floyd-protests /index.html.

Reid, Eric. "Eric Reid: Why Colin Kaepernick and I Decided to Take a Knee." *New York Times*. September 25, 2017. https://www.nytimes.com/2017/09/25/opinion /colin-kaepernick-football-protests.html.

Schulman, Henry. "Giants' Sam Coonrod Cites His Faith, Issues with Black Lives Matter for Decision Not to Kneel." *San Francisco Chronicle*. July 24, 2020. https://www.sfchronicle.com/giants/article/Giants-Coonrod-cites-his-faith-issues-with-15430927.php.

Smith, Jeremy Adam, and Dacher Keltner. "The Psychology of Taking a Knee." *Scientific American*. September 29, 2017. https://blogs.scientificamerican.com/voices/the-psychology-of-taking-a-knee/.

State of Minnesota vs. Thomas Kiernan Lane. Court File No. 27-CR-20-12951. Minnesota Judicial Branch. July 7, 2020. https://www.mncourts.gov/mncourtsgov/media/High-Profile-Cases/27-CR-20-12951-TKL/Memorandum07072020.pdf.

Staunton, George. *An Authentic Account of an Embassy from the King of Great Britain to the Emperor of China*. Vol. 1. London: W. Bulmer for G. Nicol, 1797.

Staunton, George. *An Authentic Account of an Embassy from the King of Great Britain to the Emperor of China*. Vol. 2. London: W. Bulmer for G. Nicol, 1797.

Stelter, Brian. "NFL Aired Unity Ad in Prime Time on Sunday." CNN. September 25, 2017. https://money.cnn.com/2017/09/24/media/nfl-unity-ad-trump/.

Streeter, Kurt. "Kneeling, Fiercely Debated in the N.F.L., Resonates in Protests." *New York Times*. June 5, 2020. https://www.nytimes.com/2020/06/05/sports/football/george-floyd-kaepernick-kneeling-nfl-protests.html. "横扫一切牛鬼蛇神 [Sweep Away All Evils]." Editorial. *People's Daily*. June 1, 1966.

Tan, Lin, and Wei Zhang. "网友质疑2000中学生集体跪拜父母; 我们需要何种感恩教育 [2,000 Middle School Students Kowtowing to Parents Invites Critique; What Kind of Gratitude Education Do We Need]." People's Daily Online. November 23, 2018. http://www.people.com.cn/n1/2018/1123/c347407-30418728.html.

Taylor, Derrick Bryson. "George Floyd Protests: A Timeline." *New York Times*. March 28, 2021. https://www.nytimes.com/article/george-floyd-protests-timeline.html.

Tesler, Michael. "To Many Americans, Being Patriotic Means Being White." *Washington Post*. October 13, 2017. https://www.washingtonpost.com/news/monkey-cage/wp/2017/10/13/is-white-resentment-about-the-nfl-protests-about-race-or-patriotism-or-both/.

VoteVets (@votevets). "As veterans, we swore an oath to support and defend the Constitutional rights of all citizens to speak freely and protest." Twitter, November 16, 2019, 8:17 p.m. https://twitter.com/votevets/status/1195903699706634240.

Walker, Rhiannon. "One Year Later, Steve Wyche Reflects on Breaking the Colin Kaepernick Story." The Undefeated. August 28, 2017. https://theundefeated.com/features/one-year-later-steve-wyche-colin-kaepernick-story/.

Wang, Dongqing. "Representing Kowtow: Civility and Civilization in Early Sino-British Encounters." *The Eighteenth Century* 60, no. 3 (2019): 269–92. https://doi.org/10.1353/ecy.2019.0022.

Wang, Wei. 椅子改变中国 [*Chairs Changed China*]. Beijing, China: China International Radio Press, 2009.

West Virginia State Board of Education et al. v. Barnette et al. 319 U.S. 624 (1943). Legal Information Institute. https://www.law.cornell.edu/supremecourt/text/319/624.

Witz, Billy. "This Time, Colin Kaepernick Takes a Stand by Kneeling." *New York Times.* September 1, 2016. https://www.nytimes.com/2016/09/02/sports/football/colin-kaepernick-kneels-national-anthem-protest.html.

Wood, Frances. "Britain's First View of China: The Macartney Embassy 1792–1794." *RSA Journal* 142, no. 5447 (1994): 59–68.

Wright, Jo (@JoWright59). "I'm sick of these people who believe the military is disrespected by Kaepernick & others kneeling. I'm a veteran and I support the right to kneel." Twitter. September 7, 2018, 9:42 p.m. https://mobile.twitter.com/JoWright59/status/1038271217596334080?cxt=HHwWgIC8lfyK1-gcAAAA.

Wyche, Steve. "Colin Kaepernick Explains Why He Sat During National Anthem." NFL. August 27, 2016. https://www.nfl.com/news/colin-kaepernick-explains-why-he-sat-during-national-anthem-0ap3000000691077.

Xiang, Lili, and Yan Jiang. "李阳博客贴出三千学生集体跪拜老师照片 [Li Yang's Blog Posted Photos of Three Thousand Students Kneeling to their Teachers]." Sina News. September 11, 2007. http://news.sina.com.cn/c/2007-09-11/014113860064.shtml.

6. TREATMENT OR PLACEBO

Airaksinen, Olavi V., Nils Kyrklund, Kyösti Latvala, Jukka P. Kouri, Mats Grönblad, and Pertti Kolari. "Efficacy of Cold Gel for Soft Tissue Injuries: A Prospective Randomized Double-Blinded Trial." *American Journal of Sports Medicine* 31, no. 5 (2003): 680–84. https://doi.org/10.1177/03635465030310050801.

Albright, John P., John W. Powell, Walter Smith, Al Martindale, Edward Crowley, Jeff Monroe, Russ Miller, et al. "Medial Collateral Ligament Knee Sprains in College Football: Brace Wear Preferences and Injury Risk." *American Journal of Sports Medicine* 22, no. 1 (1994): 2–11. https://doi.org/10.1177/036354659402200102.

——. "Medial Collateral Ligament Knee Sprains in College Football: Effectiveness of Preventive Braces." *American Journal of Sports Medicine* 22, no. 1 (1994): 12–18. https://doi.org/10.1177/036354659402200103.

Algafly, Amin A., and Keith P. George. "The Effect of Cryotherapy on Nerve Conduction Velocity, Pain Threshold and Pain Tolerance." *British Journal of Sports Medicine* 41, no. 6 (2007): 365–69. https://doi.org/10.1136/bjsm.2006.031237.

Allen, Frederick M. "Refrigeration Anesthesia for Limb Operations." *Anesthesiology* 4, no. 1 (1943): 12–16. https://doi.org/10.1097/00000542-194301000-00003.

American Academy of Pediatrics Committee on Sports Medicine. "Knee Brace Use by Athletes." *Pediatrics* 85, no. 2 (1990): 228.

Anderson, George, Stuart C. Zeman, and Robert T. Rosenfeld. "The Anderson Knee Stabler." *Physician and Sportsmedicine* 7, no. 6 (1979): 125–27. https://doi.org/10.1080/00913847.1979.11710882.

Barrett, D. S., A. G. Cobb, and G. Bentley. "Joint Proprioception in Normal, Osteoarthritic and Replaced Knees." *Journal of Bone and Joint Surgery* 73, no. 1 (1991): 53–56. https://doi.org/10.1302/0301-620X.73B1.1991775.

Beck, Charles, David Drez Jr., John Young, W. Dilworth Cannon Jr., and Mary Lou Stone. "Instrumented Testing of Functional Knee Braces." *American Journal of Sports Medicine* 14, no. 4 (1986): 253–56. https://doi.org/10.1177/036354658601400401.

Beynnon, B. D., M. H. Pope, C. M. Wertheimer, R. J. Johnson, B. C. Fleming, C. E. Nichols, and J. G. Howe. "The Effect of Functional Knee-Braces on Strain on the Anterior Cruciate Ligament in Vivo." *Journal of Bone and Joint Surgery.* 74, no. 9 (1992): 1298–312. https://doi.org/10.2106/00004623-199274090-00003.

Bleakley, Chris M., François Bieuzen, Gareth W. Davison, and Joseph T. Costello. "Whole-Body Cryotherapy: Empirical Evidence and Theoretical Perspectives." *Open Access Journal of Sports Medicine* 5 (2014): 25–36. https://doi.org/10.2147/OAJSM.S41655.

Bleakley, Chris M., and Joseph T. Costello. "Do Thermal Agents Affect Range of Movement and Mechanical Properties in Soft Tissues? A Systematic Review." *Archives of Physical Medicine and Rehabilitation* 94, no. 1 (2013): 149–63. https://doi.org/10.1016/j.apmr.2012.07.023.

Bleakley, Chris, Suzanne McDonough, and Domhnall MacAuley. "The Use of Ice in the Treatment of Acute Soft-Tissue Injury: A Systematic Review of Randomized Controlled Trials." *American Journal of Sports Medicine* 32, no. 1 (2004): 251–61. https://doi.org/10.1177/0363546503260757.

Borden, Sam. "Colleges Swear by Football Knee Braces. Not All Players and Experts Do." *New York Times.* January 8, 2017. https://www.nytimes.com/2017/01/08/sports/ncaafootball/college-football-playoff-alabama-clemson-knee-braces.html.

Bordes, P., E. Laboute, A. Bertolotti, J. F. Dalmay, P. Puig, P. Trouve, E. Verhaegue, et al. "No Beneficial Effect of Bracing After Anterior Cruciate Ligament Reconstruction in a Cohort of 969 Athletes Followed in Rehabilitation." *Annals of Physical and Rehabilitation Medicine* 60, no. 4 (2017): 230–36. https://doi.org/10.1016/j.rehab.2017.02.001.

Branch, Thomas, Robert Hunter, and Peter Reynolds. "Controlling Anterior Tibial Displacement Under Static Load: A Comparison of Two Braces." *Orthopedics* 11, no. 9 (1988): 1249–52.

Broatch, James R., Aaron Petersen, and David J. Bishop. "Postexercise Cold Water Immersion Benefits Are Not Greater Than the Placebo Effect." *Medicine and Science in Sports and Exercise* 46, no. 11 (2014): 2139–47. https://doi.org/10.1249/MSS.0000000000000348.

Bruneau, Jacques, and Peter Heinbecker. "Effects of Cooling on Experimentally Infected Tissues." *Annals of Surgery* 120, no. 5 (1944): 716–26.

Bugaj, Ronald "The Cooling, Analgesic, and Rewarming Effects of Ice Massage on Localized Skin." *Physical Therapy* 55, no. 1 (1975): 11–19. https://doi.org/10.1093/ptj/55.1.11.

Callaghan, Michael J., James Selfe, Pam J. Bagley, and Jacqueline A. Oldham. "The Effects of Patellar Taping on Knee Joint Proprioception." *Journal of Athletic Training* 37, no. 1 (2002): 19–24.

Chandler, Anne, Joanne Preece, and Sara Lister. "Using Heat Therapy for Pain Management." *Nursing Standard* 17, no. 9 (2002): 40–42. https://doi.org/10.7748/ns2002.11.17.9.40.c3297.

Chen, Shijia. "體寒百病生？ [Cold Qi Begets All Ailments?]" Yahoo Health. February 19, 2019. https://www.edh.tw/article/21075.

Chesterton, Linda S., Nadine E. Foster, and Lesley Ross. "Skin Temperature Response to Cryotherapy." *Archives of Physical Medicine and Rehabilitation* 83, no. 4 (2002): 543–49. https://doi.org/10.1053/apmr.2002.30926.

Chuang, Shih-Hung, Mao-Hsiung Huang, Tien-Wen Chen, Ming-Chang Weng, Chin-Wei Liu, and Chia-Hsin Chen. "Effect of Knee Sleeve on Static and Dynamic Balance in Patients with Knee Osteoarthritis." *Kaohsiung Journal of Medical Sciences* 23, no. 8 (2007): 405–11. https://doi.org/10.1016/S0257-5655(07)70004-4.

Ciolek, Jeffrey J. "Cryotherapy. Review of Physiological Effects and Clinical Application." *Cleveland Clinic Quarterly* 52, no. 2 (1985): 193–201. https://doi.org/10.3949/ccjm.52.2.193.

Collins, Amber T., J. Troy Blackburn, Chris W. Olcott, Jodie Miles, Joanne Jordan, Douglas R. Dirschl, and Paul S. Weinhold. "Stochastic Resonance Electrical Stimulation to Improve Proprioception in Knee Osteoarthritis." *The Knee* 18, no. 5 (2011): 317–22. https://doi.org/10.1016/j.knee.2010.07.001.

Collins, N. C. "Is Ice Right? Does Cryotherapy Improve Outcome for Acute Soft Tissue Injury?" *Emergency Medicine Journal* 25, no. 2 (2008): 65–68. https://doi.org/10.1136/emj.2007.051664.

Cooper, S. M., and R. P. R. Dawber. "The History of Cryosurgery." *Journal of the Royal Society of Medicine* 94, no. 4 (2001): 196–201. https://doi.org/10.1177/014107680109400416.

Crossman, Lyman Weeks, Frederick M. Allen, Vincent Hurley, Wilfred Ruggiero, and Cyrus E. Warden. "Refrigeration Anesthesia." *Anesthesia and Analgesia* 21, no. 1 (1942): 241–54. https://doi.org/10.1213/00000539-194201000-00059.

Cudejko, Tomasz, Martin van der Esch, Marike van der Leeden, Josien C. van den Noort, Leo D. Roorda, Willem Lems, Jos Twisk, et al. "The Immediate Effect of a Soft Knee Brace on Pain, Activity Limitations, Self-Reported Knee Instability, and Self-Reported Knee Confidence in Patients with Knee Osteoarthritis." *Arthritis Research and Therapy* 19, no. 1 (2017): 260–60. https://doi.org/10.1186/s13075-017-1456-0.

Daniel, Dale M., Mary Lou Stone, and Diana L. Arendt. "The Effect of Cold Therapy on Pain, Swelling, and Range of Motion after Anterior Cruciate Ligament Reconstructive Surgery." *Arthroscopy* 10, no. 5 (1994): 530–33. https://doi.org/10.1016/S0749-8063(05)80008-8.

Dantas, Lucas Ogura, Carolina Carreira Breda, Paula Regina Mendes da Silva Serrao, Francisco Aburquerque-Sendín, Ana Elisa Serafim Jorge, Jonathan Emanuel Cunha, Germanna Medeiros Barbosa, Joao Luiz Quagliotti Durigan, and Tania de Fatima Salvini. "Short-Term Cryotherapy Did Not Substantially Reduce Pain and Had Unclear Effects on Physical Function and Quality of Life in People with Knee Osteoarthritis: A Randomised Trial." *Journal of Physiotherapy* 65, no. 4 (2019): 215–21. https://doi.org/10.1016/j.jphys.2019.08.004.

Dantas, Lucas Ogura, Roberta de Fátima Carreira Moreira, Flavia Maintinguer Norde, Paula Regina Mendes Silva Serrao, Francisco Alburquerque-Sendín, and Tania Fatima Salvini. "The Effects of Cryotherapy on Pain and Function in Individuals with Knee Osteoarthritis: A Systematic Review of Randomized Controlled Trials." *Clinical Rehabilitation* 33, no. 8 (2019): 1310–19. https://doi.org/10.1177/0269215519840406.

Davison, M. H. Armstrong. "The Evolution of Anaesthesia." *British Journal of Anaesthesia* 31, no. 3 (1959): 134–37. https://doi.org/10.1093/bja/31.3.134.

DeBakey, Michael E., and Fiorindo Simeone. "Acute Battle-Incurred Arterial Injuries." In *Vascular Surgery in World War II*, ed. Daniel C. Elkin and Michael E. Debakey, 60–148. Washington, DC: Office of the Surgeon General, Department of the Army, 1955.

Denegar, Craig R., Devon R. Dougherty, Jacob E. Friedman, Maureen E. Schimizzi, James E. Clark, Brett A. Comstock, and William J. Kraemer. "Preferences for Heat, Cold, or Contrast in Patients with Knee Osteoarthritis Affect Treatment Response." *Clinical Interventions in Aging* 5 (2010): 199–206. https://doi.org/10.2147/CIA.S11431.

Draper, David O. "Comparison of Shortwave Diathermy and Microwave Diathermy." *International Journal of Athletic Therapy and Training* 18, no. 6 (2013): 13–17. https://doi.org/10.1123/ijatt.18.6.13.

Draper, David O., Shane Schulthies, Pasi Sorvisto, and Anna-Mari Hautala. "Temperature Changes in Deep Muscles of Humans During Ice and Ultrasound Therapies: An In Vivo Study." *Journal of Orthopaedic and Sports Physical Therapy* 21, no. 3 (1995): 153–57. https://doi.org/10.2519/jospt.1995.21.3.153.

Duivenvoorden, Tijs, Reinoud W. Brouwer, Tom M. van Raaij, Arianne P. Verhagen, Jan A. N. Verhaar, and Sita M. A. Bierma-Zeinstra. "Braces and Orthoses for Treating Osteoarthritis of the Knee." *Cochrane Library* 2015, no. 3 (2015): CD004020-CD20. https://doi.org/10.1002/14651858.CD004020.pub3.

Edmonds, David W., Jenny McConnell, Jay R. Ebert, Tim R. Ackland, and Cyril J. Donnelly. "Biomechanical, Neuromuscular and Knee Pain Effects Following Therapeutic Knee Taping Among Patients with Knee Osteoarthritis During Walking Gait." *Clinical Biomechanics* 39 (2016): 38–43. https://doi.org/10.1016/j.clinbiomech.2016.09.003.

France, E. Paul, and Lonnie E. Paulos. "Knee Bracing." *Journal of the American Academy of Orthopaedic Surgeons* 2, no. 5 (1994): 281–87. https://doi.org/10.5435/00124635-199409000-00006.

Giombini, Arrigo, Annalisa Di Cesare, Mariachiara Di Cesare, Maurizio Ripani, and Nicola Maffulli. "Localized Hyperthermia Induced by Microwave Diathermy in Osteoarthritis of the Knee: A Randomized Placebo-Controlled Double-Blind Clinical Trial." *Knee Surgery, Sports Traumatology, Arthroscopy* 19, no. 6 (2011): 980–87. https://doi.org/10.1007/s00167-010-1350-7.

Gohal, Chetan, Ajaykumar Shanmugaraj, Patrick Tate, Nolan S. Horner, Asheesh Bedi, Anthony Adili, and Moin Khan. "Effectiveness of Valgus Offloading Knee Braces in the Treatment of Medial Compartment Knee Osteoarthritis: A Systematic Review." *Sports Health* 10, no. 6 (2018): 500–14. https://doi.org/10.1177/1941738118763913.

Grace, Thomas G., Betty J. Skipper, James C. Newberry, Michael A. Nelson, Edward R. Sweetser, and Michael L. Rothman. "Prophylactic Knee Braces and Injury to the Lower Extremity." *Journal of Bone and Joint Surgery* 70, no. 3 (1988): 422–27. https://doi.org/10.2106/00004623-198870030-00015.

Haladik, Jeffrey A., William K. Vasileff, Cathryn D. Peltz, Terrence R. Lock, and Michael J. Bey. "Bracing Improves Clinical Outcomes but Does Not Affect the Medial Knee Joint Space in Osteoarthritic Patients During Gait." *Knee Surgery, Sports Traumatology, Arthroscopy* 22, no. 11 (2014): 2715–20. https://doi.org/10.1007/s00167-013-2596-7.

Hansen, Byron L., Jack C. Ward, and Richard C. Diehl Jr. "The Preventive Use of the Anderson Knee Stabler in Football." *Physician and Sportsmedicine* 13, no. 9 (1985): 75–81. https://doi.org/10.1080/00913847.1985.11708879.

Hassan, B. S., S. Mockett, and M. Doherty. "Influence of Elastic Bandage on Knee Pain, Proprioception, and Postural Sway in Subjects with Knee Osteoarthritis." *Annals of the Rheumatic Diseases* 61, no. 1 (2002): 24–28. https://doi.org/10.1136/ard.61.1.24.

Highgenboten, Carl L., Allen Jackson, Neil Meske, and Jimmy Smith. "The Effects of Knee Brace Wear on Perceptual and Metabolic Variables During Horizontal Treadmill Running." *American Journal of Sports Medicine* 19, no. 6 (1991): 639–43. https://doi.org/10.1177/036354659101900615.

Hippocrates. "Aphorisms." In *The Genuine Works of Hippocrates*, ed. Charles Darwin Adams. New York: Dover, 1868. https://www.chlt.org/hippocrates/Adams/page.316.a.php.

Ho, Kai-Yu, Ryan Epstein, Ron Garcia, Nicole Riley, and Szu-Ping Lee. "Effects of Patellofemoral Taping on Patellofemoral Joint Alignment and Contact Area During Weight Bearing." *Journal of Orthopaedic and Sports Physical Therapy* 47, no. 2 (2017): 115–23. https://doi.org/10.2519/jospt.2017.6936.

Hubbard, Tricia J., Stephanie L. Aronson, and Craig R. Denegar. "Does Cryotherapy Hasten Return to Participation? A Systematic Review." *Journal of Athletic Training* 39, no. 1 (2004): 88–94.

Hughes, E. S. R. "Refrigeration Anaesthesia." *British Medical Journal* 1, no. 4508 (1947): 761–64. https://doi.org/10.1136/bmj.1.4508.761.

Ichinoseki-Sekine, Noriko, Hisashi Naito, Norio Saga, Yuji Ogura, Minoru Shiraishi, Arrigo Giombini, Valentina Giovannini, and Shizuo Katamoto. "Changes in Muscle Temperature Induced by 434 MHz Microwave Hyperthermia." *British Journal of Sports Medicine* 41, no. 7 (2007): 425–29. https://doi.org/10.1136/bjsm.2006.032540.

Jones, Richard K., Christopher J. Nester, Jim D. Richards, Winston Y. Kim, David S. Johnson, Sanjiv Jari, Philip Laxton, and Sarah F. Tyson. "A Comparison of the

Biomechanical Effects of Valgus Knee Braces and Lateral Wedged Insoles in Patients with Knee Osteoarthritis." *Gait and Posture* 37, no. 3 (2012): 368–72. https://doi.org/10.1016/j.gaitpost.2012.08.002.

Komistek, Richard D., Douglas A. Dennis, Eric J. Northcut, Adam Wood, Andrew W. Parker, and Steve M. Traina. "An in Vivo Analysis of the Effectiveness of the Osteoarthritic Knee Brace During Heel-Strike of Gait." *Journal of Arthroplasty* 14, no. 6 (1999): 738–42. https://doi.org/10.1016/S0883-5403(99)90230-9.

Konrath, Gregory A., Terrence Lock, Henry T. Goitz, and Jeb Scheidler. "The Use of Cold Therapy After Anterior Cruciate Ligament Reconstruction: A Prospective, Randomized Study and Literature Review." *American Journal of Sports Medicine* 24, no. 5 (1996): 629–33. https://doi.org/10.1177/036354659602400511.

Kramer, John F., Tracy Dubowitz, Peter Fowler, Candice Schachter, and Trevor Birmingham. "Functional Knee Braces and Dynamic Performance: A Review." *Clinical Journal of Sport Medicine* 7, no. 1 (1997): 32–39. https://doi.org/10.1097/00042752-199701000-00007.

Large, A., and P. Heinbecker. "The Effect of Cooling on Wound Healing." *Annals of Surgery* 120, no. 5 (1944): 727–41. https://doi.org/10.1097/00000658-194411000-00005.

——. "Refrigeration in Clinical Surgery." *Annals of Surgery* 120, no. 5 (1944): 707–15. https://doi.org/10.1097/00000658-194411000-00003.

LaVelle, Beth Elchek, and Mariah Snyder. "Differential Conduction of Cold Through Barriers." *Journal of Advanced Nursing* 10, no. 1 (1985): 55–61. https://doi.org/10.1111/j.1365-2648.1985.tb00492.x.

Lessard, Lucy A., Roger A. Scudds, Annunziato Amendola, and Margaret D. Vaz. "The Efficacy of Cryotherapy Following Arthroscopic Knee Surgery." *Journal of Orthopaedic and Sports Physical Therapy* 26, no. 1 (1997): 14–22. https://doi.org/10.2519/jospt.1997.26.1.14.

Lin, Yijun. "伤后不用冰敷？ [No Icing after Injury?]" June 22, 2019. https://www.epochtimes.com/gb/19/6/20/n11335054.htm.

Lombardi, Giovanni, Ewa Ziemann, and Giuseppe Banfi. "Whole-Body Cryotherapy in Athletes: From Therapy to Stimulation. An Updated Review of the Literature." *Frontiers in Physiology* 8 (2017): 258–58. https://doi.org/10.3389/fphys.2017.00258.

Lu, Haiyan, Danping Huang, Noah Saederup, Israel F. Charo, Richard M. Ransohoff, and Lan Zhou. "Macrophages Recruited via CCR2 Produce Insulin-Like Growth Factor-1 to Repair Acute Skeletal Muscle Injury." *FASEB Journal* 25, no. 1 (2011): 358–69. https://doi.org/10.1096/fj.10-171579.

Mac Auley, Domhnall C. "Ice Therapy: How Good Is the Evidence?" *International Journal of Sports Medicine* 22, no. 5 (2001): 379–84. https://doi.org/10.1055/s-2001-15656.

McCarberg, W., G. Erasala, M. Goodale, J. Grender, D. Hengehold, and L. Donikyan. "Therapeutic Benefits of Continuous Low-Level Heat Wrap Therapy (CLHT) for Osteoarthritis (OA) of the Knee." *Journal of Pain* 6, no. 3 (2005): S53. https://doi.org/10.1016/j.jpain.2005.01.208.

McGorm, Hamish, Llion A. Roberts, Jeff S. Coombes, and Jonathan M. Peake. "Turning Up the Heat: An Evaluation of the Evidence for Heating to Promote Exercise Recovery, Muscle Rehabilitation and Adaptation." *Sports Medicine* 48, no. 6 (2018): 1311–28. https://doi.org/10.1007/s40279-018-0876-6.

Melick, Dermont W. "Refrigeration Anesthesia." *American Journal of Surgery* 70, no. 3 (1945): 364–68. https://doi.org/10.1016/0002-9610(45)90184-X.

Merrick, Mark A. "Secondary Injury After Musculoskeletal Trauma: A Review and Update." *Journal of Athletic Training* 37, no. 2 (2002): 209–17.

Mirkin, Gabe. "Why Ice Delays Recovery." September 16, 2015. https://www.drmirkin.com/fitness/why-ice-delays-recovery.html.

Mock, Harry E. and Harry E. Mock Jr. "Refrigeration Anesthesia in Amputations." *Journal of the American Medical Association* 123, no. 1 (1943): 81–89. https://doi.org/10.1001/jama.1943.02840360015003.

Myrer, William J., Kimberly A. Myrer, Gary J. Measom, Gilbert W. Fellingham, and Stacey L. Evers. "Muscle Temperature Is Affected by Overlying Adipose When Cryotherapy Is Administered." *Journal of Athletic Training* 36, no. 1 (2001): 32–36.

Nadaud, Matthew C., Richard D. Komistek, Mohamed R. Mahfouz, Douglas A. Dennis, and Matthew R. Anderle. "In Vivo Three-Dimensional Determination of the Effectiveness of the Osteoarthritic Knee Brace: A Multiple Brace Analysis." *Journal of Bone and Joint Surgery* 87A, Suppl. 2 (2005): 114–19. https://doi.org/10.2106/00004623-200511002-00013.

Najibi, Soheil, and John P. Albright. "The Use of Knee Braces, Part 1: Prophylactic Knee Braces in Contact Sports." *American Journal of Sports Medicine* 33, no. 4 (2005): 602–11. https://doi.org/10.1177/0363546505275128.

Newman, Barclay Moon. "Shockless Surgery." *Scientific American* 166, no. 4 (1942): 182–84. https://doi.org/10.1038/scientificamerican0442-182.

Noble, Ethel A. "Refrigeration Anaesthesia." *Anesthesiology* 10, no. 1 (1948): 121–22. https://doi.org/10.1097/00000542-194901000-00028.

Özgönenel, Levent, Ebru Aytekin, and Gulis Durmuşog-lu. "A Double-Blind Trial of Clinical Effects of Therapeutic Ultrasound in Knee Osteoarthritis." *Ultrasound in Medicine and Biology* 35, no. 1 (2008): 44–49. https://doi.org/10.1016/j.ultrasmedbio.2008.07.009.

Park, Kyue-nam, and Si-hyun Kim. "Effects of Knee Taping During Functional Activities in Older People with Knee Osteoarthritis: A Randomized Controlled Clinical Trial: Effects of Taping on Knee Osteoarthritis." *Geriatrics and Gerontology International* 18, no. 8 (2018): 1206–10. https://doi.org/10.1111/ggi.13448.

Petrofsky, Jerrold S., Michael S. Laymon, Faris S. Alshammari, and Haneul Lee. "Use of Low Level of Continuous Heat as an Adjunct to Physical Therapy Improves Knee Pain Recovery and the Compliance for Home Exercise in Patients with Chronic Knee Pain: A Randomized Controlled Trial." *Journal of Strength and Conditioning Research* 30, no. 11 (2016): 3107–15. https://doi.org/10.1519/JSC.0000000000001409.

Petrofsky, Jerrold Scott, Michael Laymon, and Haneul Lee. "Effect of Heat and Cold on Tendon Flexibility and Force to Flex the Human Knee." *Medical Science Monitor* 19 (2013): 661–67. https://doi.org/10.12659/MSM.889145.

Randall, Frank, Harold Miller, and Donald Shurr. "The Use of Prophylactic Knee Orthoses at Iowa State University." *Orthotics and Prosthetics* 37, no. 4 (1983): 54–57.

Risberg, May Arna, Inger Holm, Harald Steen, Jan Eriksson, and Arne Ekeland. "The Effect of Knee Bracing After Anterior Cruciate Ligament Reconstruction: A Prospective, Randomized Study with Two Years' Follow-Up." *American Journal of Sports Medicine* 27, no. 1 (1999): 76–83.

Roberts, Llion A., Truls Raastad, James F. Markworth, Vandre C. Figueiredo, Ingrid M. Egner, Anthony Shield, David Cameron-Smith, Jeff S. Coombes, and Jonathan M. Peake. "Post-Exercise Cold Water Immersion Attenuates Acute Anabolic Signalling and Long-Term Adaptations in Muscle to Strength Training." *Journal of Physiology* 593, no. 18 (2015): 4285–301. https://doi.org/10.1113/JP270570.

Rutjes, Anne W. S., Eveline Nüesch, Rebekka Sterchi, and Peter Jüni. "Therapeutic Ultrasound for Osteoarthritis of the Knee or Hip." *Cochrane Library* 2010, no. 1 (2010): CD003132-CD32. https://doi.org/10.1002/14651858.CD003132.pub2.

Salata, Michael J., Aimee E. Gibbs, and Jon K. Sekiya. "The Effectiveness of Prophylactic Knee Bracing in American Football: A Systematic Review." *Sports Health: A Multidisciplinary Approach* 2, no. 5 (2010): 375–79. https://doi.org/10.1177/1941738110378986.

Scarcella, Joseph B., and Bruce T. Cohn. "The Effect of Cold Therapy on the Postoperative Course of Total Hip and Knee Arthroplasty Patients." *American Journal of Orthopedics* 24, no. 11 (1995): 847–52.

Schween, Raphael, Dominic Gehring, and Albert Gollhofer. "Immediate Effects of an Elastic Knee Sleeve on Frontal Plane Gait Biomechanics in Knee Osteoarthritis." *PLoS One* 10, no. 1 (2015): e0115782-e82. https://doi.org/10.1371/journal.pone.0115782.

Seto, Hiroaki, Hiroshi Ikeda, Hidehiko Hisaoka, and Hisashi Kurosawa. "Effect of Heat- and Steam-Generating Sheet on Daily Activities of Living in Patients with Osteoarthritis of the Knee: Randomized Prospective Study." *Journal of Orthopaedic Science: Official Journal of the Japanese Orthopaedic Association* 13, no. 3 (2008): 187–91. https://doi.org/10.1007/s00776-008-1214-x.

Sitler, Michael, Jack Ryan, William Hopkinson, James Wheeler, James Santomier, Rickey Kolb, and David Polley. "The Efficacy of a Prophylactic Knee Brace to Reduce Knee Injuries in Football: A Prospective, Randomized Study at West Point." *American Journal of Sports Medicine* 18, no. 3 (1990): 310–15. https://doi.org/10.1177/036354659001800315.

Skurvydas, Albertas, Sigitas Kamandulis, Aleksas Stanislovaitis, Vytautas Streckis, Gediminas Mamkus, and Adomas Drazdauskas. "Leg Immersion in Warm Water, Stretch-Shortening Exercise, and Exercise-Induced Muscle Damage." *Journal of Athletic Training* 43, no. 6 (2008): 592–99. https://doi.org/10.4085/1062-6050-43.6.592.

Song, Chang W. "Effect of Local Hyperthermia on Blood Flow and Microenvironment: A Review." Supplement, *Cancer Research* 44, no. 10 (1984): 4721s-30s.

Takeuchi, Kousuke, Takuya Hatade, Soushi Wakamiya, Naoto Fujita, Takamitsu Arakawa, and Akinori Miki. "Heat Stress Promotes Skeletal Muscle Regeneration

After Crush Injury in Rats." *Acta Histochemica* 116, no. 2 (2014): 327–34. https://doi.org/10.1016/j.acthis.2013.08.010.

Tegner, Y., and R. Lorentzon. "Evaluation of Knee Braces in Swedish Ice Hockey Players." *British Journal of Sports Medicine* 25, no. 3 (1991): 159–61. https://doi .org/10.1136/bjsm.25.3.159.

Teitz, Carol C., Bonnie K. Hermanson, Richard A. Kronmal, and Paula H. Diehr. "Evaluation of the Use of Braces to Prevent Injury to the Knee in Collegiate Football Players." *Journal of Bone and Joint Surgery* 69, no. 1 (1987): 2–9. https:// doi.org/10.2106/00004623-198769010-00002.

Tseng, Ching-Yu, Jo-Ping Lee, Yung-Shen Tsai, Shin-Da Lee, Chung-Lan Kao, Te-Chih Liu, Cheng Hsiu Lai, M. Brennan Harris, and Chia-Hua Kuo. "Topical Cooling (Icing) Delays Recovery from Eccentric Exercise–Induced Muscle Damage." *Journal of Strength and Conditioning Research* 27, no. 5 (2013): 1354–61. https://doi.org/10.1519/JSC.0b013e318267a22c.

Usuba, Mariko, Yutaka Miyanaga, Shumpei Miyakawa, Toru Maeshima, and Yoshio Shirasaki. "Effect of Heat in Increasing the Range of Knee Motion After the Development of a Joint Contracture: An Experiment with an Animal Model." *Archives of Physical Medicine and Rehabilitation* 87, no. 2 (2006): 247–53. https://doi.org/10.1016/j.apmr.2005.10.015.

Vaile, Joanna, Shona Halson, Nicholas Gill, and Brian Dawson. "Effect of Hydrotherapy on the Signs and Symptoms of Delayed Onset Muscle Soreness." *European Journal of Applied Physiology* 103, no. 1 (2008): 121–22. https://doi.org /10.1007/s00421-007-0653-y.

Van Tiggelen, Damien, Pascal Coorevits, and Erik Witvrouw. "The Effects of a Neoprene Knee Sleeve on Subjects with a Poor Versus Good Joint Position Sense Subjected to an Isokinetic Fatigue Protocol." *Clinical Journal of Sport Medicine* 18, no. 3 (2008): 259–65. https://doi.org/10.1097/JSM.0b013e31816d78c1.

Viitasalo, J. T., K. Niemelä, R. Kaappola, T. Korjus, M. Levola, H. V. Mononen, H. K. Rusko, and T. E. S. Takala. "Warm Underwater Water-Jet Massage Improves Recovery from Intense Physical Exercise." *European Journal of Applied Physiology* 71, no. 5 (1995): 431–38. https://doi.org/10.1007/BF00635877.

Wojtys, Edward M., and Laura J. Huston. "'Custom-Fit' Versus 'Off-the-Shelf' ACL Functional Braces." *American Journal of Knee Surgery* 14, no. 3 (2001): 157–62.

Wojtys, Edward M., Sandip U. Kothari, and Laura J. Huston. "Anterior Cruciate Ligament Functional Brace Use in Sports." *American Journal of Sports Medicine* 24, no. 4 (1996): 539–46. https://doi.org/10.1177/036354659602400421.

Wu, Gloria K. H., Gabriel Y. F. Ng, and Arthur F. T. Mak. "Effects of Knee Bracing on the Functional Performance of Patients with Anterior Cruciate Ligament Reconstruction." *Archives of Physical Medicine and Rehabilitation* 82, no. 2 (2001): 282–85. https://doi.org/10.1053/apmr.2001.19020.

Wu, Yu, Shibo Zhu, Zenghui Lv, Shunli Kan, Qiuli Wu, Wenye Song, Guangzhi Ning, and Shiqing Feng. "Effects of Therapeutic Ultrasound for Knee Osteoarthritis: A Systematic Review and Meta-Analysis." *Clinical Rehabilitation* 33, no. 12 (2019): 1863–75. https://doi.org/10.1177/0269215519866494.

Yang, Xiong-gang, Jiang-tao Feng, Xin He, Feng Wang, and Yong-cheng Hu. "The Effect of Knee Bracing on the Knee Function and Stability Following Anterior Cruciate Ligament Reconstruction: A Systematic Review and Meta-Analysis of Randomized Controlled Trials." *Orthopaedics and Traumatology, Surgery and Research* 105, no. 6 (2019): 1107–14. https://doi.org/10.1016/j.otsr.2019.04.015.

Yudin, Sergei S. "Refrigeration Anesthesia for Amputations." *Anesthesia and Analgesia* 24, no. 5 (1945): 216–19. https://doi.org/10.1213/00000539-194509000-00005.

Yurtkuran, Merih, and Tuncer Kocagil. "TENS, Electroacupuncture and Ice Massage: Comparison of Treatment for Osteoarthritis of the Knee." *American Journal of Acupuncture* 27, nos. 3–4 (1999): 133–40.

7. THE HURTFUL KNEE

Associated Press. "13-Year-Old Dies After Getting KO'd in Thai Kickboxing Match." *New York Post.* November 13, 2018. https://nypost.com/2018/11/13/13-year-old-dies-after-getting-kod-in-thai-kickboxing-match/.

Ali, Md. Ayub, Teruo Uetake, and Fumio Ohtsuki. "Secular Changes in Relative Leg Length in Post-War Japan." *American Journal of Human Biology* 12, no. 3 (2000): 405–16.

Alosco, Michael L., Jesse Mez, Yorghos Tripodis, Patrick T. Kiernan, Bobak Abdolmohammadi, Lauren Murphy, Neil W. Kowall, et al. "Age of First Exposure to Tackle Football and Chronic Traumatic Encephalopathy." *Annals of Neurology* 83, no. 5 (2018): 886–901. https://doi.org/10.1002/ana.25245.

Babu, Deepika, and Bruno Bordoni. "Anatomy, Bony Pelvis and Lower Limb, Medial Longitudinal Arch of the Foot." In *StatPearls*, 32965960. Treasure Island, FL: StatPearls, 2021.

Bailey, Holly. "George Floyd Died of Low Level of Oxygen, Medical Expert Testifies; Derek Chauvin Kept Knee on His Neck 'Majority of the Time.'" *Washington Post.* April 8, 2021. https://www.washingtonpost.com/nation/2021/04/08/derek-chauvin-trial-2/.

BJJ World. "BJJ Knee on Belly." October 4, 2020. https://bjj-world.com/bjj-knee-on-belly/.

Birrer, R. B., and S. P. Halbrook. "Martial Arts Injuries: The Results of a Five Year National Survey." *American Journal of Sports Medicine* 16, no. 4 (1988): 408–10. https://doi.org/10.1177/036354658801600418.

Bogel-Burroughs, Nicholas. "George Floyd Showed Signs of a Brain Injury 4 Minutes Before Derek Chauvin Lifted His Knee, a Doctor Testifies." *New York Times.* April 8, 2021. https://www.nytimes.com/2021/04/08/us/george-floyd-knee-on-neck.html.

Burke, David, Samir al-Adawi, Daniel Burke, Paolo Bonato, and Casey Leong. "The Kicking Process in Tae Kwon Do: A Biomechanical Analysis." *International Physical Medicine and Rehabilitation Journal* 1, no. 1 (2017). https://doi.org/10.15406/ipmrj.2017.01.00002.

Carter, Bob. "The Violent World." ESPN Classic. n.d. https://www.espn.com/classic/biography/s/Huff_Sam.html.

Chan, Theodore C., Gary M. Vilke, Tom Neuman, and Jack L. Clausen. "Restraint Position and Positional Asphyxia." *Annals of Emergency Medicine* 30, no. 5 (1997): 578–86. https://doi.org/10.1016/S0196-0644(97)70072-6.

Chappell, Bill. "Chauvin's Restraint on Floyd's Neck Isn't Taught by Police, Use-of-Force Trainer Says." NPR. April 6, 2021. https://www.npr.org/sections/trial-over-killing-of-george-floyd/2021/04/06/984717386/watch-live-minneapolis-police-crisis-intervention-trainer-testifies-in-chauvin-t.

Clifton, Daniel R., James A. Onate, Eric Schussler, Aristarque Djoko, Thomas P. Dompier, and Zachary Y. Kerr. "Epidemiology of Knee Sprains in Youth, High School, and Collegiate American Football Players." *Journal of Athletic Training* 52, no. 5 (2017): 464–73. https://doi.org/10.4085/1062-6050-52.3.09.

Corbí, Alberto, and Olga Santos. "Myshikko: Modelling Knee Walking in Aikido Practice." In *Proceedings for the 26th Conference on User Modeling, Adaptation and Personalization*, 217–18. New York: Association for Computing Machinery, 2018.

Cornish, Dean. "'We Will Never Have Champions If Fighters Can't Start Young.' Getting to Know Thailand's Child Fighters." Dateline. May 21, 2019. https://www.sbs.com.au/news/dateline/we-will-never-have-champions-if-fighters-can-t-start-young-getting-to-know-thailand-s-child-fighters.

Cruz, Guilherme. "Photos: 'Cyborg' Santos Before and After Surgery." MMA Fighting. July 28, 2016. https://www.mmafighting.com/2016/7/28/12319768/photos-cyborg-santos-before-and-after-surgery.

Daniel, Ray W., Steven Rowson, and Stefan M. Duma. "Head Impact Exposure in Youth Football." *Annals of Biomedical Engineering* 40, no. 4 (2012): 976–81. https://doi.org/10.1007/s10439-012-0530-7.

Dawson, Alan. "Yes, You Can Make Millions as a Professional Fighter—but Only 19 out of 21,000 Successfully Manage It." Business Insider. June 9, 2017. https://www.businessinsider.com/mcgregor-mayweather-fight-economics-ufc-mma-bellator-boxing-paul-daley-2017-6.

DeSilva, Jeremy. *First Steps: How Walking Upright Made Us Human*. New York: HarperCollins, 2021.

Dewan, Shaila, and Sheri Fink. "Does It Matter Whether Chauvin Knelt on Floyd's Neck Versus His Shoulder?" *New York Times*. April 8, 2021. https://www.nytimes.com/2021/04/08/us/does-it-matter-whether-chauvin-knelt-on-floyds-neck-versus-his-shoulder.html.

"The Different Types of Knees in Muay Thai." *Evolve Mixed Martial Arts* (blog). 2021. https://evolve-vacation.com/blog/the-different-types-of-knees-in-muay-thai/.

Dompier, Thomas P., Zachary Y. Kerr, Stephen W. Marshall, Brian Hainline, Erin M. Snook, Ross Hayden, and Janet E. Simon. "Incidence of Concussion During Practice and Games in Youth, High School, and Collegiate American Football Players." *JAMA Pediatrics* 169, no. 7 (2015): 659–65. https://doi.org/10.1001/jamapediatrics.2015.0210.

Forliti, Amy, Steve Karnowski, and Tammy Webber. "Medical Examiner Blames Police Pressure for Floyd's Death." Associated Press News. April 9, 2021. https://apnews.com /article/derek-chauvin-trial-live-updates-05458e47134a4934bc38ce28c7543ebb.

——. "Police Chief: Kneeling on Floyd's Neck Violated Policy." Associated Press News. April 5, 2021. https://apnews.com/article/derek-chauvin-trial-live-updates -c3e3fe08773cd2f012654e782e326f6e.

Grasgruber, P., M. Sebera, E. Hrazdíra, J. Cacek, and T. Kalina. "Major Correlates of Male Height: A Study of 105 Countries." *Economics and Human Biology* 21 (2016): 172–95. https://doi.org/10.1016/j.ehb.2016.01.005.

Hays, Jeffrey. "Muay Thai (Thai Kick Boxing) and Olympic and Pro Boxing in Thailand." Facts and Details. Updated May 2014. https://factsanddetails.com /southeast-asia/Thailand/sub5_8e/entry-3270.html.

Hick, John L., Stephen W. Smith, and Michael T. Lynch. "Metabolic Acidosis in Restraint-Associated Cardiac Arrest: A Case Series." *Academic Emergency Medicine* 6, no. 3 (1999): 239–43.

Homma, Gaku. "No Suwariwaza (Kneeling Techniques) at Nippon Kan." Nippon Kan Kancho. June 15, 2007. http://www.nippon-kan.org/no-suwariwaza-kneeling -techniques-at-nippon-kan/.

Ingram, Jay G., Sarah K. Fields, Ellen E. Yard, and R. Dawn Comstock. "Epidemiology of Knee Injuries Among Boys and Girls in US High School Athletics." *American Journal of Sports Medicine* 36, no. 6 (2008): 1116–22. https://doi.org /10.1177/0363546508314400.

Kaewjinda, Kaweewit. "Death of Young Thai Kickboxer Brings Focus on Dangers." *Seattle Times.* November 18, 2018. https://www.seattletimes.com/nation-world /death-of-young-thai-kickboxer-brings-focus-on-dangers/.

Karch, Steven B. "The Problem of Police-Related Cardiac Arrest." *Journal of Forensic and Legal Medicine* 41 (2016): 36–41. https://doi.org/10.1016/j.jflm.2016.04.008.

Kerr, Zachary Y., Gary B. Wilkerson, Shane V. Caswell, Dustin W. Currie, Lauren A. Pierpoint, Erin B. Wasserman, Sarah B. Knowles, et al. "The First Decade of Web-Based Sports Injury Surveillance: Descriptive Epidemiology of Injuries in United States High School Football (2005–2006 Through 2013–2014) and National Collegiate Athletic Association Football (2004–2005 Through 2013–2014)." *Journal of Athletic Training* 53, no. 8 (2018): 738–51. https://doi.org/10.4085 /1062-6050-144-17.

Kim, Young Kwan, Yoon Hyuk Kim, and Shin Ja Im. "Inter-Joint Coordination in Producing Kicking Velocity of Taekwondo Kicks." *Journal of Sports Science and Medicine* 10, no. 1 (2011): 31–38.

Koerth, Maggie. "The Two Autopsies of George Floyd Aren't as Different as They Seem." FiveThirtyEight. June 8, 2020. https://fivethirtyeight.com/features/the-two -autopsies-of-george-floyd-arent-as-different-as-they-seem/.

Kroll, Mark W., Michael A. Brave, Scott R. Kleist, Mollie B. Ritter, Darrell L. Ross, and Steven B. Karch. "Applied Force During Prone Restraint: Is Officer Weight a Factor?" *American Journal of Forensic Medicine and Pathology* 40, no. 1 (2019): 1–7. https://doi.org/10.1097/PAF.0000000000000457.

Kroll, Mark W., G. Keith Still, Tom S. Neuman, Michael A. Graham, and Lanny V. Griffin. "Acute Forces Required for Fatal Compression Asphyxia: A Biomechanical Model and Historical Comparisons." *Medicine, Science, and the Law* 57, no. 2 (2017): 61–68. https://doi.org/10.1177/0025802417695711.

Kurland, Harvey. "A Comparison of Judo and Aikido Injuries." *Physician and Sportsmedicine* 8, no. 6 (1980): 71–74. https://doi.org/10.1080/00913847.1980.11948618.

Laothamatas, Jiraporn, Adisak Plitponkarnpim, Onousa Sangfai, Thirawat Suparatpriyakon, Mattana Pongsopon, Daochompu Nakawiro, Chakrit Sukying, Anannit Visudtibhan, and Witaya Sungkarat. "Child Muaythai Boxing: Conflict of Health and Culture." *Injury Prevention* 24, no. Suppl. 1 (2018): A126. https://doi.org/10.1136/injuryprevention-2018-safety.349.

Mash, Deborah C. "Excited Delirium and Sudden Death: A Syndromal Disorder at the Extreme End of the Neuropsychiatric Continuum." *Frontiers in Physiology* 7 (2016): 435–35. https://doi.org/10.3389/fphys.2016.00435.

McPherson, Mark, and William Pickett. "Characteristics of Martial Art Injuries in a Defined Canadian Population: A Descriptive Epidemiological Study." *BMC Public Health* 10, no. 1 (2010): 795–95. https://doi.org/10.1186/1471-2458-10-795.

Michalewicz, Betty A., Theodore C. Chan, Gary M. Vilke, Susan S. Levy, Tom S. Neuman, and Fred W. Kolkhorst. "Ventilatory and Metabolic Demands During Aggressive Physical Restraint in Healthy Adults." *Journal of Forensic Sciences* 52, no. 1 (2007): 171–75. https://doi.org/10.1111/j.1556-4029.2006.00296.x.

"Muay Thai." *Human Weapon*, Season 1 Episode 1. Created by Terry Bullman. History Channel. 2007.

"Muay Thai Children Fighting for Cash." Channel 4, British Public Broadcast Service. 2014. https://www.youtube.com/watch?v=u2ueOF7tm1k.

O'Halloran, Ronald L., and Janice G. Frank. "Asphyxial Death During Prone Restraint Revisited: A Report of 21 Cases." *American Journal of Forensic Medicine and Pathology* 21, no. 1 (2000): 39–52. https://doi.org/10.1097/00000433-200003000-00007.

Paterson, B., P. Bradley, C. Stark, D. Saddler, D. Leadbetter, and D. Allen. "Deaths Associated with Restraint Use in Health and Social Care in the UK: The Results of a Preliminary Survey." *Journal of Psychiatric and Mental Health Nursing* 10, no. 1 (2003): 3–15. https://doi.org/10.1046/j.1365-2850.2003.00523.x.

Perawongmetha, Athit, and Jiraporn Kuhakan. "Punching out of Poverty: Despite Risks, 9-Year-Old Thai Fighter Eager to Return to Ring." Reuters. April 6, 2021. https://www.reuters.com/article/us-thailand-muaythai-wideimage/punching-out-of-poverty-despite-risks-9-year-old-thai-fighter-eager-to-return-to-ring-idUSKBN2BT31S.

Pieter, F., and W. Pieter. "Speed and Force in Selected Taekwondo Techniques." *Biology of Sport* 12, no. 4 (1995): 257–66.

Pollanen, Michael S., David A. Chiasson, James T. Cairns, and James G. Young. "Unexpected Death Related to Restraint for Excited Delirium: A Retrospective

Study of Deaths in Police Custody and in the Community." *Canadian Medical Association Journal* 158, no. 12 (1998): 1603–7.

Power, Julie, and Kate Geraghty. "Inside Muay Thai: Where Culture and Children's Well-Being Collide." *Sydney Morning Herald.* November 24, 2018. https://www .smh.com.au/sport/boxing/inside-muay-thai-where-culture-and-children-s -well-being-collide-20181123-p50htq.html.

Preuschl, Emanuel, Michaela Hassmann, and Arnold Baca. "A Kinematic Analysis of the Jumping Front-Leg Axe-Kick in Taekwondo." *Journal of Sports Science and Medicine* 15, no. 1 (2016): 92–101.

Rachnavy, P, T Khaothin, and W Rittiwat. "Kinematics Analysis of Muay Thai Knee Techniques." International Conference of Sport Science-AESA. 2018. https:// journal.aesasport.com/index.php/AESA-Conf/article/view/84.

Raguse, Lou. "MPD Training Materials Show Knee-to-Neck Restraint Similar to the One Used on Floyd." KARE11. July 8, 2020. https://www.kare11.com/article /news/local/george-floyd/minneapolis-police-training-materials-show -knee-to-neck-restraint-similar-to-used-on-george-floyd/89-9f002e3f-972a -4410-86cb-50a1237fc496.

Reay, Donald T., and John W. Eisele. "Death from Law Enforcement Neck Holds." *American Journal of Forensic Medicine and Pathology* 3, no. 3 (1982): 253–58. https://doi.org/10.1097/00000433-198209000-00012.

Reay, Donald T., Corinne L. Fligner, Allan D. Stilwell, and Judy Arnold. "Positional Asphyxia During Law-Enforcement Transport." *American Journal of Forensic Medicine and Pathology* 13, no. 2 (1992): 90–97. https://doi.org/10.1097/00000433 -199206000-00002.

Ross, Darrell L., and Michael H. Hazlett. "A Prospective Analysis of the Outcomes of Violent Prone Restraint Incidents in Policing." *Forensic Research and Criminology International Journal* 2, no. 1 (2016): 16–24. https://doi.org/10.15406/frcij .2016.02.00040.

Rowe, Marc, and Lee Wedlake. "The Carotid Choke: To Sleep, Perchance to Die?" *Journal of Asian Martial Arts* 18, no. 3 (2009): 50–69.

Schlosser, Michael. "Unlocking the Confusion Around Chokeholds." *Police* 43, no. 3 (2019): 36.

Schroll, Rebecca, Alison Smith, Norman E. McSwain Jr., John Myers, Kristin Rocchi, Kenji Inaba, Stefano Siboni, et al. "A Multi-Institutional Analysis of Prehospital Tourniquet Use." *Journal of Trauma and Acute Care Surgery* 79, no. 1 (2015): 10–14. https://doi.org/10.1097/TA.0000000000000689.

Serina, E. R., and D. K. Lieu. "Thoracic Injury Potential of Basic Competition Taekwondo Kicks." *Journal of Biomechanics* 24, no. 10 (1991): 951–60. https://doi .org/10.1016/0021-9290(91)90173-K.

Slevin, John P., Cierra Harrison, Eric Da Silva, and Nathan J. White. "Martial Arts Technique for Control of Severe External Bleeding." *Emergency Medicine Journal* 36, no. 3 (2019): 154–58. https://doi.org/10.1136/emermed-2018-207966.

Sloane, Christian, Theodore C. Chan, Fred Kolkhorst, Tom Neuman, Edward M. Castillo, and Gary M. Vilke. "Evaluation of the Ventilatory Effects of the Prone

Maximum Restraint (PMR) Position on Obese Human Subjects." *Forensic Science International* 237 (2014): 86–89. https://doi.org/10.1016/j.forsciint.2014.01.017.

Solomon, Ben. "'Destroying Our Children for Sport': Thailand May Limit Underage Boxing." *New York Times*. December 23, 2018. https://www.nytimes.com/2018/12/23/world/asia/thailand-children-muay-thai.html.

Stamm, Julie M., Alexandra P. Bourlas, Christine M. Baugh, Nathan G. Fritts, Daniel H. Daneshvar, Brett M. Martin, Michael D. McClean, Yorghos Tripodis, and Robert A. Stern. "Age of First Exposure to Football and Later-Life Cognitive Impairment in Former NFL Players." *Neurology* 84, no. 11 (2015): 1114–20. https://doi.org/10.1212/WNL.0000000000001358.

State of Minnesota v. Derek Michael Chauvin, Tou Thao, J. Alexander Kueng, Thomas Kiernan Lane, Hennepin County Court File No. 27-CR-12646, 27-CR-20-12949, 27-CR-20-12953, 27-CR-20-12951. Minnesota Judicial Branch. 2020. https://www.mncourts.gov/mncourtsgov/media/High-Profile-Cases/27-CR-20-12646/ExhibitMtD08282020.pdf.

"Stealth Fighters." *Fight Science*, Season 1, Episode 5. Created by Michael Stern. National Geographic. 2008.

Steinberg, Alon. "Prone Restraint Cardiac Arrest: A Comprehensive Review of the Scientific Literature and an Explanation of the Physiology." *Medicine, Science and the Law* 61, no. 3 (2021): 215–26. https://doi.org/10.1177/0025802420988370.

Stratton, Samuel J., Christopher Rogers, Karen Brickett, and Ginger Grunzinski. "Factors Associated with Sudden Death of Individuals Requiring Restraint for Excited Delirium." *American Journal of Emergency Medicine* 19, no. 3 (2001): 187–91. https://doi.org/10.1053/ajem.2001.22665.

Surowiecki, James. "Beautiful. Violent. American. The N.F.L. At 100." *New York Times*. December 19, 2019. https://www.nytimes.com/2019/12/19/sports/football/nfl-100-violence-american-culture.html.

"Thailand's Child Fighters." Al Jazeera. October 3, 2019. https://www.youtube.com/watch?v=bh2pviMAB9I.

Thibordee, Sutima, and Orawan Prasartwuth. "Effectiveness of Roundhouse Kick in Elite Taekwondo Athletes." *Journal of Electromyography and Kinesiology* 24, no. 3 (2014): 353–58. https://doi.org/10.1016/j.jelekin.2014.02.002.

Tolan, Casey. "Two-Thirds of People Put in Neck Restraints by Minneapolis Police Were Black, Department Data Shows." CNN. June 2, 2020. https://www.cnn.com/2020/06/02/us/mn-minneapolis-police-neck-restraints-george-floyd-invs/index.html.

U.S. Department of Labor. "Child Labor and Forced Labor Reports: Thailand." Bureau of International Labor Affairs. 2018. https://www.dol.gov/sites/dolgov/files/ILAB/child_labor_reports/tda2018/Thailand.pdf.

——. "Child Labor and Forced Labor Reports: Thailand." Bureau of International Labor Affairs. 2020. https://www.dol.gov/sites/dolgov/files/ILAB/child_labor_reports/tda2020/Thailand.pdf.

von Krenner, Walther G., Damon Apodaca, and Ken Jeremiah. *Aikido Ground Fighting: Grappling and Submission Techniques*. Berkeley, CA: Blue Snake Books, 2013.

Wamsley, Laurel, and Vanessa Romo. "Defense Medical Expert: Floyd's Manner of Death 'Undetermined', Not 'Homicide'." NPR. April 14, 2021. https://www .npr.org/sections/trial-over-killing-of-george-floyd/2021/04/14/987134841 /watch-live-defense-testimony-resumes-in-derek-chauvins-trial.

Wąsik, Jacek. "Kinematics and Kinetics of Taekwon-Do Side Kick." *Journal of Human Kinetics* 30 (2011): 13–20. https://doi.org/10.2478/v10078-011-0068-z.

Webber, James T., and David A. Raichlen. "The Role of Plantigrady and Heel-Strike in the Mechanics and Energetics of Human Walking with Implications for the Evolution of the Human Foot." *Journal of Experimental Biology* 219 (2016): 3729–37. https://doi.org/10.1242/jeb.138610.

Winter, Deena. "Dr. Andrew Baker, Key Witness, Stands by Homicide Determination in Chauvin Trial." *Florida Phoenix*. April 10, 2021. https://floridaphoenix .com/2021/04/10/dr-andrew-baker-key-witness-stands-by-homicide-determination -in-chauvin-trial/.

Wilson, William Scott, trans. *Ideals of the Samurai: Writings of Japanese Warriors*. Burbank, CA: Ohara, 1982.

Xiong, Chao, Paul Walsh, and Rochelle Olson. "Cardiac Arrest and Drugs, Not Low Oxygen, Caused Floyd's Death, Defense Expert Says." *Star Tribune*. April 15, 2021. https://www.startribune.com/cardiac-arrest-and-drugs-not-low-oxygen -caused-floyd-s-death-defense-expert-says/600046038/.

Zetaruk, M. N., M. A. Violán, D. Zurakowski, and L. J. Micheli. "Injuries in Martial Arts: A Comparison of Five Styles." *British Journal of Sports Medicine* 39, no. 1 (2005): 29–33. https://doi.org/10.1136/bjsm.2003.010322.

8. RACE AND MONEY

Akgun, M., O. Araz, I. Akkurt, A. Eroglu, F. Alper, L. Saglam, A. Mirici, M. Gorguner, and B. Nemery. "An Epidemic of Silicosis Among Former Denim Sandblasters." *European Respiratory Journal* 32, no. 5 (2008): 1295–303. https://doi.org/10.1183 /09031936.00093507.

Akgun, Metin, Metin Gorguner, Mehmet Meral, Atila Turkyilmaz, Fazli Erdogan, Leyla Saglam, and Arzu Mirici. "Silicosis Caused by Sandblasting of Jeans in Turkey: A Report of Two Concomitant Cases." *Journal of Occupational Health* 47, no. 4 (2005): 346–49. https://doi.org/10.1539/joh.47.346.

Akgun, Metin, Arzu Mirici, Elif Yilmazel Ucar, Mecit Kantarci, Omer Araz, and Metin Gorguner. "Silicosis in Turkish Denim Sandblasters." *Occupational Medicine* 56, no. 8 (2006): 554–58. https://doi.org/10.1093/occmed/kql094.

Amin, Shreyasee, Joyce Goggins, Jingbo Niu, Ali Guermazi, Mikayel Grigoryan, David J. Hunter, Harry K. Genant, and David T. Felson. "Occupation-Related Squatting, Kneeling, and Heavy Lifting and the Knee Joint: A Magnetic Resonance Imaging-Based Study in Men." *Journal of Rheumatology* 35, no. 8 (2008): 1645–49.

Atif, U., A. Philip, J. Aponte, E. M. Woldu, S. Brady, V. B. Kraus, J. M. Jordan, et al. "Absence of Association of Asporin Polymorphisms and Osteoarthritis

Susceptibility in US Caucasians." *Osteoarthritis and Cartilage* 16, no. 10 (2008): 1174–77. https://doi.org/10.1016/j.joca.2008.03.007.

Bachtiar, Maulana, and Caroline G. L. Lee. "Genetics of Population Differences in Drug Response." *Current Genetic Medicine Reports* 1, no. 3 (2013): 162–70. https://doi.org/10.1007/s40142-013-0017-3.

Bass, Anne R., Kelly McHugh, Kara Fields, Rie Goto, Michael L. Parks, and Susan M. Goodman. "Higher Total Knee Arthroplasty Revision Rates Among United States Blacks Than Whites: A Systematic Literature Review and Meta-Analysis." *Journal of Bone and Joint Surgery* 98, no. 24 (2016): 2103–8. https://doi.org/10.2106/JBJS.15.00976.

Brooks, Andrew. *Clothing Poverty: The Hidden World of Fast Fashion and Second-Hand Clothes.* London: Zed Books, 2015.

Buxton, Paul, Christopher Edwards, Charles W. Archer, and Philippa Francis-West. "Growth/Differentiation Factor-5 (GDF-5) and Skeletal Development." *Journal of Bone and Joint Surgery* 83-A, Suppl. 1 (2001): S23-S30.

Cavanaugh, A. M., M. J. Rauh, C. A. Thompson, J. Alcaraz, W. M. Mihalko, C. E. Bird, C. B. Eaton, et al. "Racial and Ethnic Disparities in Utilization of Total Knee Arthroplasty Among Older Women." *Osteoarthritis and Cartilage* 27, no. 12 (2019): 1746–54. https://doi.org/10.1016/j.joca.2019.07.015.

Clean Clothes Campaign. *Deadly Denim: Sandblasting in the Bangladesh Garment Industry.* 2012. https://archive.cleanclothes.org/resources/publications/ccc-deadly-denim.pdf/view.

Center for Disease Control and Prevention. "Osteoarthritis." July 27, 2020. https://www.cdc.gov/arthritis/basics/osteoarthritis.htm.

Chaturvedi, Nishi, and Yoav Ben-Shlomo. "From the Surgery to the Surgeon: Does Deprivation Influence Consultation and Operation Rates?" *British Journal of General Practice* 45, no. 392 (1995): 127–31.

Cho, Hyung Joon, Vivek Morey, Jong Yeal Kang, Ki Woong Kim, and Tae Kyun Kim. "Prevalence and Risk Factors of Spine, Shoulder, Hand, Hip, and Knee Osteoarthritis in Community-Dwelling Koreans Older Than Age 65 Years." *Clinical Orthopaedics and Related Research* 473, no. 10 (2015): 3307–14. https://doi.org/10.1007/s11999-015-4450-3.

Coggon, David, Peter Croft, Samantha Kellingray, David Barrett, Magnus McLaren, and Cyrus Cooper. "Occupational Physical Activities and Osteoarthritis of the Knee." *Arthritis and Rheumatism* 43, no. 7 (2000): 1443–49.

Cunningham, Bill. "'Distressed' Jeans Fad: When Tatty Is Natty." *New York Times.* September 8, 1987.

Dalrymple, Theodore. "Torn Jeans: The Politics of a Fashion Statement." *City Journal.* Autumn 2004. https://www.city-journal.org/html/torn-jeans-12831.html.

Demura, Shinichi, and Masanobu Uchiyama. "Effect of Japanese Sitting Style (Seiza) on the Center of Foot Pressure after Standing." *Journal of Physiological Anthropology and Applied Human Science* 24, no. 2 (2005): 167–73. https://doi.org/10.2114/jpa.24.167.

Dunlop, Dorothy D., Larry M. Manheim, Jing Song, Min-Woong Sohn, Joseph M. Feinglass, Huan J. Chang, and Rowland W. Chang. "Age and Racial/Ethnic

Disparities in Arthritis-Related Hip and Knee Surgeries." *Medical Care* 46, no. 2 (2008): 200–8. https://doi.org/10.1097/MLR.0b013e31815cecd8.

Evangelou, Evangelos, Kay Chapman, Ingrid Meulenbelt, Fotini B. Karassa, John Loughlin, Andrew Carr, Michael Doherty, et al. "Large-Scale Analysis of Association Between *GDF5* and *FRZB* Variants and Osteoarthritis of the Hip, Knee, and Hand." *Arthritis and Rheumatism* 60, no. 6 (2009): 1710–21. https://doi.org/10.1002/art.24524.

Feldman, Candace H., Yan Dong, Jeffrey N. Katz, Laurel A. Donnell-Fink, and Elena Losina. "Association Between Socioeconomic Status and Pain, Function and Pain Catastrophizing at Presentation for Total Knee Arthroplasty." *BMC Musculoskeletal Disorders* 16, no. 1 (2015). https://doi.org/10.1186/s12891-015-0475-8.

Freburger, Janet K., George M. Holmes, Li-Jung E. Ku, Malcolm P. Cutchin, Kendra Heatwole-Shank, and Lloyd J. Edwards. "Disparities in Post–Acute Rehabilitation Care for Joint Replacement." *Arthritis Care and Research (2010)* 63, no. 7 (2011): 1020–30. https://doi.org/10.1002/acr.20477.

Fukunaga, Michihiko, and Kentaro Morimoto. "Calculation of the Knee Joint Force at Deep Squatting and Kneeling." *Journal of Biomechanical Science and Engineering* 10, no. 4 (2015): 15-00452-15-52. https://doi.org/10.1299/jbse.15-00452.

Goodman, Susan M., Lisa A. Mandl, Michael L. Parks, Meng Zhang, Kelly R. McHugh, Yuo-Yu Lee, Joseph T. Nguyen, et al. "Disparities in TKA Outcomes: Census Tract Data Show Interactions Between Race and Poverty." *Clinical Orthopaedics and Related Research* 474, no. 9 (2016): 1986–95. https://doi.org/10.1007/s11999-016-4919-8.

Haga, Susanne B. "Impact of Limited Population Diversity of Genome-Wide Association Studies." *Genetics in Medicine* 12, no. 2 (2010): 81–84. https://doi.org/10.1097/GIM.0b013e3181ca2bbf.

Hall, Howard. " 'Distressed' Look Not Well-Dressed but Distressing." *Bradenton Herald* (Bradenton, FL). September 11, 1987.

Han, Hyuk-Soo, and Seung-Baik Kang. "Does High-Flexion Total Knee Arthroplasty Allow Deep Flexion Safely in Asian Patients?" *Clinical Orthopaedics and Related Research* 471, no. 5 (2013): 1492–97. https://doi.org/10.1007/s11999-012-2628-5.

Hefzy, Mohamed Samir, Brian P. Kelly, T. Derek V. Cooke, Abdel Mohsen Al-Baddah, and Laurie Harrison. "Knee Kinematics in-Vivo of Kneeling in Deep Flexion Examined by Bi-Planar Radiographs." *Biomedical Sciences Instrumentation* 33 (1997): 453–58.

Heiden, David. "Clothes, Poverty, and the Global Economy." *Western Journal of Medicine* 175, no. 1 (2001): 72–72. https://doi.org/10.1136/ewjm.175.1.72.

Heran, Milad. *Man in Brown Jacket and Blue Denim Jeans.* March 27, 2020. Photograph. Pexels. https://www.pexels.com/photo/man-in-brown-jacket-and-blue-denim-jeans-4069117/.

Hodge, W. Andrew, Melinda K. Harman, and Scott A. Banks. "Patterns of Knee Osteoarthritis in Arabian and American Knees." *Journal of Arthroplasty* 24, no. 3 (2009): 448–53. https://doi.org/10.1016/j.arth.2007.12.012.

Huang, Hsuan-Ti, Jiing Yuan Su, and Gwo-Jaw Wang. "The Early Results of High-Flex Total Knee Arthroplasty: A Minimum of 2 Years of Follow-Up."

Journal of Arthroplasty 20, no. 5 (2005): 674–79. https://doi.org/10.1016/j.arth.2004.09.053.

Ibrahim, Said A., Laura A. Siminoff, Christopher J. Burant, and C. Kent Kwoh. "Differences in Expectations of Outcome Mediate African American/White Patient Differences in 'Willingness' to Consider Joint Replacement." *Arthritis and Rheumatism* 46, no. 9 (2002): 2429–35. https://doi.org/10.1002/art.10494.

——. "Variation in Perceptions of Treatment and Self-Care Practices in Elderly with Osteoarthritis: A Comparison Between African American and White Patients." *Arthritis and Rheumatism* 45, no. 4 (2001): 340–45.

Jiang, Qing, Dongquan Shi, Long Yi, Shiro Ikegawa, Yong Wang, Takahiro Nakamura, Di Qiao, Cheng Liu, and Jin Dai. "Replication of the Association of the Aspartic Acid Repeat Polymorphism in the Asporin Gene with Knee-Osteoarthritis Susceptibility in Han Chinese." *Journal of Human Genetics* 51, no. 12 (2006): 1068–72. https://doi.org/10.1007/s10038-006-0065-6.

Jordan, Joanne M., Charles G. Helmick, Jordan B. Renner, Gheorghe Luta, Anca D. Dragomir, Janice Woodard, Fang Fang, et al. "Prevalence of Knee Symptoms and Radiographic and Symptomatic Knee Osteoarthritis in African Americans and Caucasians: The Johnston County Osteoarthritis Project." *Journal of Rheumatology* 34, no. 1 (2007): 172–80.

Kim, T. H., D. H. Lee, and S. I. Bin. "The NexGen LPS-Flex to the Knee Prosthesis at a Minimum of Three Years." *Journal of Bone and Joint Surgery* 90, no. 10 (2008): 1304–10. https://doi.org/10.1302/0301-620X.90B10.21050.

Kim, Young Hoo, Shuichi Matsuda, and Tae Kyun Kim. "Clinical Faceoff: Do We Need Special Strategies for Asian Patients with TKA?" *Clinical Orthopaedics and Related Research* 474, no. 5 (2016): 1102–7. https://doi.org/10.1007/s11999-016-4716-4.

Kizawa, Hideki, Ikuyo Kou, Aritoshi Iida, Akihiro Sudo, Yoshinari Miyamoto, Akira Fukuda, Akihiko Mabuchi, et al. "An Aspartic Acid Repeat Polymorphism in Asporin Inhibits Chondrogenesis and Increases Susceptibility to Osteoarthritis." *Nature Genetics* 37, no. 2 (2005): 138–44. https://doi.org/10.1038/ng1496.

Kronebusch, Karl, Bradford H. Gray, and Mark Schlesinger. "Explaining Racial/Ethnic Disparities in Use of High-Volume Hospitals: Decision-Making Complexity and Local Hospital Environments." *Inquiry: Journal of Health Care Organization, Provision, and Financing* 51, no. 1 (2014): 1–21. https://doi.org/10.1177/0046958014545575.

Kurosaka, Masahiro, Shinichi Yoshiya, Kiyonori Mizuno, and Tetsuji Yamamoto. "Maximizing Flexion After Total Knee Arthroplasty: The Need and the Pitfalls." *Journal of Arthroplasty* 17, no. 4 (2002): 59–62. https://doi.org/10.1054/arth.2002.32688.

Lawrence, Reva C., David T. Felson, Charles G. Helmick, Lesley M. Arnold, Hyon Choi, Richard A. Deyo, Sherine Gabriel, et al. "Estimates of the Prevalence of Arthritis and Other Rheumatic Conditions in the United States: Part II." *Arthritis and Rheumatism* 58, no. 1 (2008): 26–35. https://doi.org/10.1002/art.23176.

Lee, Bum-Sik, Jong-Won Chung, Jong-Min Kim, Kyung-Ah Kim, and Seong-Il Bin. "High-Flexion Prosthesis Improves Function of TKA in Asian Patients Without Decreasing Early Survivorship." *Clinical Orthopaedics and Related Research* 471, no. 5 (2012): 1504–11. https://doi.org/10.1007/s11999-012-2661-4.

Marsh, Graham, and Paul Trynka. *Denim: From Cowboys to Catwalks: A Visual History of the World's Most Legendary Fabric.* London: Aurum, 2002.

Mat, Sumaiyah, Mohamad Hasif Jaafar, Chin Teck Ng, Sargunan Sockalingam, Jasmin Raja, Shahrul Bahyah Kamaruzzaman, Ai-Vyrn Chin, et al. "Ethnic Differences in the Prevalence, Socioeconomic and Health Related Risk Factors of Knee Pain and Osteoarthritis Symptoms in Older Malaysians." *PLoS One* 14, no. 11 (2019): e0225075.

Miyamoto, Yoshinari, Akihiko Mabuchi, Dongquan Shi, Toshikazu Kubo, Yoshio Takatori, Susumu Saito, Mikihiro Fujioka, et al. "A Functional Polymorphism in the 5` UTR of GDF5 Is Associated with Susceptibility to Osteoarthritis." *Nature Genetics* 39, no. 4 (2007): 529–33. https://doi.org/10.1038/2005.

Mulholland, Susan J., and URS P. Wyss. "Activities of Daily Living in Non-Western Cultures: Range of Motion Requirements for Hip and Knee Joint Implants." *International Journal of Rehabilitation Research* 24, no. 3 (2001): 191–98. https://doi.org/10.1097/00004356-200109000-00004.

Murphy, Michael, Simon Journeaux, and Trevor Russell. "High-Flexion Total Knee Arthroplasty: A Systematic Review." *International Orthopaedics* 33, no. 4 (2009): 887–93. https://doi.org/10.1007/s00264-009-0774-5.

Mustafa, Zehra, Barbara Dowling, Kay Chapman, Janet S. Sinsheimer, Andrew Carr, and John Loughlin. "Investigating the Aspartic Acid (D) Repeat of Asporin as a Risk Factor for Osteoarthritis in a UK Caucasian Population." *Arthritis and Rheumatism* 52, no. 11 (2005): 3502–6. https://doi.org/10.1002/art.21399.

Neame, R. L., K. Muir, S. Doherty, and M. Doherty. "Genetic Risk of Knee Osteoarthritis: A Sibling Study." *Annals of the Rheumatic Diseases* 63, no. 9 (2004): 1022–27. https://doi.org/10.1136/ard.2003.014498.

Peat, G., R. McCarney, and P. Croft. "Knee Pain and Osteoarthritis in Older Adults: A Review of Community Burden and Current Use of Primary Health Care." *Annals of the Rheumatic Diseases* 60, no. 2 (2001): 91–97. https://doi.org/10.1136/ard.60.2.91.

Popejoy, Alice B., and Stephanie M. Fullerton. "Genomics Is Failing on Diversity." *Nature* 538, no. 7624 (2016): 161–64. https://doi.org/10.1038/538161a.

Raichlen, David A., Herman Pontzer, Theodore W. Zderic, Jacob A. Harris, Audax Z. P. Mabulla, Marc T. Hamilton, and Brian M. Wood. "Sitting, Squatting, and the Evolutionary Biology of Human Inactivity." *Proceedings of the National Academy of Sciences* 117, no. 13 (2020): 7115–21. https://doi.org/10.1073/pnas.1911868117.

Rainey, Sarah. "Why Is Everyone Wearing Ripped Jeans . . . And Why Do They Cost More Than Ones with No Holes?" *Daily Mail.* July 24, 2017. https://www.dailymail.co.uk/femail/article-4726702/Why-wearing-ripped-jeans.html.

Riddselius, Christopher. *Fashion Victims: A Report on Sand Blasted Denim.* Swedish Fair Trade Centre and the Clean Clothes Campaign. 2010. https://archive

.cleanclothes.org/resources/national-cccs/fashion-victims-a-report-on-sandblasted
-denim/view.

Rodrigues, Wesner. *Woman Looking Up.* July 17, 2018. Photograph. Pexels. https://
www.pexels.com/photo/woman-looking-up-1892779/.

Rosquist, Catrin. *Still Fashion Victims? Monitoring a Ban on Sandblasted Denim.* Fair
Trade Center. 2012. https://fairaction.se/wp-content/uploads/2015/06/English
-version_Still-fashion-victims-Monitoring-a-ban-on-sandblasted-denim-2012_0.pdf.

Saha, Somnath, Jose J. Arbelaez, and Lisa A. Cooper. "Patient-Physician Relationships and Racial Disparities in the Quality of Health Care." *American Journal
of Public Health* 93, no. 10 (2003): 1713–19. https://doi.org/10.2105/AJPH.93
.10.1713.

Salazar, James B. "Fashioning the Historical Body: The Political Economy of
Denim." *Social Semiotics* 20, no. 3 (2010): 293–308. https://doi.org/10.1080
/10350331003722851.

Scharff, Darcell P., Katherine J. Mathews, Pamela Jackson, Jonathan Hoffsuemmer,
Emeobong Martin, and Dorothy Edwards. "More Than Tuskegee: Understanding Mistrust About Research Participation." *Journal of Health Care for the Poor
and Underserved* 21, no. 3 (2010): 879–97. https://doi.org/10.1353/hpu.0.0323.

Sirugo, Giorgio, Scott M. Williams, and Sarah A. Tishkoff. "The Missing Diversity in
Human Genetic Studies." *Cell* 177, no. 1 (2019): 26–31. https://doi.org/10.1016/j
.cell.2019.02.048.

Skinner, Jonathan, Weiping Zhou, and James Weinstein. "The Influence of Income
and Race on Total Knee Arthroplasty in the United States." *Journal of Bone and
Joint Surgery* 88, no. 10 (2006): 2159–66. https://doi.org/10.2106/JBJS.E.00271.

Somerville, Heather. "Retailer Sandblasting Bans Have Changed Little in the Garment Industry." *Mercury News.* October 28, 2013. https://www.mercurynews
.com/2013/10/28/retailer-sandblasting-bans-have-changed-little-in-the-garment
-industry-2/.

Spector, Tim D., Flavia Cicuttini, Juliet Baker, John Loughlin, and Deborah Hart.
"Genetic Influences on Osteoarthritis in Women: A Twin Study." *BMJ* 312,
no. 7036 (1996): 940–43. https://doi.org/10.1136/bmj.312.7036.940.

Steel, Nicholas, Allan Clark, Iain A. Lang, Robert B. Wallace, and David Melzer.
"Racial Disparities in Receipt of Hip and Knee Joint Replacements Are Not
Explained by Need: The Health and Retirement Study 1998–2004." *Journals of
Gerontology. Series A, Biological Sciences and Medical Sciences* 63, no. 6 (2008):
629–34. https://doi.org/10.1093/gerona/63.6.629.

Suarez-Almazor, Maria E., Julianne Souchek, P. Adam Kelly, Kimberly O'Malley,
Margaret Byrne, Marsha Richardson, and Chong Pak. "Ethnic Variation in Knee
Replacement: Patient Preferences or Uninformed Disparity?" *Archives of Internal
Medicine* 165, no. 10 (2005): 1117–24. https://doi.org/10.1001/archinte.165.10.1117.

Sullivan, James. *Jeans: A Cultural History of an American Icon.* New York: Gotham
Books, 2006.

Tanavalee, Aree, Srihatach Ngarmukos, Saran Tantavisut, and Arak Limtrakul. "High-
Flexion TKA in Patients with a Minimum of 120 Degrees of Pre-Operative

Knee Flexion: Outcomes at Six Years of Follow-Up." *International Orthopaedics* 35, no. 9 (2010): 1321–26. https://doi.org/10.1007/s00264-010-1140-3.

Tangtrakulwanich, Boonsin, Virasakdi Chongsuvivatwong, and Alan F. Geater. "Habitual Floor Activities Increase Risk of Knee Osteoarthritis." *Clinical Orthopaedics and Related Research* 454 (2007): 147–54. https://doi.org/10.1097/01 .blo.0000238808.72164.1d.

Thompson, Kathryn A., Ellen L. Terry, Kimberly T. Sibille, Ethan W. Gossett, Erin N. Ross, Emily J. Bartley, Toni L. Glover, et al. "At the Intersection of Ethnicity/Race and Poverty: Knee Pain and Physical Function." *Journal of Racial and Ethnic Health Disparities* 6, no. 6 (2019): 1131–43. https://doi.org/10.1007/s40615-019-00615-7.

Tsumaki, Noriyuki, Kazuhiro Tanaka, Eri Arikawa-Hirasawa, Takanobu Nakase, Tomoatsu Kimura, J. Terrig Thomas, Takahiro Ochi, Frank P. Luyten, and Yoshihiko Yamada. "Role of CDMP-1 in Skeletal Morphogenesis: Promotion of Mesenchymal Cell Recruitment and Chondrocyte Differentiation." *Journal of Cell Biology* 144, no. 1 (1999): 161–73. https://doi.org/10.1083/jcb.144.1.161.

Valdes, Ana M., and Tim D. Spector. "Genetic Epidemiology of Hip and Knee Osteoarthritis." *Nature Reviews: Rheumatology* 7, no. 1 (2011): 23–32. https:// doi.org/10.1038/nrrheum.2010.191.

Virayavanich, Warapat, Hamza Alizai, Thomas Baum, Lorenzo Nardo, Michael C. Nevitt, John A. Lynch, Charles E. McCulloch, and Thomas M. Link. "Association of Frequent Knee Bending Activity with Focal Knee Lesions Detected with 3T Magnetic Resonance Imaging: Data from the Osteoarthritis Initiative." *Arthritis Care and Research (2010)* 65, no. 9 (2013): 1441–48. https://doi.org /10.1002/acr.22017.

Wallace, Ian J., Steven Worthington, David T. Felson, Robert D. Jurmain, Kimberly T. Wren, Heli Maijanen, Robert J. Woods, and Daniel E. Lieberman. "Knee Osteoarthritis Has Doubled in Prevalence Since the Mid-20th Century." *Proceedings of the National Academy of Sciences* 114, no. 35 (2017): 9332–36. https:// doi.org/10.1073/pnas.1703856114.

Williams, David R., Selina A. Mohammed, Jacinta Leavell, and Chiquita Collins. "Race, Socioeconomic Status, and Health: Complexities, Ongoing Challenges, and Research Opportunities." *Annals of the New York Academy of Sciences* 1186, no. 1 (2010): 69–101. https://doi.org/10.1111/j.1749-6632.2009.05339.x.

Woolf, Jake. "Everything You Need to Know Before Buying Ripped Jeans." *GQ.* March 24, 2016. https://www.gq.com/story/ripped-jeans-guide-wearing-gq.

Wright, Nicole C., Gail Kershner Riggs, Jeffrey R. Lisse, and Zhao Chen. "Self-Reported Osteoarthritis, Ethnicity, Body Mass Index, and Other Associated Risk Factors in Postmenopausal Women—Results from the Women's Health Initiative." *Journal of the American Geriatrics Society (JAGS)* 56, no. 9 (2008): 1736–43. https://doi.org/10.1111/j.1532-5415.2008.01812.x.

Yu, Wei, Melinda Clyne, Muin J. Khoury, and Marta Gwinn. "Phenopedia and Genopedia: Disease-Centered and Gene-Centered Views of the Evolving Knowledge of Human Genetic Associations." *Bioinformatics* (2021). https://phgkb.cdc.gov /PHGKB/startPagePhenoPedia.action.

Zengini, Eleni, Konstantinos Hatzikotoulas, Ioanna Tachmazidou, Julia Steinberg, Fernando P. Hartwig, Lorraine Southam, Sophie Hackinger, et al. "Genome-Wide Analyses Using UK Biobank Data Provide Insights Into the Genetic Architecture of Osteoarthritis." *Nature Genetics* 50, no. 4 (2018): 549–58. https://doi.org/10.1038/s41588-018-0079-y.

Zhang, Yuqing, David J. Hunter, Michael C. Nevitt, Ling Xu, Jingbo Niu, Li-Yung Lui, Wei Yu, Piran Aliabadi, and David T. Felson. "Association of Squatting with Increased Prevalence of Radiographic Tibiofemoral Knee Osteoarthritis: The Beijing Osteoarthritis Study." *Arthritis and Rheumatism* 50, no. 4 (2004): 1187–92. https://doi.org/10.1002/art.20127.

Zhang, Yuqing, Ling Xu, Michael C. Nevitt, Piran Aliabadi, Wei Yu, Mingwei Qin, Li-Yung Lui, and David T. Felson. "Comparison of the Prevalence of Knee Osteoarthritis Between the Elderly Chinese Population in Beijing and Whites in the United States: The Beijing Osteoarthritis Study." *Arthritis and Rheumatism* 44, no. 9 (2001): 2065–71.

Zimmer Biomet. "NexGen®CR-Flex and LPS-Flex Knees: Design Rationale." 2016. https://www.zimmerbiomet.com/content/dam/zimmer-biomet/medical-professionals/knee/nexgen-complete-knee-solution-legacy-knee-posterior-stabilized/nexgen-cr-flex-and-lps-flex-knees-design-rationale.pdf.

Index

Page numbers in *italics* represent illustrations.